Birkhäuser

Progress in Mathematics
Volume 293

Marco Mazzucchelli

Critical Point Theory for Lagrangian Systems

Birkhäuser

Marco Mazzucchelli
Penn State University
Department of Mathematics
University Park, PA 16802
USA

ISBN 978-3-0348-0162-1 e-ISBN 978-3-0348-0163-8
DOI 10.1007/978-3-0348-0163-8
Springer Basel Dordrecht Heidelberg London New York

Library of Congress Control Number: 2011941493

Mathematics Subject Classification (2010): 70S05, 37J45, 58E05

Printed on acid-free paper

Springer Basel AG is part of Springer Science + Business Media
(www.birkhauser-science.com)

To Ana

Contents

Preface

This monograph is devoted to presenting in detail a few selected applications of critical point theory, and in particular Morse theory, to Lagrangian dynamics. A Lagrangian system is defined by a configuration space M, which has the structure of a smooth manifold, and a Lagrangian function $\mathscr{L}$ defined on the tangent bundle of M and, in the non-conservative case, depending upon time as well. In Joseph-Louis Lagrange's reformulation of classical mechanics, the Lagrangian function is given by the difference of kinetic and potential energy. The motion of the associated mechanical system is described by the Euler-Lagrange equations, a system of second-order ordinary differential equations that involves the Lagrangian. The principle of least action, which in different settings is even anterior to Lagrange's work, states that the curves that are solutions of the Euler-Lagrange equations admit a variational characterization: they are extremal points of a functional, the action, associated to the Lagrangian. The development of critical point theory in the nineteenth and twentieth century provided a powerful machinery to investigate dynamical questions in Lagrangian systems, such as existence, multiplicity or uniqueness of solutions of the Euler-Lagrange equations with prescribed boundary conditions.

In this monograph, we will consider closed configuration spaces M and we will focus on the class of so-called Tonelli Lagrangians: these are smooth Lagrangian functions $\mathscr{L} : \mathbb{R} \times \mathrm{T}M \to \mathbb{R}$ that, when restricted to the fibers of $\mathrm{T}M$, have positive definite Hessian and superlinear growth. We will normally restrict ourselves to those Tonelli Lagrangians $\mathscr{L}$ whose solutions of the Euler-Lagrange equations are defined for all times, a condition that is always fulfilled when the time-derivative of the Lagrangian is suitably bounded. The importance of the Tonelli class is twofold. From the variational point of view, the Tonelli assumptions imply existence and regularity of action minimizing curves joining given points in the configuration space. From a symplectic point of view, Tonelli Lagrangians constitute the broadest family of fiberwise convex Lagrangian functions for which the Lagrangian-Hamiltonian duality holds. These generalities on Tonelli Lagrangians, together with a brief introduction to the Lagrangian and Hamiltonian formalism, will be the subject of **Chapter 1**.

If a Tonelli Lagrangian is 1-periodic in time, namely if it is a function of the form $\mathscr{L} : \mathbb{R}/\mathbb{Z} \times \mathrm{T}M \to \mathbb{R}$, then one can look for periodic solutions (with integer period) of the associated Euler-Lagrange equations. Finding a lower bound for the number of periodic orbits with prescribed period is one of the main themes in Lagrangian dynamics. The least action principle characterizes the n-periodic orbits as the extremal points of the Lagrangian action functional defined on the space of smooth (say C^2) n-periodic curves. In view of this, one is tempted to study the multiplicity of n-periodic orbits by means of critical point theory: the "richer" the topology of the space of n-periodic curves, the larger the minimal number of n-periodic orbits. More precisely, one expects a lower bound for the number of n-periodic orbits to be given by the cup-length of the space of n-periodic curves. Another celebrated question is whether the Euler-Lagrange systems admit infinitely many periodic solutions (without prescribing their period). The main difficulty here is to recognize when an n-periodic orbit found by abstract methods is the iteration of another periodic orbit of lower period. The Morse index of periodic orbits helps us with this: a periodic orbit that is obtained by homological techniques in a certain degree d will have Morse index close to d. In the 1950s, Bott investigated the behavior of the Morse index of periodic orbits under iteration. In **Chapter 2**, we will introduce the notion of Morse index and we will present in detail Bott's iteration theory. We will also mention a symplectic interpretation of the Morse index as a Maslov index. This latter index, which can be associated to periodic orbits of more general Hamiltonian systems, was independently introduced and investigated by many people among whom are Gel'fand, Lidskiĭ, Maslov, Conley, Zehnder, Long, Robbin and Salamon.

In order to be able to apply the abstract results of critical point theory to the Lagrangian action, we need a suitable functional setting: a space of sufficiently smooth n-periodic curves with the structure of a (possibly infinite-dimensional) manifold, over which the action is regular, say at least C^1, and has sublevels that are "sufficiently compact". For the subclass of Tonelli Lagrangians with quadratic growth (which we will synthetically call "convex quadratic-growth"), a suitable choice is given by the Hilbert manifold of n-periodic curves with $W^{1,2}$ regularity. As proved by Benci, with this functional setting the action is $C^{1,1}$ and satisfies the Palais-Smale condition, a "weak compactness" condition on its sublevels that is enough for critical point theory. In **Chapter 3**, we will introduce this functional setting for the case of periodic curves and of curves with prescribed endpoints. After studying the properties of the action of convex quadratic-growth Lagrangians in this setting, we will derive a few elementary results on the existence of action minimizing orbits and of periodic orbits with prescribed period.

Even though the $C^{1,1}$ regularity of the action of convex quadratic-growth Lagrangians is sufficient for most of the results of critical point theory, all the abstract statements involving the Morse index require at least the C^2 regularity. However, this requirement turns out to be not necessary for the Lagrangian action functional. Indeed, one can equivalently work in a finite-dimensional functional setting in which the action is C^∞ and has compact sublevels. In this setting, one considers the space of continuous and piecewise broken n-periodic solutions of the Euler-Lagrange equations. This space turns out to be a smooth finite-dimensional submanifold of the $W^{1,2}$ loop space, and it may be regarded as a homotopic approximation of this latter space. In particular, all the indices and invariants coming from critical point theory are the same in the $W^{1,2}$ setting and in the finite-dimensional one: Morse index and nullity, local homology of periodic orbits, relative homology of sublevels of the action and so forth. **Chapter 4** will be devoted to introducing this finite-dimensional functional setting, and to proving a few multiplicity results for periodic orbits with prescribed period.

Bott's iteration theory can be pushed one step further by investigating the behavior of the local homology of periodic orbits under iterations. This problem was first studied by Gromoll and Meyer, and further by Long. It turns out that the behavior of local homology is sometimes dictated by the Morse index and nullity: if these indices do not change by iteration, then the local homology does not change as well. This seemingly technical result turns out to be crucial in the study of the multiplicity of periodic orbits with unprescribed period. In **Chapter 5**, we will deduce this theorem from an analogous abstract result: the local homology of a critical point of a function does not change when restricted to a submanifold which is invariant under the gradient flow of the function, provided the Morse index and nullity do not change as well.

For a general Tonelli Lagrangian with global Euler-Lagrange flow, a functional setting in which the action is both regular and satisfies the Palais-Smale condition is not known. However, one can still apply abstract results from critical point theory to a suitable modification of the Lagrangian, that coincides with the original one in a neighborhood of the zero section of the tangent bundle and it is fiberwise quadratic at infinity. This idea is due to Abbondandolo and Figalli, who showed that, for a fixed period n and a fixed action value a, all the n-periodic orbits of the modified Lagrangian with action less than a are also periodic orbits of the original Tonelli Lagrangian, provided the modification was performed sufficiently far from the zero section. In **Chapter 6** we will discuss this idea, and we will use it to extend the validity of the multiplicity results of Chapter 4 to the Tonelli case. The second part of the chapter will be devoted to proving that Tonelli Lagrangians with global Euler-Lagrange flow always admit infinitely many periodic orbits. This result, first established by Long for mechanical Lagrangians on the torus and further extended by Lu and by the author, is based upon the iteration theory for periodic orbits discussed in Chapters 2 and 5, and upon an important technique developed by Bangert and Klingenberg in the setting of closed geodesics.

We have tried to make this monograph accessible even to non-specialists, and in particular to students from the graduate level onward. The **Appendix** collects all the background from Morse theory that is needed for our applications. The interested reader can find more material together with the proofs in [Cha93]. As we have already remarked, we have only presented a few selected applications of critical point theory to Lagrangian dynamics. A topic which is very close to ours and that we did not touch at all is the existence and multiplicity of periodic orbits with prescribed energy in autonomous Lagrangian systems. For a summary of the recent state of the art of this problem we refer the reader to [Con06] and the bibliography therein. It is impossible to mention here all the other applications of critical point theory to Lagrangian and Hamiltonian dynamics. The interested reader can find an account of some of these topics in the excellent textbooks [MW89, Eke90, HZ94, CI99, Abb01, Lon02, Fat08].

Acknowledgment

I wish to thank Alberto Abbondandolo for introducing me to the realm of critical point theory a few years ago, when I was a graduate student in Pisa. I owe a lot to Gonzalo Contreras, Albert Fathi and Eduard Zehnder who, with their classes, advice or indirectly with their books, taught me the formalism of Lagrangian and Hamiltonian dynamics. Many people provided me with feedback on preliminary versions of the manuscript. I thank Will Merry for his detailed review, for his encouragement and for his valuable comments. I thank the anonymous referee for pointing out several inaccuracies in the text. Among the others who contributed to this monograph, special thanks go to Ana Lecuona, Barney Bramham, Joana Santos, José Fernando and Pierre Pageault. Finally, it is my pleasure to acknowledge the University of Pisa, the Max Planck Institute for Mathematics in the Sciences of Leipzig, and the École Normale Supérieure of Lyon for their support during the preparation of this work.

Lyon, April 2011 Marco Mazzucchelli

Chapter 1

Lagrangian and Hamiltonian Systems

This chapter is devoted to giving an informal introduction to the subject of Lagrangian and Hamiltonian dynamics, from the point of view that is relevant in this monograph. In Section 1.1 we give the basic definitions and we briefly review the duality between Lagrangian and Hamiltonian systems. Even though these topics come from mathematical physics, we make almost no mention of this. We rather emphasize the variational and dynamical systems flavors of these theories. In Section 1.2 we introduce the important class of Tonelli Lagrangians and Tonelli Hamiltonians. The main results about the existence and multiplicity of periodic orbits, that we will give in Chapter 6, are valid for Tonelli systems. Here, we try to motivate the importance of the Tonelli assumptions, as they naturally give broad families of Lagrangians and Hamiltonians for which the above-mentioned duality occurs. In Section 1.3 we introduce the action minimizing curves associated to a Tonelli Lagrangian system, and we give detailed proofs of classical results concerning their existence, uniqueness and regularity.

1.1 The formalism of classical mechanics

Classical mechanics describes the motion of mechanical systems. In this section we briefly review its language, in particular we introduce the Lagrangian and Hamiltonian formalisms. Of course, we do not attempt to give a comprehensive introduction to the subject. For such a purpose, we refer the reader to one of the many textbooks of mathematical physics (e.g., [Arn78, AM78]), dynamical systems (e.g., [Fat08, HZ94, KH95]), calculus of variations (e.g., [BGH98, GH96, MW89]) or symplectic geometry (e.g., [CdS01, MS98]).

Lagrangian mechanics provides a description of the motion of a mechanical system constrained on a configuration space. For us, the **configuration space** will always be a closed manifold M of dimension $m \geq 1$. A **Lagrangian** on the configuration space is a smooth function $\mathscr{L} : \mathbb{R} \times \mathrm{T}M \to \mathbb{R}$. In general, the points in the domain $\mathbb{R} \times \mathrm{T}M$ will be denoted by (t, q, v), i.e., $t \in \mathbb{R}$, $v \in \mathrm{T}_q M$. The $\mathbb{R}$-factor in the domain of $\mathscr{L}$ must be interpreted as a time dependence. If the Lagrangian happens to be independent of t it is called **autonomous**. A point (q, v) in the tangent bundle $\mathrm{T}M$ of the configuration space is interpreted in the following way: q gives the position of the mechanical system, while the vector v in the tangent space of q gives the velocity of the mechanical system.

Consider a bounded real interval $[t_0, t_1] \subset \mathbb{R}$. We define the **action functional** $\mathscr{A}^{t_0,t_1}$ associated to the Lagrangian $\mathscr{L}$ as

$$\mathscr{A}^{t_0,t_1}(\gamma) = \int_{t_0}^{t_1} \mathscr{L}(t, \gamma(t), \dot{\gamma}(t))\, \mathrm{d}t, \tag{1.1}$$

where $\gamma : [t_0, t_1] \to M$. For the moment, we do not discuss a functional setting for the action $\mathscr{A}^{t_0,t_1}$. We just consider it as defined on some space of curves γ such that the function $t \mapsto \mathscr{L}(t, \gamma(t), \dot{\gamma}(t))$ is integrable on the interval $[t_0, t_1]$, for instance we can consider $\mathscr{A}^{t_0,t_1}$ to be defined on the space of C^2 curves $\gamma : [t_0, t_1] \to M$. A C^2 map $\Sigma : (-\varepsilon, \varepsilon) \times [t_0, t_1] \to M$ is called a **variation** of γ when $\Sigma(0, \cdot) = \gamma$, $\Sigma(\cdot, t_0) \equiv \gamma(t_0)$ and $\Sigma(\cdot, t_1) \equiv \gamma(t_1)$. The curve γ is a **motion** of the Lagrangian system defined by $\mathscr{L}$ when, for each variation Σ of γ, we have

$$\frac{\mathrm{d}}{\mathrm{d}s}\bigg|_{s=0} \mathscr{A}^{t_0,t_1}(\Sigma(s, \cdot)) = 0.$$

Namely, a curve on M is a motion when it is an **extremal** of the action functional.

Now, let us fix a finite atlas $\mathfrak{U} = \{\phi_\alpha : U_\alpha \to \mathbb{R}^m \mid \alpha = 0, \ldots, u\}$ for the closed manifold M, and consider the induced atlases for the tangent and cotangent bundles of M. These are the atlases $\mathrm{T}\mathfrak{U} = \{\mathrm{T}\phi_\alpha : TU_\alpha \to \mathbb{R}^m \times \mathbb{R}^m \mid \alpha = 0, \ldots, u\}$ and $\mathrm{T}^*\mathfrak{U} = \{\mathrm{T}^*\phi_\alpha : \mathrm{T}^*U_\alpha \to \mathbb{R}^m \times (\mathbb{R}^m)^* \mid \alpha = 0, \ldots, u\}$ respectively, where

$$\begin{aligned} \mathrm{T}\phi_\alpha(q, v) &= (\phi_\alpha(q), \mathrm{d}\phi_\alpha(q)v), && \forall q \in U_\alpha,\ v \in \mathrm{T}_q M, \\ \mathrm{T}^*\phi_\alpha(q, p) &= (\phi_\alpha(q), p \circ \mathrm{d}\phi_\alpha^{-1}(\phi_\alpha(q))), && \forall q \in U_\alpha,\ p \in \mathrm{T}_q^* M. \end{aligned}$$

We denote the components of the introduced charts by

$$\begin{aligned} \phi_\alpha &= (q_\alpha^1, \ldots, q_\alpha^m), \\ \mathrm{T}\phi_\alpha &= (q_\alpha^1, \ldots, q_\alpha^m, v_\alpha^1, \ldots, v_\alpha^m), \\ \mathrm{T}^*\phi_\alpha &= (q_\alpha^1, \ldots, q_\alpha^m, p_{\alpha,1}, \ldots, p_{\alpha,m}). \end{aligned}$$

We define the **fiberwise derivative** of the Lagrangian $\mathscr{L} : \mathbb{R} \times \mathrm{T}M \to \mathbb{R}$ at (t, q, v) as the covector $\partial_v \mathscr{L}(t, q, v) \in \mathrm{T}_q^* M$ given in local coordinates by

$$\partial_v \mathscr{L}(t, q, v) = \sum_{j=1}^{m} \frac{\partial \mathscr{L}}{\partial v_\alpha^j}(t, q, v)\, \mathrm{d}q_\alpha^j.$$

Notice that this definition does not depend on the chosen local coordinates, in fact, if $q \in U_\alpha \cap U_\beta$, then

$$\begin{aligned}
&\sum_{j=1}^{m} \frac{\partial \mathscr{L}}{\partial v_\alpha^j}(t,q,v)\, \mathrm{d}q_\alpha^j(q) \\
&\quad = \sum_{j=1}^{m}\sum_{h=1}^{m}\Bigg(\frac{\partial \mathscr{L}}{\partial v_\beta^h}(t,q,v) \underbrace{\frac{\partial v_\beta^h}{\partial v_\alpha^j}(q,v)}_{=\frac{\partial q_\beta^h}{\partial q_\alpha^j}(q)} \mathrm{d}q_\alpha^j(q) + \frac{\partial \mathscr{L}}{\partial q_\beta^h}(t,q,v) \underbrace{\frac{\partial q_\beta^h}{\partial v_\alpha^j}(q,v)}_{=0} \mathrm{d}q_\alpha^j(q)\Bigg) \\
&\quad = \sum_{h=1}^{m} \frac{\partial \mathscr{L}}{\partial v_\beta^h}(t,q,v)\, \mathrm{d}q_\beta^h(q).
\end{aligned}$$

Extremal curves of the action functional can be characterized as follows. Assume that $\gamma : [t_0,t_1] \to M$ is a C^2 extremal curve, and consider a subdivision $t_0 = r_0 < r_1 < \cdots < r_n = t_1$ such that the support $\gamma([r_k,r_{k+1}])$ is contained in some coordinate domain U_{α_k} of the atlas $\mathfrak{U}$, for each $k = 0,\ldots,n-1$. For each variation Σ of γ, if we denote by σ the section of $\gamma^*\mathrm{T}M$ given by

$$\sigma(t) = \frac{\partial \Sigma}{\partial s}(0,t), \qquad \forall t \in [t_0,t_1],$$

we have

$$\begin{aligned}
0 &= \frac{\mathrm{d}}{\mathrm{d}s}\bigg|_{s=0} \mathscr{A}^{t_0,t_1}(\Sigma(s,\cdot)) \\
&= \sum_{k=0}^{n-1}\sum_{j=1}^{m} \int_{r_k}^{r_{k+1}} \left(\frac{\partial \mathscr{L}}{\partial q_{\alpha_k}^j}(t,\gamma,\dot\gamma)\, \sigma_{\alpha_k}^j(t) + \frac{\partial \mathscr{L}}{\partial v_{\alpha_k}^j}(t,\gamma,\dot\gamma)\, \dot\sigma_{\alpha_k}^j(t) \right) \mathrm{d}t \\
&= \sum_{k=0}^{n-1}\sum_{j=1}^{m} \int_{r_k}^{r_{k+1}} \left(\frac{\partial \mathscr{L}}{\partial q_{\alpha_k}^j}(t,\gamma,\dot\gamma) - \frac{\mathrm{d}}{\mathrm{d}t}\frac{\partial \mathscr{L}}{\partial v_{\alpha_k}^j}(t,\gamma,\dot\gamma) \right) \sigma_{\alpha_k}^j(t)\mathrm{d}t \\
&\quad + \sum_{k=0}^{n-1}\Big(\partial_v\mathscr{L}(r_{k+1},\gamma(r_{k+1}),\dot\gamma(r_{k+1}))\,\sigma(r_{k+1}) - \partial_v\mathscr{L}(r_k,\gamma(r_k),\dot\gamma(r_k))\,\sigma(r_k) \Big) \\
&= \sum_{k=0}^{n-1}\sum_{j=1}^{m} \int_{r_k}^{r_{k+1}} \left(\frac{\partial \mathscr{L}}{\partial q_{\alpha_k}^j}(t,\gamma,\dot\gamma) - \frac{\mathrm{d}}{\mathrm{d}t}\frac{\partial \mathscr{L}}{\partial v_{\alpha_k}^j}(t,\gamma,\dot\gamma) \right) \sigma_{\alpha_k}^j(t)\mathrm{d}t,
\end{aligned}$$

where we have adopted the common notation $\sigma_\alpha(t) := \mathrm{d}\phi_\alpha(\gamma(t))\sigma(t)$. By the fundamental lemma of the calculus of variations, the above expression is zero for each variation Σ of the curve γ (that is, for each C^1 section σ of the vector bundle $\gamma^*\mathrm{T}M$) if and only if γ satisfies in local coordinates

$$\frac{\mathrm{d}}{\mathrm{d}t}\frac{\partial \mathscr{L}}{\partial v^j}(t,\gamma(t),\dot\gamma(t)) - \frac{\partial \mathscr{L}}{\partial q^j}(t,\gamma(t),\dot\gamma(t)) = 0, \qquad \forall j = 1,\ldots,m. \tag{1.2}$$

The above system of second-order ordinary differential equations is known as the **Euler-Lagrange system** associated to the Lagrangian $\mathscr{L}$. Hence, we have a second characterization of motion curves as solutions of the Euler-Lagrange system (1.2).

Now, for each $(t,q,v) \in \mathbb{R} \times \mathrm{T}M$, consider the quadratic form on T_qM given by

$$w \mapsto \sum_{j,h=1}^{m} \frac{\partial^2 \mathscr{L}}{\partial v^j \partial v^h}(t,q,v) w^j w^h, \qquad \forall w = \sum_{j=1}^{m} w^j \frac{\partial}{\partial q^j} \in \mathrm{T}_qM.$$

Notice that this quadratic form is independent of the local coordinates used to define it (by the same argument that we gave to show that the fiberwise derivative is intrinsically defined). We say that the Lagrangian $\mathscr{L}$ is **non-degenerate** when the above quadratic form is non-degenerate on the whole domain of $\mathscr{L}$, i.e., for each $(t,q,v) \in \mathbb{R} \times \mathrm{T}M$ and for each nonzero $w \in \mathrm{T}_qM$ there exists $z \in \mathrm{T}_qM$ such that

$$\sum_{j,h=1}^{m} \frac{\partial^2 \mathscr{L}}{\partial v^j \partial v^h}(t,q,v) w^j z^h \neq 0.$$

Equivalently, the Lagrangian $\mathscr{L}$ is non-degenerate when the $m \times m$ real matrix

$$\left(\frac{\partial^2 \mathscr{L}}{\partial v^j \partial v^h}(t,q,v) \right)_{j,h=1,\dots,m}$$

is invertible for each $(t,q,v) \in \mathbb{R} \times \mathrm{T}M$. If this condition is fulfilled we can put the Euler-Lagrange system (1.2) in normal form as

$$\ddot{\gamma}^j = \sum_{h=1}^{m} \left[\frac{\partial^2 \mathscr{L}}{\partial v \partial v}(t,\gamma,\dot{\gamma}) \right]^{-1}_{j,h} \left(\frac{\partial \mathscr{L}}{\partial q^h}(t,\gamma,\dot{\gamma}) - \frac{\partial^2 \mathscr{L}}{\partial t \partial v^h}(t,\gamma,\dot{\gamma}) - \sum_{l=1}^{m} \frac{\partial^2 \mathscr{L}}{\partial q^l \partial v^h}(t,\gamma,\dot{\gamma})\dot{\gamma}^l \right), \qquad \forall j=1,\dots,m.$$

In other words, the non-degeneracy condition allows us to define a smooth time-dependent vector field $X_{\mathscr{L}}$ on $\mathrm{T}M$ as

$$X_{\mathscr{L}} = \sum_{j=1}^{m} \left(v^j \frac{\partial}{\partial q^j} + \sum_{h=1}^{m} \left[\frac{\partial^2 \mathscr{L}}{\partial v\, \partial v} \right]^{-1}_{j,h} \left(\frac{\partial \mathscr{L}}{\partial q^h} - \frac{\partial^2 \mathscr{L}}{\partial t\, \partial v^h} - \sum_{l=1}^{m} \frac{\partial^2 \mathscr{L}}{\partial q^l \partial v^h} v^l \right) \frac{\partial}{\partial v^j} \right).$$

This vector field is called **Euler-Lagrange vector field** associated to $\mathscr{L}$, and its integral curves are precisely the solutions of the Euler-Lagrange system associated to $\mathscr{L}$. By the Cauchy-Lipschitz theorem, $X_{\mathscr{L}}$ can be locally uniquely integrated: there exists a continuous function $\varepsilon : \mathbb{R} \times \mathrm{T}M \to (0,\infty)$ such that, for every $(t,q,v) \in \mathbb{R} \times \mathrm{T}M$ and for every $t_0, t_1 \in \mathbb{R}$ with $0 < t_1 - t_0 < \varepsilon(t,q,v)$, there exists a unique smooth solution $\gamma : [t_0,t_1] \to M$ of the Euler-Lagrange system (1.2) with $\gamma(t) = q$ and $\dot{\gamma}(t) = v$. This defines a partial flow $\Phi_{\mathscr{L}}$ on $\mathrm{T}M$, the **Euler-Lagrange**

flow associated to $\mathscr{L}$, as

$$\Phi_{\mathscr{L}}^{t_1,t_0}(\gamma(t_0),\dot\gamma(t_0)) = (\gamma(t_1),\dot\gamma(t_1)),$$

where $\gamma : [t_0, t_1] \to M$ is a solution of the Euler-Lagrange system. We say that $\Phi_{\mathscr{L}}$ is **global** when, for each $t_0 \in \mathbb{R}$, it defines a function

$$\Phi_{\mathscr{L}}^{\cdot,t_0} : \mathbb{R} \times \mathrm{T}M \to \mathrm{T}M,$$
$$(t, q, v) \mapsto \Phi_{\mathscr{L}}^{t,t_0}(q, v).$$

So far we have recalled the language of the Lagrangian formulation of classical mechanics. As we already mentioned at the beginning of this section, there is another point of view for describing classical mechanics: the Hamiltonian formulation. Quoting Vladimir Arnold [Arn78, page 161], Hamiltonian mechanics is *"geometry in phase space"*. The **phase space** is the ambient space of the considered mechanical system, and it has the structure of a symplectic manifold. We will only consider phase spaces that are cotangent bundles over a smooth closed manifold M. In these cases, the Hamiltonian formulation is, in some sense, dual to the Lagrangian one. A **Hamiltonian** on the cotangent bundle T^*M is a smooth function $\mathscr{H} : \mathbb{R} \times \mathrm{T}^*M \to \mathbb{R}$. In general, the points in the domain $\mathbb{R} \times \mathrm{T}^*M$ will be denoted by (t, q, p), i.e., $t \in \mathbb{R}$, $p \in \mathrm{T}^*_qM$. As for the Lagrangian case, the $\mathbb{R}$-factor in the domain of $\mathscr{H}$ must be interpreted as a time dependence, and if this dependence is missing the Hamiltonian is called **autonomous**. A point (q, p) in the cotangent bundle T^*M is interpreted in the following way: q still gives the position of the mechanical system, while the covector p in the cotangent space of q gives the momentum of the mechanical system.

The cotangent bundle T^*M has a **canonical symplectic structure**[1] ω, which can be defined as follows. First of all, we define the **Liouville form** of T^*M, that is a one-form λ on T^*M given in local coordinates by

$$\lambda = \sum_{j=1}^{m} p_j \mathrm{d}q^j.$$

It is easy to verify that λ is well defined by this expression (i.e., it is independent of the chosen local coordinates). Indeed, the Liouville form can also be characterized as the unique one-form λ on T^*M such that, for each one-form μ on M, we have $\mu^*\lambda = \mu$. The canonical symplectic form of T^*M is then defined as $\omega = -\mathrm{d}\lambda$. In local coordinates we have

$$\omega = \sum_{j=1}^{m} \mathrm{d}q^j \wedge \mathrm{d}p_j,$$

and it is immediate to verify that the above expression gives a non-degenerate two-form, that is clearly closed (being exact by its definition). A Hamiltonian $\mathscr{H}$

[1]We recall that a **symplectic structure** on a manifold W is a two-form ω on W that is closed (i.e., $\mathrm{d}\omega = 0$) and non-degenerate (i.e., $\omega(v,\cdot) \neq 0$ for each non-zero $v \in \mathrm{T}W$).

as above defines a smooth time-dependent vector field $X_{\mathscr{H}}$ on T^*M given by

$$X_{\mathscr{H}}(t,q,p) \lrcorner \omega = \mathrm{d}\,(\mathscr{H}(t,\cdot))\,(q,p), \qquad \forall (t,q,p) \in \mathbb{R} \times \mathrm{T}^*M,$$

where "$\lrcorner$" stands for the interior product between vectors and forms, i.e., $X_{\mathscr{H}} \lrcorner \omega = \omega(X_{\mathscr{H}}, \cdot)$. In local coordinates we have

$$X_{\mathscr{H}} = \frac{\partial \mathscr{H}}{\partial p_j} \frac{\partial}{\partial q^j} - \frac{\partial \mathscr{H}}{\partial q^j} \frac{\partial}{\partial p_j}.$$

By the Cauchy-Lipschitz theorem, this vector field can be locally integrated around any point of $\mathbb{R} \times \mathrm{T}^*M$, and therefore it defines a partial flow $\Phi_{\mathscr{H}}$ on T^*M as

$$\Phi_{\mathscr{H}}^{t_1,t_0}(\Gamma(t_0)) = \Gamma(t_1),$$

where $\Gamma : [t_0, t_1] \to \mathrm{T}^*M$ is an integral curve of $X_{\mathscr{H}}$. If we write Γ as (γ, ρ), where $\gamma : [t_0, t_1] \to M$ and ρ is a section of $\gamma^*\mathrm{T}^*M$, then in local coordinates (γ, ρ) satisfies the **Hamilton system**

$$\dot\gamma^j(t) = \frac{\partial \mathscr{H}}{\partial p_j}(t,\gamma(t),\rho(t)), \qquad \dot\rho^j(t) = -\frac{\partial \mathscr{H}}{\partial q^j}(t,\gamma(t),\rho(t)), \qquad \forall j = 1,\ldots,m. \tag{1.3}$$

We call $X_{\mathscr{H}}$ the **Hamiltonian vector field** associated to $\mathscr{H}$, and $\Phi_{\mathscr{H}}$ the corresponding **Hamiltonian flow**. The integral curves of $X_{\mathscr{H}}$, or rather their projection onto the base manifold M, are the motions of the Hamiltonian mechanical system defined by $\mathscr{H}$.

Now, consider a Lagrangian function $\mathscr{L} : \mathbb{R} \times \mathrm{T}M \to \mathbb{R}$. We define the **Legendre transform** associated to $\mathscr{L}$ as the map $\mathrm{Leg}_{\mathscr{L}} : \mathbb{R} \times \mathrm{T}M \to \mathbb{R} \times \mathrm{T}^*M$ given by

$$\mathrm{Leg}_{\mathscr{L}}(t,q,v) = (t,q,\partial_v \mathscr{L}(t,q,v)), \qquad \forall (t,q,v) \in \mathbb{R} \times \mathrm{T}M.$$

Let us assume that, for the considered Lagrangian $\mathscr{L}$, the Legendre transform is a diffeomorphism of $\mathbb{R} \times \mathrm{T}M$ onto $\mathbb{R} \times \mathrm{T}^*M$ (conditions on $\mathscr{L}$ under which this is true will be discussed in the next section). Notice that this assumption implies that $\mathscr{L}$ is non-degenerate. Then we can define the Hamiltonian $\mathscr{H} : \mathbb{R} \times \mathrm{T}^*M \to \mathbb{R}$ **Legendre-dual** to the Lagrangian $\mathscr{L}$ as

$$\mathscr{H} \circ \mathrm{Leg}_{\mathscr{L}}(t,q,v) := \partial_v \mathscr{L}(t,q,v)v - \mathscr{L}(t,q,v), \qquad \forall (t,q,v) \in \mathbb{R} \times \mathrm{T}M.$$

Actually, this sets up a duality between the Lagrangian system of $\mathscr{L}$ and the Hamiltonian system of $\mathscr{H}$. In fact, one can show that

$$\mathrm{d}(\pi_2 \circ \mathrm{Leg}_{\mathscr{L}})(t,q,v)\,(X_{\mathscr{L}}(t,q,v)) = X_{\mathscr{H}} \circ \mathrm{Leg}_{\mathscr{L}}(t,q,v), \qquad \forall (t,q,v) \in \mathbb{R} \times \mathrm{T}M,$$

where $\pi_2 : \mathbb{R}\times \mathrm{T}^*M \to \mathrm{T}^*M$ is the projection onto the second factor of $\mathbb{R}\times \mathrm{T}^*M$. Therefore the Lagrangian and Hamiltonian flows are conjugated by the Legendre transform. In other words, a curve $\gamma : [t_0, t_1] \to M$ is a solution of the Euler-Lagrange system (1.2) if and only if the curve $\Gamma = (\gamma, \rho) : [t_0, t_1] \to \mathrm{T}^*M$, where $\rho(t) := \partial_v \mathscr{L}(t, \gamma(t), \dot\gamma(t))$, is a solution of the Hamilton system (1.3).

As in the Lagrangian case, the motion curves of the Hamiltonian system associated to a smooth $\mathscr{H} : \mathbb{R}\times \mathrm{T}^*M \to \mathbb{R}$ admit a variational characterization. If $\Gamma : [t_0, t_1] \to \mathrm{T}^*M$ is a C^2 curve, we define its **Hamiltonian action** as

$$\mathscr{A}^{t_0,t_1}(\Gamma) = \int_{t_0}^{t_1} \Big(\Gamma^*\lambda - \mathscr{H}(t, \Gamma(t))\Big)\, \mathrm{d}t.$$

An easy computation shows that the solutions of the Hamilton system of $\mathscr{H}$ are precisely the extremal curves of $\mathscr{A}^{t_0,t_1}$. Moreover, this Hamiltonian action is related to the Lagrangian one of equation (1.1) in the following way: if $\mathscr{H}$ and $\mathscr{L}$ are Legendre-dual and $\Gamma = (\gamma, \rho) : [t_0, t_1] \to \mathrm{T}^*M$ is a solution of the Hamilton system of $\mathscr{H}$ (so that $\gamma : [t_0, t_1] \to M$ is a solution of the Euler-Lagrange system of $\mathscr{L}$), then the Hamiltonian action of Γ coincides with the Lagrangian action of γ, for

$$\begin{aligned}\int_{t_0}^{t_1} \Big(\Gamma^*\lambda - \mathscr{H}(t, \Gamma(t))\Big)\, \mathrm{d}t &= \int_{t_0}^{t_1} \Big(\rho(t)[\dot\gamma(t)] - \mathscr{H}(t, \gamma(t), \rho(t))\Big)\, \mathrm{d}t \\ &= \int_{t_0}^{t_1} \mathscr{L}(t, \gamma(t), \dot\gamma(t))\, \mathrm{d}t.\end{aligned}$$

1.2 Tonelli systems

Let us fix, once for all, a Riemannian metric $\langle\cdot,\cdot\rangle_\cdot$ on the closed manifold M, so that for each $q \in M$ we will denote by $\langle\cdot,\cdot\rangle_q$ the Riemannian inner product on T_qM and by $|\cdot|_q$ the corresponding norm. This metric induces a Riemannian distance $\mathrm{dist} : M \times M \to [0, \infty)$ that turns M into a complete metric space.

We say that a smooth Lagrangian $\mathscr{L} : \mathbb{R}\times \mathrm{T}M \to \mathbb{R}$ is **Tonelli** when it satisfies the following two conditions:

(T1) the fiberwise Hessian of $\mathscr{L}$ is positive-definite, i.e.,

$$\sum_{j,h=1}^{m} \frac{\partial^2 \mathscr{L}}{\partial v^j \partial v^h}(t, q, v) w^j w^h > 0,$$

for all $(t, q, v) \in \mathbb{R}\times \mathrm{T}M$ and $w = \sum_{j=1}^m w^j \frac{\partial}{\partial q^j} \in \mathrm{T}_qM$ with $w \neq 0$;

(T2) $\mathscr{L}$ is fiberwise superlinear, i.e.,

$$\lim_{|v|_q\to\infty} \frac{\mathscr{L}(t, q, v)}{|v|_q} = \infty,$$

for all $(t, q) \in \mathbb{R}\times M$.

Now, let us consider a fiberwise convex (but not necessarily Tonelli) smooth Lagrangian $\mathscr{L} : \mathbb{R} \times \mathrm{T}M \to \mathbb{R}$. We want to show that the Legendre transform $\mathrm{Leg}_{\mathscr{L}}$ is a diffeomorphism if and only if $\mathscr{L}$ is Tonelli. First of all, notice that the Legendre transform $\mathrm{Leg}_{\mathscr{L}}$ is a fiber-preserving smooth map between $\mathbb{R} \times \mathrm{T}M$ and $\mathbb{R} \times \mathrm{T}^*M$. Assuming that $\mathrm{Leg}_{\mathscr{L}}$ is a diffeomorphism is equivalent to assuming that, for each $(t, q) \in \mathbb{R} \times M$, its fiberwise restriction

$$\partial_v \mathscr{L}(t, q, \cdot) = \mathrm{d}(\mathscr{L}|_{\{t\} \times \mathrm{T}_q M}) : \mathrm{T}_q M \to \mathrm{T}^*_q M$$

is a diffeomorphism. Hence, all we have to do is characterize convex functions on $\mathbb{R}^m$ whose differential is a diffeomorphism onto $(\mathbb{R}^m)^*$.

Proposition 1.2.1. *Let $L : \mathbb{R}^m \to \mathbb{R}$ be a convex smooth function. Then its differential $\mathrm{d}L : \mathbb{R}^m \to (\mathbb{R}^m)^*$ is a diffeomorphism if and only if L is superlinear, meaning*

$$\lim_{|v| \to \infty} \frac{L(v)}{|v|} = \infty,$$

and the Hessian $\mathrm{Hess}L$ is positive-definite at any point.

Proof. Assume that L is superlinear with positive-definite Hessian. Consider an arbitrary $p_0 \in (\mathbb{R}^m)^*$ and define $L^{p_0} : \mathbb{R}^m \to \mathbb{R}$ as $L^{p_0}(v) = L(v) - p_0(v)$. This function is superlinear, as well as L, hence it reaches its minimum at some $v_0 \in \mathbb{R}^m$. In particular $\mathrm{d}L^{p_0}(v_0) = 0$, and $\mathrm{d}L(v_0) = p_0$. This shows that $\mathrm{d}L$ is surjective. Moreover, every $v \in \mathbb{R}^m$ such that $\mathrm{d}L(v) = p_0$ must be a critical point of L^{p_0}. Since the Hessian of L^{p_0} is positive-definite, the function is strictly convex and it cannot have critical points other than v_0, and therefore $\mathrm{d}L$ is bijective. By the positivity of the Hessian, we can apply the inverse function theorem to assert that $\mathrm{d}L$ is a bijective local diffeomorphism, i.e., a global diffeomorphism.

Conversely, assume that L is convex and $\mathrm{d}L$ is a diffeomorphism. Hence the Hessian of L must be non-degenerate and, by the convexity assumption on L, even positive-definite. Moreover, since L is convex, we have

$$L(v) - L(v_0) \geq \mathrm{d}L(v_0)[v - v_0], \qquad \forall v_0, v \in \mathbb{R}^m. \tag{1.4}$$

For each real constant $k > 0$ we define the compact set

$$S_k := \{v \in \mathbb{R}^m \mid |\mathrm{d}L(v)| = k\}.$$

For each $v \in \mathbb{R}^m$, there exists a unique $v_0 = v_0(v) \in \mathbb{R}^m$ such that

$$\mathrm{d}L(v_0) = \frac{k}{|v|} \langle v, \cdot \rangle,$$

where we have denoted by $\langle \cdot, \cdot \rangle$ the standard inner product in $\mathbb{R}^m$. Notice that $\mathrm{d}L(v_0) \in S_k$ and $\mathrm{d}L(v_0)v = k|v|$. Hence, by (1.4), for each $v \in \mathbb{R}^m$ we have

$$\begin{aligned} L(v) &\geq \mathrm{d}L(v_0(v))v + L(v_0(v)) - \mathrm{d}L(v_0(v))v_0(v) \\ &\geq k|v| + \inf_{w \in S_k} \{L(w) - \mathrm{d}L(w)w\}. \end{aligned}$$

This shows that L is superlinear. □

For a smooth function L as in the above statement we define the **Legendre-dual** smooth function $H : (\mathbb{R}^m)^* \to \mathbb{R}$ by

$$H \circ \mathrm{d}L(v) = \mathrm{d}L(v)v - L(v).$$

Proposition 1.2.2. *Consider a smooth function $L : \mathbb{R}^m \to \mathbb{R}$ that is superlinear with positive-definite Hessian, and its Legendre-dual function $H : (\mathbb{R}^m)^* \to \mathbb{R}$. Then*

(i) *L and H satisfy the* **Fenchel relation** *$L(v) + H(p) \geq p(v)$, and the equality holds true if and only if $p = \mathrm{d}L(v)$;*

(ii) *$\mathrm{d}H = (\mathrm{d}L)^{-1} : (\mathbb{R}^m)^* \to \mathbb{R}^m \simeq (\mathbb{R}^m)^{**}$;*

(iii) *H is superlinear with positive-definite Hessian.*

Proof. Consider arbitrary $v \in \mathbb{R}^m$ and $p \in (\mathbb{R}^m)^*$. By the assumptions on L, $\mathrm{d}L$ is a diffeomorphism and in particular $p = \mathrm{d}L(w)$ for some $w \in \mathbb{R}^m$. Therefore

$$\begin{aligned} L(v) + H(p) - p(v) &= L(v) - H(\mathrm{d}L(w)) - \mathrm{d}L(w)v \\ &= L(v) - L(w) - \mathrm{d}L(w)(v - w) \\ &\geq 0, \end{aligned}$$

where the last inequality follows by the convexity of L. Moreover, since L is strictly convex, equality holds if and only if $v = w$, that is if and only if $p = \mathrm{d}L(v)$. This proves (i). By the Fenchel relation, for each $p_0, p_1 \in (\mathbb{R}^m)^*$, we get

$$\begin{aligned} H((1-\lambda)p_0 + \lambda p_1) &= \max_{v \in \mathbb{R}^m} \{(1-\lambda)p_0(v) + \lambda p_1(v) - L(v)\} \\ &\leq \max_{v \in \mathbb{R}^m} \{(1-\lambda)p_0(v) - (1-\lambda)L(v)\} \\ &\quad + \max_{v \in \mathbb{R}^m} \{\lambda p_1(v) - \lambda L(v)\} \\ &\leq (1-\lambda)H(p_0) + \lambda H(p_1), \end{aligned}$$

therefore the function H is convex. Now, let us fix an arbitrary $p_0 \in (\mathbb{R}^m)^*$. For $v_0 = (\mathrm{d}L)^{-1}(p_0)$, we have $p_0(v_0) = H(p_0) + L(v_0)$. By the Fenchel relation, for any $p \in (\mathbb{R}^m)^*$, we have

$$p(v_0) \leq H(p) + L(v_0) = H(p) - H(p_0) + p_0(v_0).$$

Hence, for any $p \in (\mathbb{R}^m)^*$ we have $(p - p_0)v_0 \leq H(p) - H(p_0)$, and this is possible if and only if $v_0 = \mathrm{d}H(p_0)$ (here, we are making the canonical identification between $(\mathbb{R}^m)^{**}$ and $\mathbb{R}^m$). This proves (ii), and in particular that $\mathrm{d}H$ is a diffeomorphism. Applying Proposition 1.2.1 to H in place of L we obtain that H is superlinear with positive-definite Hessian, hence (iii) holds. □

We say that a Hamiltonian $\mathscr{H} : \mathbb{R} \times \mathrm{T}^*M \to \mathbb{R}$ is **Tonelli** when it satisfies the following two conditions:

(T1′) the fiberwise Hessian of $\mathscr{H}$ is positive-definite, i.e.,

$$\sum_{j,h=1}^{m} \frac{\partial^2 \mathscr{H}}{\partial p_j \partial p_h}(t,q,p) r_j r_h > 0,$$

for all $(t,q,p) \in \mathbb{R} \times \mathrm{T}^*M$ and $r = \sum_{j=1}^{m} r_j \mathrm{d}q^j \in \mathrm{T}_q^*M$ with $r \neq 0$;

(T2′) $\mathscr{H}$ is fiberwise superlinear, i.e.,

$$\lim_{|p|_q \to \infty} \frac{\mathscr{H}(t,q,p)}{|p|_q} = \infty,$$

for all $(t,q) \in \mathbb{R} \times M$ (here, by a common abuse of notation, we write $|\cdot|_q$ also for the norm in T_q^*M induced by the Riemannian metric of M).

By Proposition 1.2.2, the Tonelli Hamiltonians are precisely those Hamiltonian functions that are dual to Tonelli Lagrangians. Namely, the Legendre duality sets up a one-to-one correspondence

$$\left\{ \begin{array}{c} \mathscr{L} : \mathbb{R} \times \mathrm{T}M \to \mathbb{R} \\ \text{Tonelli} \end{array} \right\} \stackrel{1:1}{\longleftrightarrow} \left\{ \begin{array}{c} \mathscr{H} : \mathbb{R} \times \mathrm{T}^*M \to \mathbb{R} \\ \text{Tonelli} \end{array} \right\}.$$

This discussion should have motivated the importance of the Tonelli class in the study of Lagrangian and Hamiltonian systems.

Remark 1.2.1 (Uniform fiberwise superlinearity). It turns out that the Tonelli assumptions, both for Lagrangians and Hamiltonians, imply that the fiberwise superlinearity of conditions **(T2)** and **(T2′)** are uniform over compact subsets of $\mathbb{R} \times M$, which means that, for each compact interval $[t_0, t_1] \subset \mathbb{R}$, the limits in **(T2)** and **(T2′)** are satisfied uniformly in $(t,q) \in [t_0,t_1] \times M$. In fact, if $\mathscr{L}$ is a Tonelli Lagrangian with dual Tonelli Hamiltonian $\mathscr{H}$, by the Fenchel relation (Proposition 1.2.2(i)) we have

$$\begin{aligned} \mathscr{L}(t,q,v) &\geq \max_{|p|_q \leq k} \{p(v) - \mathscr{H}(t,q,p)\} \\ &\geq \max_{|p|_q \leq k} \{p(v)\} - \max_{|p|_q \leq k} \{\mathscr{H}(t,q,p)\} \\ &\geq k|v|_q - \max\left\{\mathscr{H}(t',q',p') \,|\, (t',q',p') \in [t_0,t_1] \times \mathrm{T}^*M,\ |p'|_{q'} \leq k\right\}, \end{aligned}$$

for each $k \in \mathbb{N}$ and $(t,q,v) \in [t_0,t_1] \times \mathrm{T}M$. Analogously

$$\mathscr{H}(t,q,p) \geq k|p|_q - \max\left\{\mathscr{L}(t',q',v') \,|\, (t',q',v') \in [t_0,t_1] \times \mathrm{T}M,\ |v'|_{q'} \leq k\right\},$$

for each $k \in \mathbb{N}$ and $(t,q,p) \in [t_0,t_1] \times \mathrm{T}^*M$. □

1.3 Action minimizers

Let us consider a Tonelli Lagrangian $\mathscr{L} : \mathbb{R} \times \mathrm{T}M \to \mathbb{R}$. The Tonelli assumptions **(T1)** and **(T2)** imply that $\mathscr{L}$ is bounded from below, and therefore each absolutely continuous[2] curve $\gamma : [t_0, t_1] \to M$ has a well-defined action

$$\mathscr{A}^{t_0,t_1}(\gamma) = \int_{t_0}^{t_1} \mathscr{L}(t, \gamma(t), \dot\gamma(t))\, dt \in \mathbb{R} \cup \{+\infty\}.$$

An absolutely continuous curve $\gamma : [t_0, t_1] \to M$ is an **action minimizer** with respect to $\mathscr{L}$ when every other absolutely continuous curve $\zeta : [t_0, t_1] \to M$ with the same endpoints of γ satisfies $\mathscr{A}^{t_0,t_1}(\gamma) \leq \mathscr{A}^{t_0,t_1}(\zeta)$. The Tonelli assumptions guarantee the existence of action minimizers, as stated by the following fundamental result that is due to Tonelli [Ton34]. Here, we give a modern proof following Mather [Mat91, Appendix 1], Contreras and Iturriaga [CI99, Section 3.1] and Fathi [Fat08, Section 3.3].

Theorem 1.3.1 (Tonelli). *Let $\mathscr{L} : \mathbb{R} \times \mathrm{T}M \to \mathbb{R}$ be a Tonelli Lagrangian. For each real interval $[t_0, t_1] \subset \mathbb{R}$ and for all $q_0, q_1 \in M$ there exists an action minimizer (with respect to $\mathscr{L}$) $\gamma : [t_0, t_1] \to M$ with $\gamma(t_0) = q_0$ and $\gamma(t_1) = q_1$.*

In order to prove the Tonelli theorem, we first need two preliminary technical lemmas.

Lemma 1.3.2. *Consider a Tonelli Lagrangian $\mathscr{L} : [t_0, t_1] \times \mathrm{T}K \to \mathbb{R}$, where K is the closure of a bounded open set of $\mathbb{R}^m$. For each $R > 0$ and $\varepsilon > 0$ there exists $\delta = \delta(R, \varepsilon) > 0$ such that, for each $t \in [t_0, t_1]$ and $(q, v), (q', v') \in \mathrm{T}K \simeq K \times \mathbb{R}^m$ with $|q - q'| \leq \delta$ and $|v| \leq R$, we have*

$$\mathscr{L}(t, q', v') \geq \mathscr{L}(t, q, v) + \langle \partial_v \mathscr{L}(t, q, v), v' - v\rangle - \varepsilon.$$

Proof. Let us fix $R > 0$ and $\varepsilon > 0$ arbitrarily. We define

$$\ell_1 := \max\Big\{ \mathscr{L}(t, q, v) - \langle \partial_v \mathscr{L}(t, q, v), v\rangle \ \Big|\ (t, q, v) \in [t_0, t_1] \times \mathrm{T}K,\ |v| \leq R \Big\},$$
$$\ell_2 := \max\Big\{ |\partial_v \mathscr{L}(t, q, v)| \ \Big|\ (t, q, v) \in [t_0, t_1] \times \mathrm{T}K,\ |v| \leq R \Big\},$$

and, for each $R' > 0$,

$$k(R') := \min\left\{ \frac{\mathscr{L}(t, q, v)}{|v|} \ \middle|\ (t, q, v) \in [t_0, t_1] \times \mathrm{T}K,\ |v| \geq R' \right\}.$$

Notice that these are indeed real constants. In fact, ℓ_1 and ℓ_2 are maximums of continuous functions over a compact set, while $k(R')$ is finite due to the fiberwise superlinearity **(T2)** of the Tonelli Lagrangian $\mathscr{L}$, which also implies $k(R') \to +\infty$

[2] We recall that the absolutely continuous curves can be characterized as those curves having integrable weak derivative.

as $R' \to +\infty$. Let us fix $R' > 1$ such that $k(R') \geq \ell_1 + \ell_2$. For each $t \in [t_0, t_1]$ and $(q, v), (q', v') \in \mathrm{T}K \simeq K \times \mathbb{R}^m$ with $|v| \leq R$ and $|v'| \geq R'$, we have

$$\mathscr{L}(t, q', v') \geq k(R')\,|v'| \geq \ell_1 + \ell_2\,|v'| \geq \mathscr{L}(t, q, v) + \langle \partial_v \mathscr{L}(t, q, v), v' - v \rangle\,.$$

Now, since the fiberwise Hessian of $\mathscr{L}$ is positive-definite (by the Tonelli assumption **(T1)**), for each $t \in [t_0, t_1]$ and $(q, v), (q', v') \in \mathrm{T}K \simeq K \times \mathbb{R}^m$ with $|v| \leq R$ and $|v'| \leq R'$, we have

$$\mathscr{L}(t, q, v') \geq \mathscr{L}(t, q, v) + \langle \partial_v \mathscr{L}(t, q, v), v' - v \rangle\,.$$

Therefore, if $|q - q'|$ is sufficiently small (uniformly in t, v and v' as above), we further have

$$\mathscr{L}(t, q', v') \geq \mathscr{L}(t, q, v) + \langle \partial_v \mathscr{L}(t, q, v), v' - v \rangle - \varepsilon. \qquad \square$$

We recall that a space $\mathscr{U}$ of maps of the form $h : [t_0, t_1] \to \mathbb{R}^m$ is called **uniformly integrable** when for each $\varepsilon > 0$ there exists $\delta = \delta(\varepsilon) > 0$ such that, for each Borel subset $I \subset [t_0, t_1]$ with Lebesgue measure $\mu_{\mathrm{Leb}}(I) < \delta$, we have

$$\int_I |h(t)|\, \mathrm{d}t < \varepsilon, \qquad \forall h \in \mathscr{U}.$$

Lemma 1.3.3. *Let I be a Borel subset of a compact interval $[t_0, t_1]$, $f : I \to \mathbb{R}^m$ an L^∞ map, and $\{g_n : [t_0, t_1] \to \mathbb{R}^m \,|\, n \in \mathbb{N}\}$ a sequence of absolutely continuous maps such that $g_n \to \mathbf{0}$ uniformly as $n \to \infty$ and the sequence of weak derivatives $\{\dot{g}_n \,|\, n \in \mathbb{N}\}$ is uniformly integrable. Then*

$$\lim_{n \to \infty} \int_I \langle f(t), \dot{g}_n(t) \rangle\, \mathrm{d}t = 0. \tag{1.5}$$

Proof. For each $J \subseteq [t_0, t_1]$ which is a finite union of closed intervals, i.e.,

$$J = [r_1, r_1'] \cup \cdots \cup [r_N, r_N'],$$

we have

$$\lim_{n \to \infty} \int_J \dot{g}_n(t)\, \mathrm{d}t = \lim_{n \to \infty} \sum_{i=1}^{N} [g_n(r_i') - g_n(r_i)] = \mathbf{0}.$$

Now, for each Borel set I as in the statement, we can always find a sequence $\{J_k \subseteq [t_0, t_1] \,|\, k \in \mathbb{N}\}$ such that

$$\lim_{k \to \infty} \mu_{\mathrm{Leb}}(J_k) = \mu_{\mathrm{Leb}}(I) \tag{1.6}$$

and, for each $k \in \mathbb{N}$, J_k is a finite union of closed intervals. Let us consider an arbitrary $\varepsilon > 0$, and let $\delta = \delta(\varepsilon) > 0$ be the corresponding constant given by the uniform integrability of $\{\dot{g}_n \,|\, n \in \mathbb{N}\}$. By (1.6), there exists $\bar{k} = \bar{k}(\delta) \in \mathbb{N}$ such that, for every integer $k \geq \bar{k}$, we have $\mu_{\mathrm{Leb}}(I \setminus J_k) < \delta$ and therefore

$$\begin{aligned}\limsup_{n\to\infty}\left|\int_I \dot{g}_n(t)\,\mathrm{d}t\right| &\leq \underbrace{\limsup_{n\to\infty}\left|\int_{J_k} \dot{g}_n(t)\,\mathrm{d}t\right|}_{=0} + \limsup_{n\to\infty}\left|\int_{I\setminus J_k} \dot{g}_n(t)\,\mathrm{d}t\right| \\ &\leq \limsup_{n\to\infty}\int_{I\setminus J_k} |\dot{g}_n(t)|\;\mathrm{d}t < \varepsilon.\end{aligned}$$

Since ε can be taken arbitrarily small, we obtain

$$\lim_{n\to\infty}\int_I \dot{g}_n(t)\,\mathrm{d}t = \mathbf{0}.$$

This proves equation (1.5) for the special case in which f is a **simple function**, i.e., when f has the form

$$f(t) = \sum_{i=1}^{N} \chi_i(t)\, w_i, \qquad \forall t \in I,$$

where $N \in \mathbb{N}$, χ_i is the characteristic function of a Borel subset $I_i \subseteq [t_0, t_1]$ and $w_i \in \mathbb{R}^m$ for each $i \in \{1, \ldots, N\}$.

In the general case, given an L^∞ function f as in the statement, for each $\delta > 0$ we can find a simple function $f_\delta : I \to \mathbb{R}^m$ such that $\|f_\delta\|_{L^\infty} \leq 2\|f\|_{L^\infty}$ and $\|f - f_\delta\|_{L^1} \leq \delta^2$. We introduce the Borel set

$$I_\delta = \big\{t \in I \,\big|\, |f_\delta(t) - f(t)| < \delta\big\}$$

so that, by our choice f_δ, we have

$$\mu_{\mathrm{Leb}}(I \setminus I_\delta) = \int_{I\setminus I_\delta} 1\,\mathrm{d}t \leq \frac{1}{\delta}\int_{I\setminus I_\delta} |f_\delta(t) - f(t)|\,\mathrm{d}t \leq \frac{1}{\delta}\|f_\delta - f\|_{L^1} \leq \delta.$$

Now, let us consider an arbitrary $\varepsilon > 0$ and the associated $\delta = \delta(\varepsilon) > 0$ given by the uniform integrability of $\{\dot{g}_n \,|\, n \in \mathbb{N}\}$. If we set $\sigma := \min\{\varepsilon, \delta\} > 0$, we obtain

$$\begin{aligned}&\limsup_{n\to\infty}\left|\int_I \langle f(t), \dot{g}_n(t)\rangle\;\mathrm{d}t\right| \\ &\quad\leq \underbrace{\limsup_{n\to\infty}\left|\int_I \langle f_\sigma(t), \dot{g}_n(t)\rangle\;\mathrm{d}t\right|}_{=0} + \limsup_{n\to\infty}\left|\int_I \langle f(t) - f_\sigma(t), \dot{g}_n(t)\rangle\;\mathrm{d}t\right| \\ &\quad\leq \limsup_{n\to\infty}\left|\int_{I_\sigma} \langle f(t) - f_\sigma(t), \dot{g}_n(t)\rangle\;\mathrm{d}t\right| + \limsup_{n\to\infty}\left|\int_{I\setminus I_\sigma} \langle f(t) - f_\sigma(t), \dot{g}_n(t)\rangle\;\mathrm{d}t\right|\end{aligned}$$

$$\le \sigma \limsup_{n\to\infty} \int_{I_\sigma} |\dot{g}_n(t)|\,\mathrm{d}t + \|f - f_\sigma\|_{L^\infty} \limsup_{n\to\infty} \underbrace{\int_{I\setminus I_\sigma} |\dot{g}_n(t)|\,\mathrm{d}t}_{<\varepsilon}$$

$$\le \sigma \limsup_{n\to\infty} \|\dot{g}_n\|_{L^1} + 3\varepsilon\|f\|_{L^\infty} \le \varepsilon\Big(\underbrace{\limsup_{n\to\infty} \|\dot{g}_n\|_{L^1} + 3\|f\|_{L^\infty}}_{(*)}\Big).$$

Notice that, by the uniform integrability of $\{\dot{g}_n \mid n \in \mathbb{N}\}$, the quantity $(*)$ is finite, and since ε can be taken arbitrarily small we readily obtain (1.5). □

Proof of Theorem 1.3.1. For each $c \in \mathbb{R}$, let $\mathscr{U}(c)$ be the set of absolutely continuous curves $\zeta : [t_0, t_1] \to M$ such that $\zeta(t_0) = q_0$, $\zeta(t_1) = q_1$ and $\mathscr{A}^{t_0,t_1}(\zeta) \le c$. All we have to do in order to prove the theorem is to show that $\mathscr{U}(c)$ is compact in the topology of uniform convergence. In fact, once this is established, we can conclude as follows. First of all, we set

$$c_0 := \inf\{c \in \mathbb{R} \mid \mathscr{U}(c) \neq \varnothing\}. \tag{1.7}$$

Notice that c_0 is finite, since the action functional $\mathscr{A}^{t_0,t_1}$ is bounded from below by the real constant

$$(t_1 - t_0) \min\{\mathscr{L}(t, q, v) \mid (t, q, v) \in [t_0, t_1] \times \mathrm{T}M\}$$

over the space of absolutely continuous curves defined on $[t_0, t_1]$. Moreover, the infimum in (1.7) is actually a minimum since $\mathscr{U}(c_0)$ is equal to the intersection of nested nonempty compact sets

$$\mathscr{U}(c_0 + 1) \supseteq \mathscr{U}(c_0 + \tfrac{1}{2}) \supseteq \mathscr{U}(c_0 + \tfrac{1}{3}) \supseteq \cdots \supseteq \mathscr{U}(c_0 + \tfrac{1}{n}) \supseteq \cdots$$

and therefore it is itself nonempty. This implies that any curve $\gamma : [t_0, t_1] \to M$ that belongs to $\mathscr{U}(c_0)$ is an action minimizer joining q_0 and q_1.

In the remaining of the proof we will establish the compactness of $\mathscr{U}(c)$, for every $c \in \mathbb{R}$. Let us fix $c \in \mathbb{R}$ such that $\mathscr{U}(c)$ is nonempty (otherwise there is nothing to prove). Without loss of generality, we can assume that

$$\mathscr{L}(t, q, v) \ge 0, \qquad \forall (t, q, v) \in [t_0, t_1] \times \mathrm{T}M.$$

By the uniform fiberwise superlinearity of $\mathscr{L}$ (Remark 1.2.1), for each $k > 0$ there exists $C_k = C_k(\mathscr{L}, t_0, t_1) > 0$ such that

$$\mathscr{L}(t, q, v) \ge k|v|_q - C_k, \qquad \forall (t, q, v) \in [t_0, t_1] \times \mathrm{T}M.$$

Now, consider an arbitrarily small $\varepsilon > 0$ and fix two real constants k and δ such that

$$k > \frac{c}{\varepsilon}, \qquad 0 < \delta < \frac{k\varepsilon - c}{C_k}.$$

For each $\zeta \in \mathscr{U}(c)$ and for each Borel subset $I \subseteq [t_0, t_1]$ having Lebesgue measure $\mu_{\mathrm{Leb}}(I) < \delta$, we have

$$\int_I |\dot\zeta(t)|_{\zeta(t)} \mathrm{d}t \leq \int_I \frac{1}{k} \left[\mathscr{L}(t, \zeta(t), \dot\zeta(t)) + C_k \right] \mathrm{d}t \leq \frac{1}{k}(c + \delta\, C_k) < \varepsilon. \tag{1.8}$$

In other words, the space $\mathscr{U}'(c) := \{\dot\zeta \,|\, \zeta \in \mathscr{U}(c)\}$ is uniformly integrable. This implies that the space $\mathscr{U}(c)$ is absolutely equicontinuous, since if we take the Borel set I to be a finite union of intervals $[r_1, r_1'] \cup \cdots \cup [r_N, r_N']$, we have

$$\sum_{i=1}^N \mathrm{dist}(\zeta(r_i), \zeta(r_i')) \leq \sum_{i=1}^N \int_{r_i}^{r_i'} |\dot\zeta(t)|_{\zeta(t)} \mathrm{d}t = \int_I |\dot\zeta(t)|_{\zeta(t)} \mathrm{d}t < \varepsilon.$$

By the Arzelà-Ascoli theorem, the equicontinuity of $\mathscr{U}(c)$ implies that each sequence $\{\gamma_n \,|\, n \in \mathbb{N}\} \subseteq \mathscr{U}(c)$ admits a uniformly converging subsequence, and since we also have absolute equicontinuity all the limit points γ are actually absolutely continuous curves. Hence, in order to conclude the proof of the compactness of $\mathscr{U}(c)$ we only have to show that it is closed in the topology induced by the uniform convergence. Namely, for an arbitrary sequence $\{\gamma_n \,|\, n \in \mathbb{N}\} \subseteq \mathscr{U}(c)$ that converges uniformly to (an absolutely continuous curve) γ we must show that $\mathscr{A}^{t_0,t_1}(\gamma) \leq c$.

Let us fix a finite atlas $\mathfrak{U} = \{(U_\alpha, \phi_\alpha) \,|\, \alpha = 1, \ldots, u\}$ on the closed manifold M, and consider a subdivision $t_0 = s_0 < s_1 < \cdots < s_N = t_1$ such that the support $\gamma([s_i, s_{i+1}])$ is contained in some coordinate domain U_{α_i}, for each $i = 0, \ldots, N-1$. This implies that, for each $n \in \mathbb{N}$ sufficiently large (say, $n \geq \bar n$) and for each $i = 0, \ldots, N-1$, the support $\gamma_n([s_i, s_{i+1}])$ is contained in U_{α_i} as well. In order to conclude the proof, it is enough to show that

$$\mathscr{A}^{s_i,s_{i+1}}(\gamma|_{[s_i,s_{i+1}]}) \leq \liminf_{n\to\infty} \mathscr{A}^{s_i,s_{i+1}}(\gamma_n|_{[s_i,s_{i+1}]}), \qquad \forall i = 0, \ldots, N-1, \tag{1.9}$$

since this would readily give

$$\begin{aligned} \mathscr{A}^{t_0,t_1}(\gamma) &= \sum_{i=0}^{N-1} \mathscr{A}^{s_i,s_{i+1}}(\gamma|_{[s_i,s_{i+1}]}) \leq \sum_{i=0}^{N-1} \liminf_{n\to\infty} \mathscr{A}^{s_i,s_{i+1}}(\gamma_n|_{[s_i,s_{i+1}]}) \\ &\leq \liminf_{n\to\infty} \sum_{i=0}^{N-1} \mathscr{A}^{s_i,s_{i+1}}(\gamma_n|_{[s_i,s_{i+1}]}) = \liminf_{n\to\infty} \mathscr{A}^{t_0,t_1}(\gamma_n) \leq c. \end{aligned}$$

Hence, let us fix $i \in \{0, \ldots, N-1\}$. From now on, we will make the identification $U_{\alpha_i} \equiv \phi_{\alpha_i}(U_{\alpha_i})$ and all the expressions in local coordinates will be understood with respect to the coordinate chart $\phi_{\alpha_i} : U_{\alpha_i} \to \mathbb{R}^m$. Let us consider arbitrary real numbers $\varepsilon > 0$ and $R > 0$, and define

$$I_R := \left\{ t \in [s_i, s_{i+1}] \;\middle|\; |\dot\gamma(t)| \leq R \right\}.$$

Notice that $\mu_{\mathrm{Leb}}(I_R) \to \mu_{\mathrm{Leb}}([s_i, s_{i+1}]) = s_{i+1} - s_i$ as $R \to \infty$, where μ_{Leb} denotes the Lebesgue measure. By Lemma 1.3.2, for each integer $n \geq \bar{n}$ sufficiently large, we have

$$\begin{aligned}\int_{I_R} \big[\mathscr{L}(t, \gamma(t), \dot\gamma(t)) + \langle \partial_v \mathscr{L}(t, \gamma(t), \dot\gamma(t)), \dot\gamma_n(t) - \dot\gamma(t)\rangle - \varepsilon\big]\,\mathrm{d}t \\ \leq \int_{I_R} \mathscr{L}(t, \gamma_n(t), \dot\gamma_n(t))\,\mathrm{d}t \leq \int_{s_i}^{s_{i+1}} \mathscr{L}(t, \gamma_n(t), \dot\gamma_n(t))\,\mathrm{d}t.\end{aligned} \tag{1.10}$$

As we have already showed previously (see the paragraph before (1.8)), the sequence $\{\dot\gamma_n \,|\, n \in \mathbb{N}\}$ is uniformly integrable, which implies that the sequence $\{(\dot\gamma_n - \dot\gamma)|_{[s_i,s_{i+1}]} \,|\, n \geq \bar{n}\}$ is uniformly integrable as well. Applying Lemma 1.3.3 with $I = I_R$, $f(t) = \partial_v \mathscr{L}(t, \gamma(t), \dot\gamma(t))$ for each $t \in I_R$, and $g_n = (\gamma_n - \gamma)|_{[s_i,s_{i+1}]}$ for each integer $n \geq \bar{n}$, we obtain

$$\lim_{n\to\infty} \int_{I_R} \langle \partial_v \mathscr{L}(t, \gamma(t), \dot\gamma(t)), \dot\gamma_n(t) - \dot\gamma(t)\rangle \,\mathrm{d}t = 0.$$

This, together with (1.10), implies

$$\int_{I_R} \mathscr{L}(t, \gamma(t), \dot\gamma(t))\,\mathrm{d}t - \varepsilon\,\mu_{\mathrm{Leb}}(I_R) \leq \liminf_{n\to\infty} \int_{s_i}^{s_{i+1}} \mathscr{L}(t, \gamma_n(t), \dot\gamma_n(t))\,\mathrm{d}t.$$

Finally, taking the limits for $R \to \infty$ and $\varepsilon \to 0$ in this inequality, we obtain the estimate (1.9). □

So far we have proved the existence of action minimizers, which is part of Tonelli's theory of calculus of variations. An older result, that goes back to Weierstrass, states that each sufficiently short minimizer is **unique**, meaning that it is the only curve joining its endpoints that minimizes the action, and moreover it is a smooth solution of the Euler-Lagrange system. In modern language, the precise statement goes as follows.

Theorem 1.3.4 (Weierstrass). *Let $\mathscr{L} : \mathbb{R} \times \mathrm{T}M \to \mathbb{R}$ be a Tonelli Lagrangian. For each $C_0 > 0$ and $T_0 > 0$ there exists $\varepsilon_0 = \varepsilon_0(C_0, T_0) > 0$ such that, for each interval $[t_0, t_1] \subset \mathbb{R}$ and for all $q_0, q_1 \in M$ such that $|t_0| \leq T_0$, $0 < t_1 - t_0 \leq \varepsilon_0$ and $\mathrm{dist}(q_0, q_1) \leq C_0(t_1 - t_0)$, there is a unique action minimizer (with respect to $\mathscr{L}$) $\gamma_{q_0,q_1} : [t_0, t_1] \to M$ with $\gamma_{q_0,q_1}(t_0) = q_0$ and $\gamma_{q_0,q_1}(t_1) = q_1$. Moreover γ_{q_0,q_1} is a (smooth) solution of the Euler-Lagrange system of $\mathscr{L}$.*

Following Mañé [Mn91], the idea of the proof of this theorem consists in reducing the setting to the case of a very special class of Lagrangians, for which the stationary curves are solutions of the Euler-Lagrange system. Then one can conclude by means of the following statement. For an alternative approach based on the Hamilton-Jacobi equation we refer the reader to Fathi [Fat08, Section 3.6].

Lemma 1.3.5. *Let N be a (not necessarily compact) smooth manifold without boundary and $\mathscr{L}_0 : [t_0, t_1] \times \mathrm{T}N \to \mathbb{R}$ a smooth Lagrangian such that*

(i) *$\mathscr{L}_0(t, q, v) > \mathscr{L}_0(t, q, 0)$ for all $(t, q, v) \in [t_0, t_1] \times \mathrm{T}N$ with $v \neq 0$,*
(ii) *all the stationary curves (i.e., those $\gamma : [t_0, t_1] \to N$ such that $\gamma(t) \equiv q$ for some $q \in N$) are solutions of the Euler-Lagrange system of $\mathscr{L}_0$.*

Then, for each point $q \in N$, the unique action minimizer $\gamma : [t_0, t_1] \to N$ such that $\gamma(t_0) = \gamma(t_1) = q$ is the stationary curve at q.

Proof. Assumptions (i) and (ii) imply that, for each $t \in [t_0, t_1]$, the function $q \mapsto \mathscr{L}_0(t, q, 0)$ is constant. Indeed, by (i), in any system of local coordinates we have

$$\frac{\partial \mathscr{L}_0}{\partial v^j}(t, q, 0) = 0, \qquad \forall (t, q) \in [t_0, t_1] \times N,\ j = 1, \ldots, \dim(N),$$

and, by (ii), we further have

$$0 = -\frac{\mathrm{d}}{\mathrm{d}t}\frac{\partial \mathscr{L}_0}{\partial v^j}(t, q, 0) + \frac{\partial \mathscr{L}_0}{\partial q^j}(t, q, 0) = \frac{\partial \mathscr{L}_0}{\partial q^j}(t, q, 0),$$
$$\forall (t, q) \in [t_0, t_1] \times N,\ j = 1, \ldots, \dim(N).$$

Now, let $\gamma : [t_0, t_1] \to N$ be a stationary curve, i.e., $\gamma(t) \equiv q$ for some $q \in N$, and let $\zeta : [t_0, t_1] \to N$ be any non-stationary absolutely continuous curve such that $\zeta(t_0) = \zeta(t_1) = q$. There exists a subset $I \subseteq [t_0, t_1]$ of positive measure such that, for each $t \in I$, the derivative $\dot\zeta(t)$ exists and it is nonzero. This implies that

$$\mathscr{L}_0(t, \zeta(t), \dot\zeta(t)) > \mathscr{L}_0(t, \zeta(t), 0) = \mathscr{L}_0(t, q, 0), \qquad \forall t \in I,$$

and we conclude that

$$\begin{aligned}
\int_{t_0}^{t_1} \mathscr{L}_0(t, \zeta(t), \dot\zeta(t))\, \mathrm{d}t &= \int_{[t_0,t_1]\setminus I} \mathscr{L}_0(t, \zeta(t), \dot\zeta(t))\, \mathrm{d}t + \int_I \mathscr{L}_0(t, \zeta(t), \dot\zeta(t))\, \mathrm{d}t \\
&> \int_{[t_0,t_1]\setminus I} \mathscr{L}_0(t, \zeta(t), \dot\zeta(t))\, \mathrm{d}t + \int_I \mathscr{L}_0(t, q, 0)\, \mathrm{d}t \\
&= \int_{[t_0,t_1]\setminus I} \mathscr{L}_0(t, \zeta(t), 0)\, \mathrm{d}t + \int_I \mathscr{L}_0(t, q, 0)\, \mathrm{d}t \\
&= \int_{[t_0,t_1]\setminus I} \mathscr{L}_0(t, q, 0)\, \mathrm{d}t + \int_I \mathscr{L}_0(t, q, 0)\, \mathrm{d}t \\
&= \int_{t_0}^{t_1} \mathscr{L}_0(t, \gamma(t), \dot\gamma(t))\, \mathrm{d}t.
\end{aligned}$$

□

Before going to the proof of the Weierstrass theorem, we need a preliminary remark about the Euler-Lagrange flow $\Phi_{\mathscr{L}}$ of a Tonelli Lagrangian $\mathscr{L} : \mathbb{R} \times \mathrm{T}M \to \mathbb{R}$ (actually, the same remark holds for any second-order flow on a closed manifold). We denote by $Q_{\mathscr{L}}$ the projection of the Euler-Lagrange flow onto M, i.e., $Q_{\mathscr{L}}^{t_1,t_0} = \tau \circ \Phi_{\mathscr{L}}^{t_1,t_0}$, where $\tau : \mathrm{T}M \to M$ is the projection of the tangent bundle onto its base. For any arbitrary $C_0 > 0$, $T_0 > 0$ and $\delta_0 > 0$, we define a compact set

$$K_0(C_0, T_0, \delta_0) := \big\{(t_0, q_0, t_1, q_1) \in \mathbb{R} \times M \times \mathbb{R} \times M \,\big|\, |t_0| \leq T_0,\ 0 < t_1 - t_0 \leq \delta_0,\ \mathrm{dist}(q_0, q_1) \leq C_0(t_1 - t_0)\big\}.$$

We denote by $\pi_2 : K_0(C_0, T_0, \delta_0) \to M$ the projection

$$\pi_2(t_0, q_0, t_1, q_1) = q_0, \qquad \forall (t_0, q_0, t_1, q_1) \in K_0(C_0, T_0, \delta_0),$$

and we consider the pull-back bundle $\pi_2^* \mathrm{T}M \to K_0(C_0, T_0, \delta_0)$.

Lemma 1.3.6. *For any $C_1 > C_0 > 0$ and $T_0 > 0$ there exist $\delta_0 = \delta_0(C_0, C_1, T_0) > 0$ and a smooth section $w_0 : K_0(C_0, T_0, \delta_0) \to \pi_2^* \mathrm{T}M$ such that*

$$Q_{\mathscr{L}}^{t_1,t_0}(q_0, w_0(t_0, q_0, t_1, q_1)) = q_1, \qquad |w_0(t_0, q_0, t_1, q_1)|_{q_0} \leq C_1, \qquad \forall (t_0, q_0, t_1, q_1) \in K_0(C_0, T_0, \delta_0).$$

Proof. First of all, there exists $\delta' > 0$ such that, for each $(t_0, t_1, q_0, v_0) \in \mathbb{R} \times \mathbb{R} \times \mathrm{T}M$ with $|t_0| \leq T_0$, $|t_1 - t_0| \leq \delta'$ and $|v_0|_{q_0} \leq C_1$, $Q_{\mathscr{L}}^{t_1,t_0}(q_0, v_0)$ is well defined and its distance from q_0 is less than the injectivity radius of M. This immediately implies that there is a unique $\phi^{t_1,t_0}(q_0, v_0) \in \mathrm{T}_{q_0}M$ such that

$$\exp_{q_0}\big((t_1 - t_0)\phi^{t_1,t_0}(q_0, v_0)\big) = Q_{\mathscr{L}}^{t_1,t_0}(q_0, v_0),$$

and clearly $\phi^{t_1,t_0}(q_0, v_0)$ depends smoothly on (t_0, t_1, q_0, v_0). Notice that

$$\phi^{t_0,t_0}(q_0, v_0) = v_0.$$

Now, we claim that there exists $\delta'' \in (0, \delta')$ such that, for each $t_0 \in [-T_0, T_0]$, $t_1 \in [t_0 - \delta'', t_0 + \delta'']$ and $q_0 \in M$, the map

$$\phi^{t_1,t_0}(q_0, \cdot) : \big\{v \in \mathrm{T}_{q_0}M \,\big|\, |v|_{q_0} \leq C_1\big\} \to \mathrm{T}_{q_0}M$$

is a diffeomorphism onto its image. Indeed, by the implicit function theorem we obtain that, for some $\delta''' \in (0, \delta')$ and for each $t_0 \in [-T_0, T_0]$, $t_1 \in [t_0 - \delta''', t_0 + \delta''']$, $\phi^{t_1,t_0}(q_0, \cdot)$ is a local diffeomorphism. Let us assume by contradiction that there exists a sequence $\{(t_n, q_n, v_n, v_n') \,|\, n \in \mathbb{N}\}$ such that $t_n \to t_0$ and, for each $n \in \mathbb{N}$, v_n and v_n' are distinct tangent vectors in $\mathrm{T}_{q_n}M$ such that $|v_n|_{q_n} \leq C_1$, $|v_n'|_{q_n} \leq C_1$ and $\phi^{t_n,t_0}(q_n, v_n) = \phi^{t_n,t_0}(q_n, v_n')$. Since M is compact, up to passing to a subsequence we have that $q_n \to q_0$, $v_n \to v_0$ and $v_n' \to v_0'$. By continuity of $\phi^{\cdot,t_0}$ we have that $v_0 = \phi^{t_0,t_0}(q_0, v_0) = \phi^{t_0,t_0}(q_0, v_0') = v_0'$. Now, consider the sequence

$$\left\{ w_n := \frac{v_n - v_n'}{|v_n - v_n'|_{q_n}} \,\middle|\, n \in \mathbb{N} \right\}.$$

Notice that $|w_n|_{q_n} = 1$ for each $n \in \mathbb{N}$, so that, up to passing to a subsequence, $w_n \to w_0 \in \mathrm{T}_{q_0}M$ with $|w_0|_{q_0} = 1$. However, we have

$$\begin{aligned} 0 &= \frac{\phi^{t_n,t_0}(q_n, v_n) - \phi^{t_n,t_0}(q_n, v_n')}{|v_n - v_n'|_{q_n}} \\ &= \int_0^1 \mathrm{d}(\phi^{t_n,t_0}(q_n,\cdot))(s\, v_n + (1-s)v_n')w_n \,\mathrm{d}s \underset{n\to\infty}{\longrightarrow} \mathrm{d}(\phi^{t_0,t_0}(q_0,\cdot))(v_0)w_0 = w_0, \end{aligned}$$

which gives a contradiction.

Notice that, for each $t_0 \in [-T_0, T_0]$ and $q_0 \in M$, the map $\phi^{t_0,t_0}(q_0,\cdot)$ is the identity on its domain. Therefore, there exists $\delta_0 \in (0, \delta'')$ such that, for each $t_0 \in [-T_0, T_0]$ and $t_1 \in [t_0 - \delta_0, t_0 + \delta_0]$, the image of $\{v \in \mathrm{T}_{q_0}M \mid |v|_{q_0} \le C_1\}$ by the diffeomorphism $\phi^{t_1,t_0}(q_0,\cdot)$ contains $\{v \in \mathrm{T}_{q_0}M \mid |v|_{q_0} \le C_0\}$. Moreover, for this choice of (t_0, t_1, q_0), the map

$$Q_{\mathscr{L}}^{t_1,t_0}(q_0,\cdot) = \exp_{q_0}\big((t_1 - t_0)\phi^{t_1,t_0}(q_0,\cdot)\big) : \{v \in \mathrm{T}_{q_0}M \mid |v|_{q_0} \le C_1\} \to M$$

is a diffeomorphism onto its image and, by the definition of the exponential map, this image contains the Riemannian closed ball

$$\overline{B(q_0, C_0|t_1 - t_0|)} = \{q \in M \mid \mathrm{dist}(q, q_0) \le C_0|t_1 - t_0|\}.$$

Then the existence of a smooth section w_0 as in the statement readily follows. □

Proof of Theorem 1.3.4. Given the constants $C_0 > 0$ and $T_0 > 0$, let us consider two arbitrary points $q_0, q_1 \in M$ and an arbitrary real interval $[t_0, t_1] \subset \mathbb{R}$ such that

$$|t_0| \le T_0, \qquad \mathrm{dist}(q_0, q_1) \le C_0 \underbrace{(t_1 - t_0)}_{=:\varepsilon}.$$

All we need to do in order to conclude the proof is to show that there is a unique action minimizer as in the statement provided that ε is sufficiently small.

Let us fix an arbitrary real constant $\mu > 1$. By compactness, the manifold M admits a finite atlas $\mathfrak{U} = \{(U_\alpha, \phi_\alpha) \mid \alpha = 1, \dots, u\}$ and a finite open cover $\mathfrak{V} = \{V_\alpha \mid \alpha = 1, \dots, u\}$ such that, for all $\alpha \in \{1, \dots, u\}$, we have

$$\overline{V}_\alpha \subset U_\alpha, \tag{1.11}$$

and moreover, for each $q, q' \in U_\alpha$ and $v \in \mathrm{T}_qM$, we have

$$\begin{aligned} \mu^{-1} |\phi_\alpha(q) - \phi_\alpha(q')| \le \mathrm{dist}(q, q') \le \mu\, |\phi_\alpha(q) - \phi_\alpha(q')|, \\ \mu^{-1} |\mathrm{d}\phi_\alpha(q)v| \le |v|_q \le \mu\, |\mathrm{d}\phi_\alpha(q)v|. \end{aligned}$$

Notice that here we have denoted by $|\cdot|$ the standard Euclidean norm in $\mathbb{R}^m$ (and by $|\cdot|_q$ the Riemannian norm in T_qM as usual). Without loss of generality, we can further assume that $\phi_\alpha(V_\alpha)$ is a convex subset of $\mathbb{R}^m$.

Let us consider a real $C_1 > C_0$. For each $\alpha \in 1, \ldots, u$, we introduce the compact set

$$K_\alpha = K_\alpha(T_0, C_1) = \left\{(t, q, v, q') \in \mathbb{R} \times \mathrm{T}\overline{V}_\alpha \times \overline{V}_\alpha \,\middle|\, |t| \leq T_0,\ |v|_q \leq C_1\right\}.$$

Let $\pi_4 : K_\alpha \to \overline{V}_\alpha$ be the projection given by $\pi_4(t, q, v, q') = q'$ for all $(t, q, v, q') \in K_\alpha$. By the Tonelli assumptions (see Proposition 1.2.1), there is a unique smooth section

$$w_\alpha : K_\alpha \to \pi_4^* \mathrm{T}\overline{V}_\alpha \tag{1.12}$$

such that, in the local coordinates given by $\mathrm{T}\phi_\alpha = (q_\alpha^1, \ldots, q_\alpha^m, v_\alpha^1, \ldots, v_\alpha^m)$, we have

$$\begin{gathered}\frac{\partial \mathscr{L}}{\partial v_\alpha^j}(t, q', w_\alpha(t, q, v, q')) = \frac{\partial \mathscr{L}}{\partial v_\alpha^j}(t, q, v), \\ \forall (t, q, v, q') \in K_\alpha,\ j = 1, \ldots, m.\end{gathered} \tag{1.13}$$

Notice that $w_\alpha(t, q, v, q) = v$ for each $(t, q, v, q) \in K_\alpha$ and, since K_α is compact, we obtain a real constant

$$C_2 := \max\left\{|w_\alpha(t, q, v, q')|_{q'} \,\middle|\, (t, q, v, q') \in K_\alpha\right\} \geq C_1.$$

Let us introduce the compact subset

$$K = K(C_2) = \left\{(q, v) \in \mathrm{T}M \,\middle|\, |v|_q \leq C_2\right\}.$$

There exists $\delta = \delta(C_2, T_0) > 0$ such that, for all $t, t' \in \mathbb{R}$ with $|t| \leq T_0$ and $|t' - t| \leq \delta$, we have a well-defined map $\Phi_{\mathscr{L}}^{t',t} : K \to \mathrm{T}M$, and moreover

$$Q_{\mathscr{L}}^{t',t}(q, v) \in U_\alpha, \qquad \forall (q, v) \in \mathrm{T}V_\alpha \cap K. \tag{1.14}$$

Now, let $\mathrm{Leb}(\mathfrak{V})$ denote the Lebesgue number[3] of the open cover $\mathfrak{V}$. By definition of Lebesgue number, the Riemannian closed ball

$$\overline{B(q_0, \mathrm{Leb}(\mathfrak{V})/2)} = \{q \in M \mid \mathrm{dist}(q_0, q) \leq \mathrm{Leb}(\mathfrak{V})/2\}$$

is contained in a coordinate open set V_α for some $\alpha \in \{1, \ldots, u\}$. Therefore, if we require that $\varepsilon \leq \mathrm{Leb}(\mathfrak{V})/(2C_0)$, the points q_0 and q_1 lie in the same open set V_α, for

$$\mathrm{dist}(q_0, q_1) \leq C_0 \varepsilon \leq \mathrm{Leb}(\mathfrak{V})/2.$$

Let $r : [t_0, t_1] \to V_\alpha$ be the segment from q_0 to q_1 given by

$$r(t) = \phi_\alpha^{-1}\left(\frac{t_1 - t}{\varepsilon}\, \phi_\alpha(q_0) + \frac{t - t_0}{\varepsilon}\, \phi_\alpha(q_1)\right), \qquad \forall t \in [t_0, t_1].$$

[3]We recall that, for every open cover $\mathfrak{V}$ of a compact metric space, there exists a positive number $\mathrm{Leb}(\mathfrak{V}) > 0$, the **Lebesgue number** of $\mathfrak{V}$, such that every subset of the metric space of diameter less than $\mathrm{Leb}(\mathfrak{V})$ is contained in some member of the cover $\mathfrak{V}$.

The derivative of r is bounded from above by $\mu^2 C_0$, for

$$\begin{aligned}|\dot r(t)|_{r(t)} &= \left| \mathrm{d}\phi_\alpha^{-1}(\phi_\alpha(r(t)))\frac{\phi_\alpha(q_1)-\phi_\alpha(q_0)}{\varepsilon}\right|_{r(t)}\\ &\leq \mu\left|\frac{\phi_\alpha(q_1)-\phi_\alpha(q_0)}{\varepsilon}\right| \leq \mu^2\frac{\mathrm{dist}(q_0,q_1)}{\varepsilon}\\ &\leq \mu^2 C_0.\end{aligned}$$

On the compact set

$$K' = K'(T_0,C_0,\mu) := \left\{(t,q,v)\in\mathbb{R}\times \mathrm{T}M \,\middle|\, |t|\leq T_0+1,\ |v|_q\leq \mu^2C_0\right\}$$

the Lagrangian $\mathscr{L}$ is bounded from above by some constant $\bar b = \bar b(T_0,C_0,\mu)>0$, i.e., $\mathscr{L}(t,q,v)\leq \bar b$ for all $(t,q,v)\in K'$. Hence, if we require ε to be less than or equal to 1, we have

$$\mathscr{A}^{t_0,t_1}(r) = \int_{t_0}^{t_1}\mathscr{L}(t,r(t),\dot r(t))\,\mathrm{d}t \leq \bar b\,(t_1-t_0) = \bar b\,\varepsilon.$$

By the uniform fiberwise superlinearity of $\mathscr{L}$ (see Remark 1.2.1), there exists a real constant $C = C(\mathscr{L},T_0)>0$ such that

$$\mathscr{L}(t,q,v)\geq |v|_q - C, \qquad \forall (t,q,v)\in[-T_0-1,T_0+1]\times \mathrm{T}M.$$

By (1.11), we get a positive real number

$$R := \min_{\beta=1,\dots,u}\ \inf\left\{\mathrm{dist}(q,q')\,|\, q\in\partial V_\beta,\ q'\in\partial U_\beta\right\}>0.$$

Now, for each absolutely continuous curve $\zeta:[t_0,t_1]\to M$ such that $\zeta(t_0)=q_0$, $\zeta(t_1)=q_1$ and $\zeta(t_*)\notin U_\alpha$ for some $t_*\in[t_0,t_1]$, we have

$$\begin{aligned}\mathscr{A}^{t_0,t_1}(\zeta) &= \int_{t_0}^{t_1}\mathscr{L}(t,\zeta(t),\dot\zeta(t))\,\mathrm{d}t \geq \int_{t_0}^{t_1}\left[|\dot\zeta(t)|_{\zeta(t)} - C\right]\,\mathrm{d}t\\ &\geq R - C\,(t_0-t_1) = R - C\,\varepsilon \geq \bar b\,\varepsilon \geq \mathscr{A}^{t_0,t_1}(r)\end{aligned}$$

provided $\varepsilon < R/(\bar b + C)$. This implies that, with this choice of ε, all the action minimizers defined on $[t_0,t_1]$ and joining q_0 and q_1 have support inside the coordinate open set U_α.

Now, let us further impose that ε is smaller than the constants $\delta=\delta(C_2,T_0)$ introduced above and $\delta_0=\delta_0(C_0,C_1,T_0)$ given by Lemma 1.3.6. We introduce the smooth flow $\vartheta^{\cdot,t_0}:[t_0,t_1]\times V_\alpha\to U_\alpha$ given by

$$\vartheta^{t,t_0}(q) = \vartheta^{\cdot,t_0}(t,q) := Q_{\mathscr{L}}^{t,t_0}\big(q, w_\alpha(t_0,q_0,w_0(t_0,q_0,t_1,q_1),q)\big), \qquad \forall (t,q)\in[t_0,t_1]\times V_\alpha.$$

Here, $w_0 : K_0(C_0, T_0, \delta_0) \to \pi_2^*\mathrm{T}M$ is the section given by Lemma 1.3.6, while $w_\alpha : K_\alpha \to \pi_4^*\mathrm{T}\overline{V}_\alpha$ is the section introduced above in (1.12). The fact that $\vartheta^{\cdot,t_0}$ is a well-defined flow of the form $\vartheta^{\cdot,t_0} : [t_0, t_1] \times V_\alpha \to U_\alpha$ is guaranteed by (1.14). We denote by Y the time-dependent vector field that generates this flow, i.e.,

$$Y(t, \vartheta^{t,t_0}(q)) := \frac{\mathrm{d}}{\mathrm{d}t}\vartheta^{t,t_0}(q), \qquad \forall (t,q) \in [t_0, t_1] \times V_\alpha.$$

Notice that all the flow lines of $\vartheta^{\cdot,t_0}$ are solutions of the Euler-Lagrange system of $\mathscr{L}$. We claim that each portion of any of these flow lines is a unique action minimizer with respect to the restricted Lagrangian $\mathscr{L}|_{\mathbb{R}\times TU_\alpha}$. Assuming that this claim holds, consider the curve $\gamma_{q_0,q_1} : [t_0, t_1] \to U_\alpha$ given by

$$\begin{aligned}\gamma_{q_0,q_1}(t) := \vartheta^{t,t_0}(q_0) &= Q^{t,t_0}_{\mathscr{L}}(q, w_\alpha(t_0, q_0, w_0(t_0, q_0, t_1, q_1), q)) \\ &= Q^{t,t_0}_{\mathscr{L}}(q, w_0(t_0, q_0, t_1, q_1)), \qquad \forall t \in [t_0, t_1].\end{aligned}$$

The claim implies that this curve minimizes the action among all the absolutely continuous curves $\zeta : [t_0, t_1] \to U_\alpha$ with $\zeta(t_0) = \gamma_{q_0,q_1}(t_0) = q_0$ and $\zeta(t_1) = \gamma_{q_0,q_1}(t_1) = q_1$. Nevertheless, since all the action minimizers defined on $[t_0, t_1]$ and joining the points q_0 and q_1 have support inside U_α, we readily obtain that γ_{q_0,q_1} minimizes the action among all the absolutely continuous curves $\zeta : [t_0, t_1] \to M$ with $\zeta(t_0) = \gamma_{q_0,q_1}(t_0) = q_0$ and $\zeta(t_1) = \gamma_{q_0,q_1}(t_1) = q_1$, which proves the theorem.

In order to conclude the proof, we only need to establish the claim, i.e., that each portion of any of the flow lines of $\vartheta^{\cdot,t_0}$ is a unique action minimizer with respect to the restricted Lagrangian $\mathscr{L}|_{\mathbb{R}\times \mathrm{T}U_\alpha}$. From now on we will identify U_α with $\phi_\alpha(U_\alpha)$, so that all the expressions in local coordinates on $\mathrm{T}U_\alpha$ will be implicitly understood with respect to the chart $\mathrm{T}\phi_\alpha$. We want to prove that there exists a smooth function $G : W \to \mathbb{R}$ such that

$$\partial_v \mathscr{L}(t, q, Y(t,q)) = \partial_q G(t,q), \qquad \forall (t,q) \in W,$$

where $W := \{(t, \vartheta^{t,t_0}(q)) \mid t \in [t_0, t_1],\ q \in V_\alpha\}$. This is easily verified as follows. For each $(t,q) \in [t_0, t_1] \times V_\alpha$ we have

$$\begin{aligned}&\frac{\mathrm{d}}{\mathrm{d}t}\left[\partial_v \mathscr{L}(t, \vartheta^{t,t_0}(q), Y(t, \vartheta^{t,t_0}(q))) \circ \mathrm{d}\vartheta^{t,t_0}(q)\right] \\ &= \left[\frac{\mathrm{d}}{\mathrm{d}t}\partial_v \mathscr{L}(t, \vartheta^{t,t_0}(q), Y(t, \vartheta^{t,t_0}(q)))\right] \circ \mathrm{d}\vartheta^{t,t_0}(q) \\ &\quad + \partial_v \mathscr{L}(t, \vartheta^{t,t_0}(q), Y(t, \vartheta^{t,t_0}(q))) \circ \left[\frac{\mathrm{d}}{\mathrm{d}t}\mathrm{d}\vartheta^{t,t_0}(q)\right] \\ &= \partial_q \mathscr{L}(t, \vartheta^{t,t_0}(q), Y(t, \vartheta^{t,t_0}(q))) \circ \mathrm{d}\vartheta^{t,t_0}(q) \\ &\quad + \partial_v \mathscr{L}(t, \vartheta^{t,t_0}(q), Y(t, \vartheta^{t,t_0}(q))) \circ \left[\frac{\mathrm{d}}{\mathrm{d}t}\mathrm{d}\vartheta^{t,t_0}(q)\right] \\ &= \partial_q \left[\mathscr{L}(t, \vartheta^{t,t_0}(q), Y(t, \vartheta^{t,t_0}(q)))\right].\end{aligned} \tag{1.15}$$

Now, since $Y(t_0,q) = w_\alpha(t_0,q_0,w_0(t_0,q_0,t_1,q_1),q)$, by equation (1.13) we have

$$\begin{aligned}\partial_v\mathscr{L}(t_0,q,Y(t_0,q)) &= \partial_v\mathscr{L}(t_0,q,w_\alpha(t_0,q_0,w_0(t_0,q_0,t_1,q_1),q)) \\ &= \partial_v\mathscr{L}(t_0,q_0,w_0(t_0,q_0,t_1,q_1)),\end{aligned}$$

which, together with (1.15), gives

$$\begin{aligned}&\partial_v\mathscr{L}(t,\vartheta^{t,t_0}(q),Y(t,\vartheta^{t,t_0}(q)))\circ \mathrm{d}\vartheta^{t,t_0}(q) \\ &= \partial_v\mathscr{L}(t_0,q,Y(t_0,q)) + \partial_q\left[\int_{t_0}^{t}\mathscr{L}(s,\vartheta^{s,t_0}(q),Y(s,\vartheta^{s,t_0}(q)))\,\mathrm{d}s\right] \\ &= \partial_v\mathscr{L}(t_0,q_0,w_0(t_0,q_0,t_1,q_1)) + \partial_q\left[\int_{t_0}^{t}\mathscr{L}(s,\vartheta^{s,t_0}(q),Y(s,\vartheta^{s,t_0}(q)))\,\mathrm{d}s\right] \\ &= \partial_q\underbrace{\left[\langle\partial_v\mathscr{L}(t_0,q_0,w_0(t_0,q_0,t_1,q_1)),q\rangle + \int_{t_0}^{t}\mathscr{L}(s,\vartheta^{s,t_0}(q),Y(s,\vartheta^{s,t_0}(q)))\,\mathrm{d}s\right]}_{=:\,\widetilde{G}(t,q)}.\end{aligned}$$

Then we can build the function G as

$$G(t,q) := \widetilde{G}(t,(\vartheta^{t,t_0})^{-1}(q)), \qquad \forall (t,q)\in[t_0,t_1]\times V_\alpha.$$

We define a smooth Lagrangian $\mathscr{L}_0 : [t_0,t_1]\times \mathrm{T}V_\alpha \to \mathbb{R}$ by

$$\begin{aligned}\mathscr{L}_0(t,q,v) = \mathscr{L}&(t,\vartheta^{t,t_0}(q),\mathrm{d}\vartheta^{t,t_0}(q)v + Y(t,\vartheta^{t,t_0}(q))) \\ &- \partial_q G(t,\vartheta^{t,t_0}(q))\left[\mathrm{d}\vartheta^{t,t_0}(q)v + Y(t,\vartheta^{t,t_0}(q))\right] \\ &- \partial_t G(t,\vartheta^{t,t_0}(q)),\end{aligned}$$

for each $(t,q,v)\in[t_0,t_1]\times \mathrm{T}V_\alpha$. We denote its associated action by $\mathscr{A}_0^{t_0,t_1}$, i.e., if $\zeta_0 : [t_0,t_1]\to V_\alpha$ is an absolutely continuous curve, we have

$$\mathscr{A}_0^{t_0,t_1}(\zeta_0) = \int_{t_0}^{t_1}\mathscr{L}_0(t,\zeta_0(t),\dot\zeta_0(t))\,\mathrm{d}t.$$

For each curve ζ_0 as above there is a corresponding absolutely continuous curve $\zeta : [t_0,t_1]\to U_\alpha$ defined by $\zeta(t) := \vartheta^{t,t_0}(\zeta_0(t))$ for all $t\in[t_0,t_1]$. Moreover,

$$\begin{aligned}\mathscr{A}_0^{t_0,t_1}(\zeta_0) &= \int_{t_0}^{t_1}\mathscr{L}_0(t,\zeta_0(t),\dot\zeta_0(t))\,\mathrm{d}t \\ &= \int_{t_0}^{t_1}\left[\mathscr{L}(t,\zeta(t),\dot\zeta(t)) - \frac{\mathrm{d}}{\mathrm{d}t}G(t,\zeta(t))\right]\mathrm{d}t \\ &= \mathscr{A}^{t_0,t_1}(\zeta) + G(t_0,\zeta(t_0)) - G(t_1,\zeta(t_1)),\end{aligned} \tag{1.16}$$

which readily implies that ζ_0 is a unique action minimizer with respect to the Lagrangian $\mathscr{L}_0$ if and only if ζ is a unique action minimizer with respect to the

Lagrangian $\mathscr{L}|_{\mathbb{R}\times \mathrm{T}U_\alpha}$. Notice that ζ is a flow line of $\vartheta^{\cdot,t_0}$ if and only if ζ_0 is a stationary curve. Therefore, in order to prove that the flow lines of $\vartheta^{\cdot,t_0}$ are unique action minimizers, we only have to show that $\mathscr{L}_0$ fulfills the hypotheses (i) and (ii) of Lemma 1.3.5. Hypothesis (ii) readily follows from the fact that, by equation (1.16), γ_0 is a solution of the Euler-Lagrange system of $\mathscr{L}_0$ if and only if γ, given by $\gamma(t) = \vartheta^{t,t_0}(\gamma_0(t))$ for each $t \in [t_0, t_1]$, is a solution of the Euler-Lagrange system of $\mathscr{L}$. Therefore if and only if γ_0 is stationary. As for hypothesis (i), it follows from the fact that the fiberwise Hessian of $\mathscr{L}_0$ is positive-definite, since

$$\begin{aligned}\partial_{vv}\mathscr{L}_0(t,q,v)[w,w] \\ &= \partial_{vv}\mathscr{L}\big(t,\vartheta^{t,t_0}(q),\mathrm{d}\vartheta^{t,t_0}(q)v + Y(t,\vartheta^{t,t_0}(q))\big)[\mathrm{d}\vartheta^{t,t_0}(q)w, \mathrm{d}\vartheta^{t,t_0}(q)w] \\ &> 0\end{aligned}$$

for all $(t,q,v) \in [t_0,t_1] \times \mathrm{T}V_\alpha$ and $w \in \mathrm{T}_qV_\alpha$ with $w \neq 0$, together with the fact that

$$\begin{aligned}\partial_v\mathscr{L}_0(t,q,0) \\ &= \partial_v\mathscr{L}\big(t,\vartheta^{t,t_0}(q), Y(t,\vartheta^{t,t_0}(q))\big)\circ \mathrm{d}\vartheta^{t,t_0}(q) - \partial_q G(t,\vartheta^{t,t_0}(q))\circ \mathrm{d}\vartheta^{t,t_0}(q) \\ &= 0\end{aligned}$$

for all $(t,q) \in [t_0,t_1] \times V_\alpha$. □

Unlike the Weierstrass theorem, the Tonelli theorem (Theorem 1.3.1) does not imply that the action minimizers, whose existence is guaranteed there, are smooth solutions of the Euler-Lagrange system of $\mathscr{L}$. Indeed, Ball and Mizel [BM85] provided several examples of Tonelli Lagrangian systems possessing action minimizers that are not C^1. However, this lack of regularity cannot occur if we deal with Tonelli Lagrangians with global Euler-Lagrange flow.

Theorem 1.3.7 (Regularity of minimizers). *Let $\mathscr{L} : \mathbb{R}\times \mathrm{T}M \to \mathbb{R}$ be a Tonelli Lagrangian with global Euler-Lagrange flow. Then every action minimizer (with respect to $\mathscr{L}$) is a smooth solution of the Euler-Lagrange system of $\mathscr{L}$.*

Proof. Let $\gamma : [t_0,t_1] \to M$ be an action minimizer with respect to $\mathscr{L}$. Since γ is absolutely continuous, its derivative $\dot\gamma(t)$ exists for almost every $t \in [t_0,t_1]$. Namely, it exists for every $t \in I$, where I is a subset of full Lebesgue measure in $[t_0,t_1]$. Let us consider an arbitrary $s_* \in I$ and let us fix a real quantity $C_0 > |\dot\gamma(s_*)|_{\gamma(s_*)}$. Notice that, for some $\varepsilon_1 > 0$, we must have

$$\mathrm{dist}(\gamma(s_0),\gamma(s_1)) \leq C_0(s_1 - s_0), \qquad \forall s_0 \in [s_* - \varepsilon_1, s_*],\ s_1 \in [s_*, s_* + \varepsilon_1].$$

By the Weierstrass theorem (Theorem 1.3.4), there exists $\varepsilon_0 = \varepsilon_0(C_0,t_1) > 0$ such that, for each $s_0 \in [s_* - \varepsilon_1, s_*)$ and $s_1 \in (s_*, s_* + \varepsilon_1]$ with $|s_1 - s_0| \leq \varepsilon_0$, there is a unique action minimizer $\gamma_{q_0,q_1} : [s_0,s_1] \to M$ such that $\gamma_{q_0,q_1}(s_0) = q_0 := \gamma(s_0)$

and $\gamma_{q_0,q_1}(s_1) = q_1 := \gamma(s_1)$. But since $\gamma|_{[s_0,s_1]}$ is an action minimizer, we must have $\gamma|_{[s_0,s_1]} = \gamma_{q_0,q_1}$. This proves that I is a full measure open subset of $[t_0, t_1]$ such that, for each interval $J \subseteq I$, the curve $\gamma|_J$ is a solution of the Euler-Lagrange system of $\mathscr{L}$.

Now, let J be a connected component of $I \subseteq [t_0, t_1]$, and let $r_* := \sup J$. We assume by contradiction that $r_* \neq t_1$. Since the Euler-Lagrange flow of $\mathscr{L}$ is global, there exists a solution $\zeta : \mathbb{R} \to M$ of the Euler-Lagrange system of $\mathscr{L}$ such that $\zeta|_J = \gamma|_J$, which implies that

$$\lim_{r\uparrow r_*} |\dot\gamma(r)|_{\gamma(r)} = |\dot\zeta(r_*)|_{\zeta(r_*)} < \infty.$$

Fix a real quantity $C_0 > |\dot\zeta(r_*)|_{\zeta(r_*)}$ and consider $\varepsilon_0 = \varepsilon_0(C_0, t_1) > 0$ given by the Weierstrass theorem (Theorem 1.3.4). We can choose $r_0 \in (r_* - \frac{\varepsilon_0}{2}, r_*)$ sufficiently close to r_* such that

$$\operatorname{dist}(\gamma(r_0), \gamma(r_*)) < C_0(r_* - r_0).$$

By the continuity of γ, we can further choose $r_1 \in (r_*, r_* + \frac{\varepsilon_0}{2})$ sufficiently close to r_* such that

$$\operatorname{dist}(\gamma(r_0), \gamma(r_1)) < C_0(r_1 - r_0).$$

By the Weierstrass theorem, there exists a unique action minimizer $\zeta : [r_0, r_1] \to M$ such that $\zeta(r_0) = \gamma(r_0)$ and $\zeta(r_1) = \gamma(r_1)$. But since $\gamma|_{[r_0,r_1]}$ is an action minimizer, we must have $\gamma|_{[r_0,r_1]} = \zeta$ and therefore $[r_0, r_1] \subset J$, contradicting the definition of r_*. This proves that $\sup J = \sup I = t_1$. By the same proof, we have $\inf J = \inf I = t_0$. □

The hypothesis of global flow is always fulfilled when the Tonelli Lagrangian under consideration is autonomous, as a consequence of the **conservation of energy** (i.e., the invariance of the Legendre-dual Hamiltonian along its Hamiltonian flow lines). In fact, let $\mathscr{L} : \mathrm{T}M \to \mathbb{R}$ be an autonomous Tonelli Lagrangian with Legendre-dual Hamiltonian $\mathscr{H} : \mathrm{T}^*M \to \mathbb{R}$. Then

$$\begin{aligned}\frac{\mathrm{d}}{\mathrm{d}t}\mathscr{H}(\Phi_{\mathscr{H}}^{t,0}(q,p)) &= \mathrm{d}\mathscr{H}(\Phi_{\mathscr{H}}^{t,0}(q,p))\, X_{\mathscr{H}}(\Phi_{\mathscr{H}}^{t,0}(q,p)) \\ &= \omega\big(X_{\mathscr{H}}(\Phi_{\mathscr{H}}^{t,0}(q,p)), X_{\mathscr{H}}(\Phi_{\mathscr{H}}^{t,0}(q,p))\big) \\ &= 0\end{aligned} \tag{1.17}$$

for each $(t, q, p) \in \mathbb{R} \times \mathrm{T}^*M$ such that $\Phi_{\mathscr{H}}^{t,0}(q,p)$ is defined. Let $\Gamma : (t_0, t_1) \to \mathrm{T}^*M$, for some $t_0 < 0$ and $t_1 > 0$, be a maximal solution of the Hamilton system of $\mathscr{H}$ with $c := \mathscr{H}(\Gamma(0))$. Since $\mathscr{H}$ is a Tonelli Hamiltonian, the fiberwise superlinearity assumption **(T2')** implies that $\mathscr{H}$ is **coercive**, i.e.,

$$\lim_{|p|_q\to\infty} \mathscr{H}(q,p) = \infty, \qquad \forall q \in M,$$

and in particular the hypersurface $\mathscr{H}^{-1}(c)$ is compact. By (1.17) we have that

$$\mathscr{H}(\Gamma(t)) = \mathscr{H}(\Phi^{t,0}_{\mathscr{H}}(\Gamma(0))) = \mathscr{H}(\Gamma(0)) = c, \qquad \forall t \in (t_0, t_1),$$

namely Γ lies on the compact hypersurface $\mathscr{H}^{-1}(c)$ and since it is a maximal solution of the Hamilton system we must have $(t_0, t_1) = \mathbb{R}$. This proves that the Hamiltonian flow of $\mathscr{H}$ (and equivalently the Euler-Lagrange flow of $\mathscr{L}$) is global.

More generally, the following statement holds.

Proposition 1.3.8. *Let $\mathscr{L} : \mathbb{R} \times \mathrm{T}M \to \mathbb{R}$ be a Tonelli Lagrangian such that, for some $C > 0$ and $D > 0$, we have*

$$\frac{\partial \mathscr{L}}{\partial t}(t,q,v) \geq C\left(\mathscr{L}(t,q,v) - \partial_v \mathscr{L}(t,q,v)v\right) - D, \qquad \forall (t,q,v) \in \mathbb{R} \times \mathrm{T}M. \tag{1.18}$$

Then the Euler-Lagrange flow of $\mathscr{L}$ is global.

Proof. By the properties of Legendre-dual functions (see Proposition 1.2.2), for each $(t,q,p) \in \mathbb{R} \times \mathrm{T}^*M$ we have

$$\begin{aligned}
\frac{\partial \mathscr{H}}{\partial t}(t,q,p) &= \frac{\partial}{\partial t}\Big[p\left(\partial_p \mathscr{H}(t,q,p)\right) - \mathscr{L}(t,q,\partial_p \mathscr{H}(t,q,p))\Big] \\
&= p\left(\frac{\partial}{\partial t}\partial_p \mathscr{H}(t,q,p)\right) - \frac{\partial \mathscr{L}}{\partial t}(t,q,\partial_p \mathscr{H}(t,q,p)) \\
&\quad - \underbrace{\partial_v \mathscr{L}(t,q,\partial_p \mathscr{H}(t,q,p))}_{=p} \frac{\partial}{\partial t}\partial_p \mathscr{H}(t,q,p) \\
&= -\frac{\partial \mathscr{L}}{\partial t}(t,q,\partial_p \mathscr{H}(t,q,p))
\end{aligned}$$

and

$$\begin{aligned}
\mathscr{H}(t,q,p) &= p\left(\partial_p \mathscr{H}(t,q,p)\right) - \mathscr{L}(t,q,\partial_p \mathscr{H}(t,q,p)) \\
&= \partial_v \mathscr{L}(t,q,\partial_p \mathscr{H}(t,q,p))\, \partial_p \mathscr{H}(t,q,p) - \mathscr{L}(t,q,\partial_p \mathscr{H}(t,q,p)).
\end{aligned}$$

Hence, the inequality (1.18) is equivalent to

$$\frac{\partial}{\partial t}\mathscr{H}(t,q,p) \leq C\,\mathscr{H}(t,q,p) + D, \qquad \forall (t,q,p) \in \mathbb{R} \times \mathrm{T}^*M. \tag{1.19}$$

Let us fix $(t_0, q_0, p_0) \in \mathbb{R} \times \mathrm{T}^*M$ and consider a smooth curve $\Gamma : (t', t'') \to \mathrm{T}^*M$, for some $t' < t_0$ and $t'' > t_0$, which is a maximal solution of the Hamilton system of $\mathscr{H}$ with initial condition $\Gamma(t_0) = (q_0, p_0)$. By (1.19), for each $t \in (t', t'')$ we have

$$\begin{aligned}
\frac{\mathrm{d}}{\mathrm{d}t}\mathscr{H}(t,\Gamma(t)) &= \frac{\partial \mathscr{H}}{\partial t}(t,\Gamma(t)) + \mathrm{d}\mathscr{H}(t,\Gamma(t)) X_{\mathscr{H}}(t,\Gamma(t)) \\
&= \frac{\partial \mathscr{H}}{\partial t}(t,\Gamma(t)) + \underbrace{\omega(X_{\mathscr{H}}(t,\Gamma(t)), X_{\mathscr{H}}(t,\Gamma(t)))}_{=0} \\
&\leq C\,\mathscr{H}(t,\Gamma(t)) + D,
\end{aligned}$$

and therefore, by the Gronwall lemma (see for instance [Ver96, Theorem 1.3]), we have

$$\mathscr{H}(t,\Gamma(t)) \leq \left(\mathscr{H}(t_0,\Gamma(t_0)) + \frac{D}{C}\right) e^{C\,|t-t_0|} - \frac{D}{C}.$$

Now, let us assume by contradiction that $t'' < +\infty$. By the coercivity of $\mathscr{H}$, the curve $\Gamma|_{[t_0,t'')}$ lies inside the compact set

$$K_{t''} = \{(q,p) \in \mathrm{T}^*M \mid \mathscr{H}(t,q,p) \leq k_{t''} \quad \forall t \in [t_0,t'']\},$$

where

$$k_{t''} := \left(\mathscr{H}(t_0,\Gamma(t_0)) + \frac{D}{C}\right) e^{C\,|t''-t_0|} - \frac{D}{C} \in \mathbb{R},$$

but this contradicts the maximality of t''. Hence $t'' = +\infty$. Analogously, we must have $t' = -\infty$. □

Chapter 2

The Morse Indices in Lagrangian Dynamics

In this chapter we introduce the Morse index and nullity for periodic orbits of Euler-Lagrange systems of Tonelli type, an orbit being regarded as an extremal point of the action functional. These indices were first introduced by Morse [Mor96] in the study of closed geodesics. As we will see in the forthcoming chapters, they play a crucial role in the proof of existence and multiplicity results of periodic orbits. In Section 2.1 we give the definition of Morse index and nullity of a periodic orbit, and we prove that they are always finite. In Section 2.2 we outline the beautiful iteration theory of Bott, which studies the behavior of the Morse indices as a periodic orbit is iterated. Finally, in Section 2.3 we describe the relation between the Morse index and the Maslov index from symplectic geometry, which is an index for periodic orbits of general Hamiltonian systems.

2.1 The Morse index and nullity

Let us consider a closed manifold M of dimension $m \geq 1$ with a fixed Riemannian metric $\langle \cdot, \cdot \rangle_\cdot$, and 1-periodic Tonelli Lagrangian $\mathscr{L} : \mathbb{R}/\mathbb{Z} \times \mathrm{T}M \to \mathbb{R}$. We denote by $\mathscr{A}$ the associated action defined on the space of 1-periodic curves $\gamma : \mathbb{R}/\mathbb{Z} \to M$ of class C^2, given by

$$\mathscr{A}(\gamma) = \int_0^1 \mathscr{L}(t, \gamma(t), \dot{\gamma}(t))\,\mathrm{d}t, \qquad \forall \gamma \in C^2(\mathbb{R}/\mathbb{Z}; M).$$

The extremal points of this functional are precisely the 1-periodic solutions of the Euler-Lagrange system of $\mathscr{L}$ (cf. Section 1.1). We want to compute the second variation of the functional $\mathscr{A}$ at an extremal γ. Hence, let us consider a smooth section σ of the pull-back bundle $\gamma^*\mathrm{T}M$. Notice that σ is 1-periodic, being a map of the form $\sigma : \mathbb{R}/\mathbb{Z} \to \gamma^*\mathrm{T}M$. We can use this section to define a homotopy $\Sigma : (-\varepsilon, \varepsilon) \times \mathbb{R}/\mathbb{Z} \to M$ of γ as

$$\Sigma(s,t) := \exp_{\gamma(t)}(s\,\sigma(t)), \qquad \forall (s,t) \in (-\varepsilon,\varepsilon) \times \mathbb{R}/\mathbb{Z},$$

where $\varepsilon > 0$ is a sufficiently small real constant and exp is the exponential map associated to a Riemannian metric on M. The section σ can be reobtained by differentiating the homotopy Σ in the s direction at $s = 0$, for

$$\left.\frac{\partial}{\partial s}\right|_{s=0} \Sigma(s,t) = \mathrm{d}\exp_{\gamma(t)}(0)(\sigma(t)) = \sigma(t) \qquad \forall t \in \mathbb{R}/\mathbb{Z}.$$

If we consider a finite atlas $\mathfrak{U} = \{\phi_\alpha : U_\alpha \to \mathbb{R}^m \mid \alpha = 0, \dots, u\}$ for M, as in section 1.1, we can find a subdivision $0 = r_0 < r_1 < \cdots < r_n = 1$ such that the support $\gamma([r_k, r_{k+1}])$ is contained in some coordinate domain U_{α_k}, for each $k = 0, \dots, n-1$. We define

$$\begin{aligned}
\mathscr{B}_\gamma(\sigma) &:= \left.\frac{\mathrm{d}^2}{\mathrm{d}s^2}\right|_{s=0} \mathscr{A}(\Sigma(s,t)) \\
&= \left.\frac{\mathrm{d}}{\mathrm{d}s}\right|_{s=0} \sum_{k=0}^{n-1}\sum_{j=1}^{m} \int_{r_k}^{r_{k+1}} \left[\frac{\partial \mathscr{L}}{\partial v^j_{\alpha_k}} \left(t, \Sigma(s,t), \frac{\partial \Sigma}{\partial t}(s,t)\right) \frac{\partial^2 \Sigma^j_{\alpha_k}}{\partial s\,\partial t}(s,t) \right. \\
&\quad \left. + \frac{\partial \mathscr{L}}{\partial q^j_{\alpha_k}} \left(t, \Sigma(s,t), \frac{\partial \Sigma}{\partial t}(s,t)\right) \frac{\partial \Sigma^j_{\alpha_k}}{\partial s}(s,t) \right] \mathrm{d}t \\
&= \sum_{k=0}^{n-1}\sum_{j,h=1}^{m} \int_{r_k}^{r_{k+1}} \left[\frac{\partial^2 \mathscr{L}}{\partial v^h_{\alpha_k} \partial v^j_{\alpha_k}} (t,\gamma,\dot\gamma) \dot\sigma^j_{\alpha_k} \dot\sigma^h_{\alpha_k} \right. \\
&\quad \left. + 2 \frac{\partial^2 \mathscr{L}}{\partial v^h_{\alpha_k} \partial q^j_{\alpha_k}} (t,\gamma,\dot\gamma) \sigma^j_{\alpha_k} \dot\sigma^h_{\alpha_k} + \frac{\partial^2 \mathscr{L}}{\partial q^h_{\alpha_k} \partial q^j_{\alpha_k}} (t,\gamma,\dot\gamma) \sigma^j_{\alpha_k} \sigma^h_{\alpha_k} \right] \mathrm{d}t \in \mathbb{R}.
\end{aligned}$$

Notice that $\mathscr{B}_\gamma(\sigma)$ is independent of the particular choice of the homotopy Σ, and for each $r \in \mathbb{R}$ we have $\mathscr{B}_\gamma(r\sigma) = r^2 \mathscr{B}_\gamma(\sigma)$. This shows that $\mathscr{B}_\gamma$ is a well-defined quadratic form, and by polarization we can define the symmetric bilinear form

$$\begin{aligned}
\mathscr{B}_\gamma(\sigma,\xi) &:= \frac{1}{4}\Big[\mathscr{B}_\gamma(\sigma+\xi) - \mathscr{B}_\gamma(\sigma-\xi)\Big] \qquad (2.1) \\
&= \sum_{k=0}^{n-1}\sum_{j,h=1}^{m} \int_{r_k}^{r_{k+1}} \left[\frac{\partial^2 \mathscr{L}}{\partial v^h_{\alpha_k} \partial v^j_{\alpha_k}} (t,\gamma,\dot\gamma) \dot\sigma^j_{\alpha_k} \dot\xi^h_{\alpha_k} + \frac{\partial^2 \mathscr{L}}{\partial v^h_{\alpha_k} \partial q^j_{\alpha_k}} (t,\gamma,\dot\gamma) \sigma^j_{\alpha_k} \dot\xi^h_{\alpha_k} \right. \\
&\quad \left. + \frac{\partial^2 \mathscr{L}}{\partial q^h_{\alpha_k} \partial v^j_{\alpha_k}} (t,\gamma,\dot\gamma) \dot\sigma^j_{\alpha_k} \xi^h_{\alpha_k} + \frac{\partial^2 \mathscr{L}}{\partial q^h_{\alpha_k} \partial q^j_{\alpha_k}} (t,\gamma,\dot\gamma) \sigma^j_{\alpha_k} \xi^h_{\alpha_k} \right] \mathrm{d}t,
\end{aligned}$$

where σ and ξ are smooth sections of $\gamma^*\mathrm{T}M$. The above expression still makes sense if we only require that σ and ξ have $W^{1,2}$ regularity[1], and actually $\mathscr{B}_\gamma$ extends to a bounded symmetric bilinear form

$$\mathscr{B}_\gamma : W^{1,2}(\gamma^*\mathrm{T}M) \times W^{1,2}(\gamma^*\mathrm{T}M) \to \mathbb{R}.$$

Here we have denoted by $W^{1,2}(\gamma^*\mathrm{T}M)$ the space of $W^{1,2}$ sections of $\gamma^*\mathrm{T}M$. This is a Hilbert space with inner product

$$\langle\!\langle \xi, \zeta \rangle\!\rangle_{W^{1,2}} := \int_0^1 \Big[\langle \xi(t), \zeta(t) \rangle_{\gamma(t)} + \langle \nabla_t \xi, \nabla_t \zeta \rangle_{\gamma(t)} \Big] \,\mathrm{d}t, \qquad \forall \xi, \zeta \in W^{1,2}(\gamma^*\mathrm{T}M),$$

where ∇_t denotes the covariant derivative of the Riemannian manifold $(M, \langle \cdot, \cdot \rangle_\cdot)$.

Before studying the properties of $\mathscr{B}_\gamma$, let us recall some definitions concerning symmetric bilinear forms over a Hilbert space. Let $\boldsymbol{E}$ be a real separable Hilbert space with inner product $\langle\!\langle \cdot, \cdot \rangle\!\rangle_{\boldsymbol{E}}$ and norm $\| \cdot \|_{\boldsymbol{E}}$. We denote by $\mathrm{Bil}(\boldsymbol{E})$ the set of bounded (or, equivalently, continuous) bilinear forms on $\boldsymbol{E}$, which is a Banach space with norm

$$\|\mathscr{B}\|_{\mathrm{Bil}(\boldsymbol{E})} = \max \left\{ \mathscr{B}(\boldsymbol{v}, \boldsymbol{w}) \mid \|\boldsymbol{v}\|_{\boldsymbol{E}} = \|\boldsymbol{w}\|_{\boldsymbol{E}} = 1 \right\}, \qquad \forall \mathscr{B} \in \mathrm{Bil}(\boldsymbol{E}).$$

A bounded symmetric bilinear form $\mathscr{B} : \boldsymbol{E} \times \boldsymbol{E} \to \mathbb{R}$ is called **Fredholm** when the unique bounded self-adjoint linear operator B on $\boldsymbol{E}$, given by

$$\mathscr{B}(\boldsymbol{v}, \boldsymbol{w}) = \langle\!\langle B\boldsymbol{v}, \boldsymbol{w} \rangle\!\rangle_{\boldsymbol{E}}, \qquad \forall \boldsymbol{v}, \boldsymbol{w} \in \boldsymbol{E}, \tag{2.2}$$

is Fredholm (see section A.2). We define the **Morse index** $\mathrm{ind}(\mathscr{B})$ as the supremum of the dimension of the vector subspaces of $\boldsymbol{E}$ on which $\mathscr{B}$ is negative-definite. Analogously, we define the **nullity** $\mathrm{nul}(\mathscr{B})$ as the dimension of the kernel of the associated operator B, and the **large Morse index** $\mathrm{ind}^*(\mathscr{B})$ as $\mathrm{ind}(\mathscr{B}) + \mathrm{nul}(\mathscr{B})$.

Lemma 2.1.1. *The Morse index [resp. the large Morse index] is lower semi-continuous [resp. upper semi-continuous] over the space of Fredholm symmetric bilinear forms on* $\boldsymbol{E}$.

Proof. Let $\mathscr{B}$ be a Fredholm symmetric bilinear form on $\boldsymbol{E}$, with associated operator B as in (2.2). By the spectral theorem, this operator induces an orthogonal splitting $\boldsymbol{E} = \boldsymbol{E}^0 \oplus \boldsymbol{E}^- \oplus \boldsymbol{E}^+$, where $\boldsymbol{E}^0$ is the kernel of B, and $\boldsymbol{E}^-$ [resp. $\boldsymbol{E}^+$] is a subspace of $\boldsymbol{E}$ over which B is negative-definite [resp. positive-definite]. Notice that the dimension of $\boldsymbol{E}^0$, $\boldsymbol{E}^-$ and $\boldsymbol{E}^+$ is respectively $\mathrm{nul}(\mathscr{B})$, $\mathrm{ind}(\mathscr{B})$ and $\mathrm{ind}(-\mathscr{B})$.

Since B is a Fredholm operator, $0 \in \mathbb{R}$ is an isolated point in the spectrum of B. In fact, the subspace $\boldsymbol{E}^0$ is finite-dimensional, so that the quotient $\boldsymbol{E}/\boldsymbol{E}^0$ is still a Hilbert space, and B induces a bounded self-adjoint linear operator $\bar{B}$ on $\boldsymbol{E}/\boldsymbol{E}^0$ that is bijective. By the inverse mapping theorem, $\bar{B}$ is an isomorphism.

[1]See Section 3.1 for the background on $W^{1,2}$ sections of the pull-back bundle $\gamma^*\mathrm{T}M$.

Therefore, since the spectrum of B is the disjoint union of the spectrum of $\bar{B}$ and $\{0\}$, the claim follows. This implies that there exists a constant $c_B > 0$ such that

$$\mathscr{B}(\boldsymbol{v},\boldsymbol{v}) \geq c_B \|\boldsymbol{v}\|_{\boldsymbol{E}}^2, \qquad \forall \boldsymbol{v} \in \boldsymbol{E}^+,$$
$$\mathscr{B}(\boldsymbol{v},\boldsymbol{v}) \leq -c_B \|\boldsymbol{v}\|_{\boldsymbol{E}}^2, \qquad \forall \boldsymbol{v} \in \boldsymbol{E}^-.$$

Now, consider an arbitrary Fredholm symmetric bilinear form $\mathscr{B}'$ such that $\|\mathscr{B}' - \mathscr{B}\|_{\mathrm{Bil}(\boldsymbol{E})} < c_B$. For each $\boldsymbol{v} \in \boldsymbol{E}^- \setminus \{\boldsymbol{0}\}$ and $\boldsymbol{w} \in \boldsymbol{E}^+ \setminus \{\boldsymbol{0}\}$, we have

$$\mathscr{B}'(\boldsymbol{v},\boldsymbol{v}) \leq \mathscr{B}(\boldsymbol{v},\boldsymbol{v}) + \|\mathscr{B}' - \mathscr{B}\|_{\mathrm{Bil}(\boldsymbol{E})} \|\boldsymbol{v}\|_{\boldsymbol{E}}^2 < (-c_B + c_B)\|\boldsymbol{v}\|_{\boldsymbol{E}}^2 = 0,$$
$$\mathscr{B}'(\boldsymbol{w},\boldsymbol{w}) \geq \mathscr{B}(\boldsymbol{w},\boldsymbol{w}) - \|\mathscr{B}' - \mathscr{B}\|_{\mathrm{Bil}(\boldsymbol{E})} \|\boldsymbol{w}\|_{\boldsymbol{E}}^2 > (c_B - c_B)\|\boldsymbol{w}\|_{\boldsymbol{E}}^2 = 0.$$

Therefore, $\mathscr{B}'$ is still negative-definite on $\boldsymbol{E}^-$ and positive-definite on $\boldsymbol{E}^+$. This implies that $\mathrm{ind}(\mathscr{B}') \geq \mathrm{ind}(\mathscr{B})$ and $\mathrm{ind}(-\mathscr{B}') \geq \mathrm{ind}(-\mathscr{B})$, which forces $\mathrm{ind}^*(\mathscr{B}') \leq \mathrm{ind}^*(\mathscr{B})$. □

Lemma 2.1.2. *Consider a bounded symmetric bilinear form* $\mathscr{B} = \mathscr{P} + \mathscr{K}$, *where:*

- $\mathscr{P} = \langle\langle P\cdot,\cdot\rangle\rangle_{\boldsymbol{E}} : \boldsymbol{E} \times \boldsymbol{E} \to \mathbb{R}$ *is a* **semipositive-definite** *symmetric bilinear Fredholm form, i.e.,* $\mathscr{P}(\boldsymbol{v},\boldsymbol{v}) \geq 0$ *for each* $\boldsymbol{v} \in \boldsymbol{E}$,
- $\mathscr{K} = \langle\langle K\cdot,\cdot\rangle\rangle_{\boldsymbol{E}} : \boldsymbol{E} \times \boldsymbol{E} \to \mathbb{R}$ *is a* **compact** *symmetric bilinear form, i.e., its associated self-adjoint operator* K *is compact.*

Then $\mathscr{B}$ *is a Fredholm bilinear form and its Morse index* $\mathrm{ind}(\mathscr{B})$ *is finite.*

Proof. First of all, notice that the self-adjoint bounded operator B associated to $\mathscr{B}$ is given by $P+K$. In particular, B is a compact perturbation of the Fredholm operator P, and therefore it is Fredholm (see, e.g., [Arv02, Section 3.3]). This proves that $\mathscr{B}$ is a Fredholm form. Now, consider the orthogonal splittings of $\boldsymbol{E}$ induced by $\mathscr{B}$, $\mathscr{P}$ and $\mathscr{K}$, i.e.,

$$\boldsymbol{E} = \boldsymbol{E}^0_{\mathscr{B}} \oplus \boldsymbol{E}^+_{\mathscr{B}} \oplus \boldsymbol{E}^-_{\mathscr{B}} = \boldsymbol{E}^0_{\mathscr{P}} \oplus \boldsymbol{E}^+_{\mathscr{P}} = \boldsymbol{E}^0_{\mathscr{K}} \oplus \boldsymbol{E}^+_{\mathscr{K}} \oplus \boldsymbol{E}^-_{\mathscr{K}}.$$

Here, $\boldsymbol{E}^0_{\mathscr{B}}$, $\boldsymbol{E}^0_{\mathscr{P}}$ and $\boldsymbol{E}^0_{\mathscr{K}}$ are the kernels of B, P and K respectively, and

$$\pm\,\mathscr{B}(\boldsymbol{v},\boldsymbol{v}) > 0 \qquad \forall \boldsymbol{v} \in \boldsymbol{E}^\pm_{\mathscr{B}} \setminus \{\boldsymbol{0}\},$$
$$\pm\,\mathscr{K}(\boldsymbol{v},\boldsymbol{v}) > 0 \qquad \forall \boldsymbol{v} \in \boldsymbol{E}^\pm_{\mathscr{K}} \setminus \{\boldsymbol{0}\},$$
$$\mathscr{P}(\boldsymbol{v},\boldsymbol{v}) \geq c_P \|\boldsymbol{v}\|_{\boldsymbol{E}}^2 \qquad \forall \boldsymbol{v} \in \boldsymbol{E}^+_{\mathscr{P}},$$

where $c_P > 0$ is a constant determined by the spectrum of the Fredholm operator P (see the proof of Lemma 2.1.1). We can further express the negative eigenspace $\boldsymbol{E}^-_{\mathscr{B}}$ as the orthogonal direct sum

$$\boldsymbol{E}^-_{\mathscr{B}} = (\boldsymbol{E}^-_{\mathscr{B}} \cap \boldsymbol{E}^0_{\mathscr{P}}) \oplus (\boldsymbol{E}^-_{\mathscr{B}} \cap \boldsymbol{E}^+_{\mathscr{P}}),$$

where the first direct summand $\boldsymbol{E}^-_{\mathscr{B}} \cap \boldsymbol{E}^0_{\mathscr{P}}$ is finite-dimensional (since $\mathscr{P}$ is a Fredholm form). Hence, in order to conclude we only need to show that the second summand $\boldsymbol{E}^-_{\mathscr{B}} \cap \boldsymbol{E}^+_{\mathscr{P}}$ is finite-dimensional as well. This is easily proved as follows. For each $\boldsymbol{v} \in (\boldsymbol{E}^-_{\mathscr{B}} \cap \boldsymbol{E}^+_{\mathscr{P}})$, we have

$$0 \geq \mathscr{B}(\boldsymbol{v},\boldsymbol{v}) = \mathscr{P}(\boldsymbol{v},\boldsymbol{v}) + \mathscr{K}(\boldsymbol{v},\boldsymbol{v}) \geq c_P \,\|\boldsymbol{v}\|^2_{\boldsymbol{E}} + \mathscr{K}(\boldsymbol{v},\boldsymbol{v}),$$

and therefore

$$\boldsymbol{E}^-_{\mathscr{B}} \cap \boldsymbol{E}^+_{\mathscr{P}} \subseteq \boldsymbol{E}^{\leq -c_P}_{\mathscr{K}} := \{\boldsymbol{v} \in \boldsymbol{E} \,|\, \mathscr{K}(\boldsymbol{v},\boldsymbol{v}) \leq -c_P \,\|\boldsymbol{v}\|^2_{\boldsymbol{E}}\}.$$

Notice that $\boldsymbol{E}^{\leq -c_P}_{\mathscr{K}}$ is the direct sum of the eigenspaces of K corresponding to the eigenvalues less than or equal to $-c_P$. Since K is a compact operator, each of these eigenspaces is finite-dimensional. Moreover, by the spectral theorem for compact self-adjoint operators (see, e.g., [Mac09, Section 4.3]), K admits only a finite number of distinct eigenvalues less than or equal to $-c_P$. This proves that $\boldsymbol{E}^{\leq -c_P}_{\mathscr{K}}$ is finite-dimensional, which in turn implies that $\boldsymbol{E}^-_{\mathscr{B}} \cap \boldsymbol{E}^+_{\mathscr{P}}$ is finite-dimensional. □

After these preliminaries, let us go back to consider the form $\mathscr{B}_\gamma$ defined in (2.1). In order to simplify the notation, we denote its Morse index $\mathrm{ind}(\mathscr{B}_\gamma)$ by $\mathrm{ind}(\gamma)$ and its nullity $\mathrm{nul}(\mathscr{B}_\gamma)$ by $\mathrm{nul}(\gamma)$. In the following chapters, we will also write them as $\mathrm{ind}(\mathscr{A},\gamma)$ and $\mathrm{nul}(\mathscr{A},\gamma)$ whenever we need to keep track of the action functional with respect to which γ is an extremal.

Remark 2.1.1. In the next chapter we will see that, for a certain subclass of Tonelli Lagrangians $\mathscr{L}$ and for a suitable functional setting for the Lagrangian action $\mathscr{A}$, the bilinear form $\mathscr{B}_\gamma$ is the Hessian (in the Gâteaux or Fréchet sense, depending on the properties of $\mathscr{L}$) of $\mathscr{A}$ at γ. Hence, in that case, the indices $\mathrm{ind}(\gamma)$ and $\mathrm{nul}(\gamma)$ are respectively the Morse index and the nullity of the functional $\mathscr{A}$ at the critical point γ (see Section A.1). By abuse of terminology, for any Tonelli Lagrangian $\mathscr{L}$, from now on we will always call $\mathrm{ind}(\gamma)$ and $\mathrm{nul}(\gamma)$ the Morse index and nullity of γ. □

Proposition 2.1.3. *The symmetric bilinear form $\mathscr{B}_\gamma$ is Fredholm and the Morse index $\mathrm{ind}(\gamma) = \mathrm{ind}(\mathscr{B}_\gamma)$ is finite.*

Proof. Consider the bounded bilinear forms

$$\begin{aligned}
\mathscr{C} &: W^{1,2}(\gamma^*\mathrm{T}M) \times W^{1,2}(\gamma^*\mathrm{T}M) \to \mathbb{R},\\
\mathscr{K}_0 &: L^2(\gamma^*\mathrm{T}M) \times L^2(\gamma^*\mathrm{T}M) \to \mathbb{R},\\
\mathscr{K}' &: L^2(\gamma^*\mathrm{T}M) \times W^{1,2}(\gamma^*\mathrm{T}M) \to \mathbb{R},\\
\mathscr{K}'' &: W^{1,2}(\gamma^*\mathrm{T}M) \times L^2(\gamma^*\mathrm{T}M) \to \mathbb{R},\\
\mathscr{K}''' &: L^2(\gamma^*\mathrm{T}M) \times L^2(\gamma^*\mathrm{T}M) \to \mathbb{R},
\end{aligned}$$

given by

$$\mathscr{C}(\sigma,\xi) = \sum_{k=0}^{n-1} \sum_{j,h=1}^{m} \int_{r_k}^{r_{k+1}} \frac{\partial^2 \mathscr{L}}{\partial v^h_{\alpha_k}\,\partial v^j_{\alpha_k}}(t,\gamma(t),\dot\gamma(t))\,\dot\sigma^j_{\alpha_k}(t)\,\dot\xi^h_{\alpha_k}(t)\,\mathrm{d}t,$$

$$\mathscr{K}_0(\zeta,\chi) = \int_0^1 \langle\zeta(t),\chi(t)\rangle_{\gamma(t)}\,\mathrm{d}t,$$

$$\mathscr{K}'(\zeta,\sigma) = \sum_{k=0}^{n-1}\sum_{j,h=1}^{m}\int_{r_k}^{r_{k+1}} \frac{\partial^2\mathscr{L}}{\partial v^h_{\alpha_k}\,\partial q^j_{\alpha_k}}(t,\gamma(t),\dot\gamma(t))\,\zeta^j_{\alpha_k}(t)\,\dot\sigma^h_{\alpha_k}(t)\,\mathrm{d}t,$$

$$\mathscr{K}''(\sigma,\zeta) = \sum_{k=0}^{n-1}\sum_{j,h=1}^{m}\int_{r_k}^{r_{k+1}} \frac{\partial^2\mathscr{L}}{\partial q^h_{\alpha_k}\,\partial v^j_{\alpha_k}}(t,\gamma(t),\dot\gamma(t))\,\dot\sigma^j_{\alpha_k}(t)\,\zeta^h_{\alpha_k}(t)\,\mathrm{d}t,$$

$$\mathscr{K}'''(\zeta,\chi) = \sum_{k=0}^{n-1}\sum_{j,h=1}^{m}\int_{r_k}^{r_{k+1}} \frac{\partial^2\mathscr{L}}{\partial q^h_{\alpha_k}\,\partial q^j_{\alpha_k}}(t,\gamma(t),\dot\gamma(t))\,\zeta^j_{\alpha_k}(t)\,\chi^h_{\alpha_k}(t)\,\mathrm{d}t,$$

for each $\sigma,\xi \in W^{1,2}(\gamma^*\mathrm{T}M)$ and $\zeta,\chi \in L^2(\gamma^*\mathrm{T}M)$. By the Rellich-Kondrachov theorem (see [AF03, page 168]), the embedding $W^{1,2}(\gamma^*\mathrm{T}M) \hookrightarrow L^2(\gamma^*\mathrm{T}M)$ is compact. This implies that $\mathscr{K}_0$, $\mathscr{K}'$, $\mathscr{K}''$ and $\mathscr{K}'''$, restricted as bilinear forms

$$W^{1,2}(\gamma^*\mathrm{T}M) \times W^{1,2}(\gamma^*\mathrm{T}M) \to \mathbb{R},$$

are compact (meaning that their associated bounded operator on $W^{1,2}(\gamma^*\mathrm{T}M)$ is compact). The symmetric bilinear form $\mathscr{C} + \mathscr{K}_0$ is Fredholm with

$$\mathrm{ind}(\mathscr{C} + \mathscr{K}_0) = \mathrm{nul}(\mathscr{C} + \mathscr{K}_0) = 0.$$

In fact, it is easy to see that $\mathscr{C} + \mathscr{K}_0$ is a positive-definite bounded symmetric bilinear form. Therefore its associated operator on $W^{1,2}(\gamma^*\mathrm{T}M)$ is an isomorphism, and in particular it is Fredholm. Notice that the forms $\mathscr{K}_0$ and $\mathscr{K}' + \mathscr{K}'' + \mathscr{K}'''$ are both symmetric and compact, which implies that $\mathscr{B}_\gamma$ is a symmetric compact perturbation of a positive-definite symmetric Fredholm form, for

$$\mathscr{B}_\gamma = (\mathscr{C} + \mathscr{K}_0) + (\mathscr{K}' + \mathscr{K}'' + \mathscr{K}''' - \mathscr{K}_0).$$

Applying the abstract Lemma 2.1.2, with $\boldsymbol{E} = W^{1,2}(\gamma^*\mathrm{T}M)$, $\mathscr{P} = \mathscr{C} + \mathscr{K}_0$ and $\mathscr{K} = \mathscr{K}' + \mathscr{K}'' + \mathscr{K}''' - \mathscr{K}_0$, we obtain the claim. □

The Morse indices are often used to distinguish among periodic orbits that have been detected by abstract methods. To this purpose, it is useful to have some a priori estimates for them, such as the one given by the following statement due to Abbondandolo and Figalli [AF07, Lemma 2.2].

Proposition 2.1.4. *Assume that the Euler-Lagrange flow $\Phi_{\mathscr{L}}$ of $\mathscr{L}$ is global (see* Section 1.1*). Then for each $a \in \mathbb{R}$ there exists $j = j(\mathscr{L}, a) \in \mathbb{N}$ such that, for each 1-periodic solution γ of the Euler-Lagrange system of $\mathscr{L}$ with $\mathscr{A}(\gamma) \leq a$, we have* $\mathrm{ind}(\gamma) + \mathrm{nul}(\gamma) \leq j$.

Proof. For each $\zeta \in C^1(\mathbb{R}/\mathbb{Z}; M)$, we can define a bilinear form $\mathscr{B}_\zeta$ as in equation (2.1). We introduce the smooth infinite-dimensional vector bundle

$$\pi : \mathscr{E} \to C^1(\mathbb{R}/\mathbb{Z}; M)$$

such that, for each $\zeta \in C^1(\mathbb{R}/\mathbb{Z}; M)$, the fiber $\pi^{-1}(\zeta)$ is the space of Fredholm symmetric bilinear forms on $W^{1,2}(\zeta^*\mathrm{T}M)$. By the dominated convergence theorem, the map $\zeta \mapsto \mathscr{B}_\zeta$ is a continuous section of this vector bundle. Now, we claim that the set of 1-periodic orbits with action less than or equal to a are compact in the C^1 topology. By the upper semi-continuity of the large Morse index (Lemma 2.1.1), the thesis of the proposition follows. All we need to do in order to conclude the proof is to establish the claim.

Consider γ as in the statement. Since $\mathscr{A}(\gamma) \leq a$, there exists $t_0 \in [0,1]$ such that

$$\mathscr{L}(t_0, \gamma(t_0), \dot\gamma(t_0)) \leq a. \tag{2.3}$$

By the uniform fiberwise superlinearity of $\mathscr{L}$ (see Remark 1.2.1) there exists a real constant $C(\mathscr{L}) > 0$ such that

$$\mathscr{L}(t,q,v) \geq |v|_q - C(\mathscr{L}), \qquad \forall (t,q,v) \in \mathbb{R}/\mathbb{Z} \times \mathrm{T}M. \tag{2.4}$$

Since we are assuming that the Euler-Lagrange flow $\Phi_{\mathscr{L}}$ is global, we can define a compact subset of $\mathrm{T}M$ by

$$K = K(a) = \left\{ \Phi_{\mathscr{L}}^{t,s}(q,v) \in \mathrm{T}M \,\middle|\, t,s \in [0,1],\ (q,v) \in \mathrm{T}M,\ |v|_q \leq a + C(\mathscr{L}) \right\}.$$

By (2.3) and (2.4), $|\dot\gamma(t_0)|_{\gamma(t_0)} \leq a + C(\mathscr{L})$ and therefore, for each $t \in [0,1]$, we have that $(\gamma(t), \dot\gamma(t)) = \Phi_{\mathscr{L}}^{t,t_0}(\gamma(t_0), \dot\gamma(t_0))$ belongs to the compact set K. By the compactness of K and the arbitrariness of the choice of γ, we obtain a uniform C^1 bound for the 1-periodic orbits with action less than or equal to a. This immediately implies also a uniform C^2 bound, since, for each γ as above, the lifted orbit $(\gamma, \dot\gamma)$ is a 1-periodic integral curve of the (smooth and 1-periodic in time) Euler-Lagrange vector field $X_{\mathscr{L}}$, i.e.,

$$\frac{\mathrm{d}}{\mathrm{d}t}(\gamma(t), \dot\gamma(t)) = X_{\mathscr{L}}(t, \gamma(t), \dot\gamma(t)), \qquad \forall t \in \mathbb{R}/\mathbb{Z}.$$

This uniform C^2 bound, together with the Arzelà-Ascoli theorem, implies that the set of 1-periodic orbits with action less than or equal to a is compact in the C^1 topology. □

2.2 Bott's iteration theory

Consider again the 1-periodic solution γ of the Euler-Lagrange system of $\mathscr{L}$. For each $n \in \mathbb{N}$, we denote by $\gamma^{[n]} : \mathbb{R}/n\mathbb{Z} \to M$ its nth **iteration** defined as the composition of γ with the n-fold covering map of the circle $\mathbb{R}/n\mathbb{Z} \to \mathbb{R}/\mathbb{Z}$, i.e.,

$$\gamma^{[n]}(k+t) = \gamma(t), \qquad \forall k \in \{0, \dots, n-1\},\ t \in [0,1].$$

Geometrically, γ and $\gamma^{[n]}$ are the same curve. However, at a formal level they are different objects, and indeed $\gamma^{[n]}$ is an extremal point of the Lagrangian action $\mathscr{A}^{[n]}$ in period n, given by

$$\mathscr{A}^{[n]}(\zeta) = \frac{1}{n}\int_0^n \mathscr{L}(t,\zeta(t),\dot{\zeta}(t))\,\mathrm{d}t.$$

Here, we consider $\mathscr{A}^{[n]}$ as a functional defined on a space of n-periodic curves $\zeta : \mathbb{R}/n\mathbb{Z} \to M$ with suitable regularity (for instance C^2). As we did in the previous section, we can introduce the bilinear form $\mathscr{B}_{\gamma^{[n]}}$ representing the second variation of $\mathscr{A}^{[n]}$ at $\gamma^{[n]}$, the Morse index $\mathrm{ind}(\gamma^{[n]})$ and the nullity $\mathrm{nul}(\gamma^{[n]})$. In the 1950s, Bott [Bot56] studied the behavior of these indices as the period n varies. Here, we wish to outline the part of his beautiful iteration theory that is relevant in Lagrangian dynamics, more specifically to the problem of the multiplicity of periodic orbits with unprescribed period (see Section 6.5)

For each integer $n \in \mathbb{N}$, consider the Hilbert space $\boldsymbol{E}^{[n]} := W^{1,2}((\gamma^{[n]})^*\mathrm{T}M)$ with inner product

$$\langle\!\langle \xi,\zeta\rangle\!\rangle_{\boldsymbol{E}^{[n]}} := \frac{1}{n}\int_0^n \Big[\langle \xi(t),\zeta(t)\rangle_{\gamma(t)} + \langle \nabla_t\xi,\nabla_t\zeta\rangle_{\gamma(t)}\Big]\,\mathrm{d}t, \qquad \forall \xi,\zeta \in \boldsymbol{E}^{[n]},$$

where ∇_t denotes the covariant derivative of the Riemannian manifold $(M,\langle\cdot,\cdot\rangle.)$. Let $B^{[n]}$ be the self-adjoint bounded operator on $\boldsymbol{E}^{[n]}$ associated to the bilinear form $\mathscr{B}_{\gamma^{[n]}}$, i.e.,

$$\mathscr{B}_{\gamma^{[n]}}(\sigma,\xi) = \langle\!\langle B^{[n]}\sigma,\xi\rangle\!\rangle_{\boldsymbol{E}^{[n]}}, \qquad \forall \sigma,\xi \in \boldsymbol{E}^{[n]}.$$

Notice that, for each 1-periodic section $\sigma \in \boldsymbol{E}^{[1]}$, the nth iteration of the curve $B^{[1]}\sigma$ is precisely $B^{[n]}\sigma^{[n]}$, i.e.,

$$(B^{[1]}\sigma)^{[n]} = B^{[n]}\sigma^{[n]}. \tag{2.5}$$

Consider the orthogonal spectral decomposition of $\boldsymbol{E}^{[n]}$ associated to the operator $B^{[n]}$. The Morse index $\mathrm{ind}(\gamma^{[n]})$ and the nullity $\mathrm{nul}(\gamma^{[n]})$ are equal to the dimension of the negative eigenspace and of the kernel of $B^{[n]}$ respectively. Bott's idea was to study the eigenvalue problems associated to $B^{[n]}$ in the complexified setting $\boldsymbol{E}^{[n]} \otimes \mathbb{C}$: since $B^{[n]}$ is a real self-adjoint operator, its eigenvalues are all real and the dimension of its eigenspaces in $\boldsymbol{E}^{[n]}$ is the same as the complex dimension of its corresponding eigenspaces in $\boldsymbol{E}^{[n]} \otimes \mathbb{C}$. As we will see in a moment, in the complexified setting, the Fourier decomposition allows to reduce the analysis in period n to the one in period 1.

Let $S^1 \subset \mathbb{C}$ be the unit circle in the complex plane. For each $z \in S^1$ we define the space of complex sections

$$\boldsymbol{E}_z^{[n]} := \Big\{\sigma \in W^{1,2}(\mathbb{R};\mathrm{T}M\otimes\mathbb{C}) \;\Big|\; \sigma(t) \in \mathrm{T}_{\gamma(t)}M\otimes\mathbb{C},\ \sigma(t+n) = z\sigma(t)\ \forall t \in \mathbb{R}\Big\}.$$

This is a complex Hilbert space with Hermitian inner product

$$\langle\!\langle \xi,\zeta\rangle\!\rangle_{\boldsymbol{E}_z^{[n]}} := \frac{1}{n}\int_0^n \left[\langle \xi(t),\overline{\zeta(t)}\rangle_{\gamma(t)} + \langle \nabla_t\xi,\overline{\nabla_t\zeta}\rangle_{\gamma(t)}\right] \mathrm{d}t, \qquad \forall \xi,\zeta\in \boldsymbol{E}_z^{[n]}.$$

In particular $\boldsymbol{E}_1^{[n]} = \boldsymbol{E}^{[n]}\otimes\mathbb{C}$, and the symmetric bilinear form $\mathscr{B}_{\gamma^{[n]}}$ can be extended as a Hermitian form on $\boldsymbol{E}_1^{[n]}$. By the same expression, for each $z\in S^1$ we can define the Hermitian bilinear form $\mathscr{B}_{\gamma^{[n]}}$ and the associated operator $B^{[n]}$ on $\boldsymbol{E}_z^{[n]}$.

Remark 2.2.1. Notice that $\boldsymbol{E}_z^{[1]}$ is a vector subspace of $\boldsymbol{E}_{z^n}^{[n]}$, hence the restriction of the operator $B^{[n]} : \boldsymbol{E}_{z^n}^{[n]} \to \boldsymbol{E}_{z^n}^{[n]}$ to $\boldsymbol{E}_z^{[1]}$ is precisely $B^{[1]} : \boldsymbol{E}_z^{[1]} \to \boldsymbol{E}_z^{[1]}$ (cf. equation (2.5)). □

For each $\lambda\in\mathbb{R}$ we consider the eigenvalue problem

$$\begin{cases} B^{[n]}\sigma = \lambda\sigma, \\ \sigma\in \boldsymbol{E}_z^{[n]}. \end{cases}$$

We denote by $\mathrm{ind}_{z,\lambda}(\gamma^{[n]})$ the complex dimension of the vector space of solutions of this eigenvalue problem, and we set

$$\begin{aligned} \mathrm{ind}_z(\gamma^{[n]}) &:= \sum_{\lambda<0} \mathrm{ind}_{z,\lambda}(\gamma^{[n]}),\\ \mathrm{nul}_z(\gamma^{[n]}) &:= \mathrm{ind}_{z,0}(\gamma^{[n]}). \end{aligned}$$

As for the standard Morse index, for each $z\in S^1$ and $\lambda\in\mathbb{R}$, the indices $\mathrm{ind}_{z,\lambda}(\gamma^{[n]})$ are finite. Moreover, for each $z\in S^1$, the set of those $\lambda\leq 0$ for which $\mathrm{ind}_{z,\lambda}(\gamma^{[n]})\neq 0$ is finite (in order to verify this, the reader can simply check that the proof of Proposition 2.1.3 goes through in the complexified setting $\boldsymbol{E}_z^{[n]}$). Notice that the standard Morse index and nullity are a particular case of these indices, as

$$\begin{aligned} \mathrm{ind}(\gamma^{[n]}) &= \mathrm{ind}_1(\gamma^{[n]}),\\ \mathrm{nul}(\gamma^{[n]}) &= \mathrm{nul}_1(\gamma^{[n]}). \end{aligned}$$

The advantage of working with these more general indices is that the function $z\mapsto \mathrm{ind}_{z,\lambda}(\gamma)$ completely determines the indices $\mathrm{ind}_{z,\lambda}(\gamma^{[n]})$ in any period n. The precise relation is described in the following statement.

Lemma 2.2.1. *For each $n\in\mathbb{N}$, $z\in S^1$ and $\lambda\leq 0$ we have*

$$\mathrm{ind}_{z,\lambda}(\gamma^{[n]}) = \sum_{w\in\sqrt[n]{z}} \mathrm{ind}_{w,\lambda}(\gamma).$$

Proof. Every section $\sigma \in \boldsymbol{E}_z^{[n]}$ admits a unique Fourier expansion

$$\sigma = \sum_{w \in \sqrt[n]{z}} \sigma_w,$$

where each σ_w belongs to the vector space $\boldsymbol{E}_w^{[1]}$ and can be computed as

$$\sigma_w(t) = \frac{1}{n} \sum_{j=0}^{n-1} w^{1-j}\, \sigma(t+j), \qquad \forall t \in \mathbb{R}.$$

Now, as we already pointed out in Remark 2.2.1, $\boldsymbol{E}_w^{[1]}$ is a vector subspace of $\boldsymbol{E}_z^{[n]}$ and

$$B^{[n]}\sigma = \sum_{w \in \sqrt[n]{z}} B^{[n]}\sigma_w = \sum_{w \in \sqrt[n]{z}} B^{[1]}\sigma_w.$$

Since $B^{[n]}\sigma$ belongs to $\boldsymbol{E}_z^{[n]}$ and each $B^{[1]}\sigma_w$ belongs to $\boldsymbol{E}_w^{[1]}$, the above expression gives the unique Fourier expansion of $B^{[n]}\sigma$. From this, we conclude that σ satisfies $B^{[n]}\sigma = \lambda\sigma$ is and only if each σ_w satisfies $B^{[1]}\sigma_w = \lambda\sigma_w$, and the lemma follows. □

Thanks to this lemma, we can now restrict ourselves to study the indices $\mathrm{ind}_{z,\lambda}(\gamma^{[n]})$ for $n = 1$. To start with, let us investigate the properties of $\mathrm{nul}_z(\gamma) = \mathrm{ind}_{z,0}(\gamma)$. Following Bott's terminology, we call $z \in S^1$ a **Poincaré point** of γ when $\mathrm{nul}_z(\gamma) \neq 0$.

Lemma 2.2.2. *The curve γ has only finitely many Poincaré points, and we have*

$$\sum_{z \in S^1} \mathrm{nul}_z(\gamma) \leq 2m,$$

where $m = \dim(M)$.

Proof. By the definition of $B^{[1]}$ and after an integration by parts in (2.1), we infer that a curve $\sigma \in \boldsymbol{E}_n^{[1]}$ is contained in the kernel of the operator $B^{[1]}$ if and only if it satisfies the linear second-order differential system that can by written in local coordinates as

$$\sum_{h=1}^{m} \Big[a_{hj}(t)\, \ddot\sigma^h(t) + \big(b_{hj}(t) - b_{jh}(t) + \dot a_{hj}(t)\big)\, \dot\sigma^h(t) + \big(\dot b_{hj}(t) - c_{hj}(t)\big)\, \sigma^h(t) \Big] = 0,$$

$$j = 1, \dots, m,$$

where

$$a_{hj}(t) = \frac{\partial^2 \mathscr{L}}{\partial v^h \partial v^j}(t, \gamma(t), \dot\gamma(t)),$$
$$b_{hj}(t) = \frac{\partial^2 \mathscr{L}}{\partial v^h \partial q^j}(t, \gamma(t), \dot\gamma(t)),$$
$$c_{hj}(t) = \frac{\partial^2 \mathscr{L}}{\partial q^h \partial q^j}(t, \gamma(t), \dot\gamma(t)).$$

Since $\mathscr{L}$ is a Tonelli Lagrangian, by condition **(T1)** in Section 1.2 the $m \times m$ matrix $a(t) = [a_{hj}(t)]$ is invertible. Hence we can put the above differential system in normal form as

$$\ddot\sigma^j(t) = \sum_{h,k=1}^{m} -a(t)^{-1}_{kj} \Big[\big(b_{hk}(t) - b_{kh}(t) + \dot a_{hk}(t)\big)\, \dot\sigma^h(t) + \big(\dot b_{hk}(t) - c_{hk}(t)\big)\, \sigma^h(t) \Big],$$
$$j = 1, \ldots, m,$$

Now, consider a curve $\sigma : \mathbb{R} \to \mathrm{T}M$ with $\sigma(t) \in \mathrm{T}_{\gamma(t)}M$ for each $t \in \mathbb{R}$, and assume that σ is a solution of this system. Then $\sigma(t)$ depends linearly on the initial conditions $(\sigma(0), \dot\sigma(0))$. Notice that the initial conditions belong to a vector space that we can identify with $\mathrm{T}_{\gamma(0)}M \oplus \mathrm{T}_{\gamma(0)}M$. Therefore the vector space of solutions of the above system has dimension at most $2m = \dim(\mathrm{T}_{\gamma(0)}M \oplus \mathrm{T}_{\gamma(0)}M)$. This proves the lemma. □

Now, let us discuss the properties of the index $\mathrm{ind}_z(\gamma)$.

Lemma 2.2.3. *Let $z_1, \ldots, z_r \in S^1$ be the Poincaré points of γ. Then the function $z \mapsto \mathrm{ind}_z(\gamma)$ is locally constant on $S^1 \setminus \{z_1, \ldots, z_r\}$ and lower semi-continuous on the whole of S^1. Moreover, the jump of this function at any Poincaré point z_j is bounded in absolute value by $\mathrm{nul}_{z_j}(\gamma)$, i.e.,*

$$\mathrm{ind}_{z_j}(\gamma) \leq \lim_{z \to z_j^{\pm}} \mathrm{ind}_z(\gamma) \leq \mathrm{ind}_{z_j}(\gamma) + \mathrm{nul}_{z_j}(\gamma).$$

Proof. For each $z \in S^1$, the real self-adjoint linear operator $B^{[1]} : \boldsymbol{E}_z^{[1]} \to \boldsymbol{E}_z^{[1]}$ is Fredholm (this can be proved by the same argument as in Proposition 2.1.3). We denote by σ_z its spectrum. By continuity, for each interval $(\alpha, \beta] \subset \mathbb{R}$ such that α and β do not belong to σ_z, there is a neighborhood of z in S^1 such that, for each z' in this neighborhood, α and β do not belong to $\sigma_{z'}$ and moreover

$$\sum_{\lambda \in (\alpha,\beta)} \mathrm{ind}_{z,\lambda}(\gamma) = \sum_{\lambda \in (\alpha,\beta)} \mathrm{ind}_{z',\lambda}(\gamma).$$

Assume that $z \in S^1$ is not a Poincaré point of γ, namely 0 does not belong to σ_z. By the above continuity property, 0 does not belong to $\sigma_{z'}$ for each z' in a sufficiently small neighborhood of z, and therefore

$$\operatorname{ind}_z(\gamma) = \sum_{\lambda<0} \operatorname{ind}_{z,\lambda}(\gamma) = \sum_{\lambda<0} \operatorname{ind}_{z',\lambda}(\gamma) = \operatorname{ind}_{z'}(\gamma).$$

Now, assume that $z \in S^1$ is a Poincaré point of γ. Since $B^{[1]} : \boldsymbol{E}_z^{[1]} \to \boldsymbol{E}_z^{[1]}$ is a Fredholm operator, 0 is an isolated point in the spectrum σ_z (see the second paragraph in the proof of Lemma 2.1.1). Therefore, we can fix a sufficiently small $\varepsilon > 0$ such that $[-\varepsilon, \varepsilon] \cap \sigma_z = \{0\}$. By applying once more the above continuity property, $-\varepsilon$ and ε do not belong to $\sigma_{z'}$ for each z' in a sufficiently small neighborhood of z, and we have

$$\begin{aligned}
\operatorname{ind}_{z'}(\gamma) &= \sum_{\lambda<-\varepsilon} \operatorname{ind}_{z',\lambda}(\gamma) + \sum_{\lambda\in(-\varepsilon,0)} \operatorname{ind}_{z',\lambda}(\gamma) \\
&= \sum_{\lambda<-\varepsilon} \operatorname{ind}_{z,\lambda}(\gamma) + \sum_{\lambda\in(-\varepsilon,0)} \operatorname{ind}_{z',\lambda}(\gamma) \\
&= \operatorname{ind}_z(\gamma) + \sum_{\lambda\in(-\varepsilon,0)} \operatorname{ind}_{z',\lambda}(\gamma).
\end{aligned}$$

This proves that $\operatorname{ind}_z(\gamma) \leq \operatorname{ind}_{z'}(\gamma)$. Finally

$$\begin{aligned}
\operatorname{ind}_{z'}(\gamma) &= \operatorname{ind}_z(\gamma) + \sum_{\lambda\in(-\varepsilon,0)} \operatorname{ind}_{z',\lambda}(\gamma) \\
&\leq \operatorname{ind}_z(\gamma) + \sum_{\lambda\in(-\varepsilon,\varepsilon)} \operatorname{ind}_{z',\lambda}(\gamma) \\
&= \operatorname{ind}_z(\gamma) + \sum_{\lambda\in(-\varepsilon,\varepsilon)} \operatorname{ind}_{z,\lambda}(\gamma) \\
&= \operatorname{ind}_z(\gamma) + \operatorname{nul}_z(\gamma).
\end{aligned}$$

□

Now, we introduce the index $\overline{\operatorname{ind}}(\gamma)$ defined by

$$\overline{\operatorname{ind}}(\gamma) = \frac{1}{2\pi} \int_0^{2\pi} \operatorname{ind}_{e^{i\vartheta}}(\gamma)\, \mathrm{d}\vartheta.$$

We call $\overline{\operatorname{ind}}(\gamma)$ the **mean Morse index** of γ, as it is obtained by averaging the indices $\operatorname{ind}_z(\gamma)$ on S^1. Notice that, by Lemma 2.2.3, $\overline{\operatorname{ind}}(\gamma)$ is always finite (it is a non-negative real number). We conclude the section by showing that the Morse index of $\gamma^{[n]}$ grows linearly in the period n, and the asymptotic slope is exactly given by the mean Morse index. For a stronger result we refer the reader to Liu and Long [LL98, LL00] (see also [Lon02, page 213]).

Proposition 2.2.4. *For each periodic solution γ of the Euler-Lagrange system of a Tonelli Lagrangian $\mathscr{L} : \mathbb{R}/\mathbb{Z} \times \mathrm{T}M \to \mathbb{R}$, the following claims hold.*

(i) *The nullity of* $\gamma^{[n]}$ *is uniformly bounded by* $2m = 2\dim(M)$, *i.e.*,

$$\mathrm{nul}(\gamma^{[n]}) \leq 2m.$$

(ii) *We have* $\overline{\mathrm{ind}}(\gamma) = 0$ *if and only if* $\mathrm{ind}(\gamma^{[n]}) = 0$ *for every* $n \in \mathbb{N}$.

(iii) *The Morse index of* $\gamma^{[n]}$ *verifies the inequality*

$$n\,\overline{\mathrm{ind}}(\gamma^{[n]}) - 2m \leq \mathrm{ind}(\gamma^{[n]}) \leq n\,\overline{\mathrm{ind}}(\gamma^{[n]}) + 2m - \mathrm{nul}(\gamma^{[n]}).$$

In particular the mean Morse index $\overline{\mathrm{ind}}(\gamma)$ *can be computed as*

$$\overline{\mathrm{ind}}(\gamma) = \lim_{n\to\infty} \frac{1}{n}\mathrm{ind}(\gamma^{[n]}).$$

Proof. Point (i) follows from Lemma 2.2.2 together with Lemma 2.2.1. By the definition of mean Morse index, $\overline{\mathrm{ind}}(\gamma) = 0$ if and only if the function $z \mapsto \mathrm{ind}_z(\gamma)$ is zero almost everywhere. By Lemma 2.2.3, this function is locally constant outside the Poincaré points and lower semi-continuous on the whole S^1. Hence $\overline{\mathrm{ind}}(\gamma) = 0$ if and only if $\mathrm{ind}_z(\gamma) = 0$ for every $z \in S^1$. This, together with Lemma 2.2.1, proves point (ii). As for point (iii), let us denote by $z_1, \dots, z_r \in S^1$ the Poincaré points of γ. By Lemma 2.2.1, the Poincaré points of the nth iteration $\gamma^{[n]}$ are precisely $z_1^n, \dots, z_r^n$. By Lemma 2.2.3, for each $w, z \in S^1$ we have

$$\mathrm{ind}_w(\gamma^{[n]}) + \mathrm{nul}_w(\gamma^{[n]}) \leq \mathrm{ind}_z(\gamma^{[n]}) + \sum_{j=1}^{r} \mathrm{nul}_{z_j^n}(\gamma^{[n]}). \tag{2.6}$$

By Lemmas 2.2.1 and 2.2.2 we have

$$\sum_{j=1}^{r} \mathrm{nul}_{z_j^n}(\gamma^{[n]}) = \sum_{j=1}^{r} \mathrm{nul}_{z_j}(\gamma) \leq 2m,$$

and, together with (2.6), we obtain

$$\mathrm{ind}_w(\gamma^{[n]}) + \mathrm{nul}_w(\gamma^{[n]}) \leq \mathrm{ind}_z(\gamma^{[n]}) + 2m. \tag{2.7}$$

Notice that, by Lemma 2.2.2, the mean nullity of γ and of $\gamma^{[n]}$ is zero, i.e.,

$$\frac{1}{2\pi}\int_0^{2\pi} \mathrm{nul}_{e^{i\vartheta}}(\gamma)\,\mathrm{d}\vartheta = \frac{1}{2\pi}\int_0^{2\pi} \mathrm{nul}_{e^{i\vartheta}}(\gamma^{[n]})\,\mathrm{d}\vartheta = 0,$$

whereas, by Lemma 2.2.1,

$$\overline{\mathrm{ind}}(\gamma^{[n]}) := \frac{1}{2\pi}\int_0^{2\pi} \mathrm{ind}_{e^{i\vartheta}}(\gamma^{[n]})\,\mathrm{d}\vartheta = n\,\overline{\mathrm{ind}}(\gamma).$$

Now, by setting $z = 1$ and integrating w on S^1 in (2.7), we get

$$n\,\overline{\mathrm{ind}}(\gamma) \leq \mathrm{ind}(\gamma^{[n]}) + 2m.$$

By setting $w = 1$ and integrating z on S^1 in (2.7), we get

$$\mathrm{ind}(\gamma^{[n]}) + \mathrm{nul}(\gamma^{[n]}) \leq n\,\overline{\mathrm{ind}}(\gamma) + 2m. \qquad \square$$

2.3 A symplectic excursion: the Maslov index

In this section we wish to provide a symplectic interpretation of the Morse index for Tonelli Lagrangian systems: the Morse index of a periodic orbit coincides with the Maslov index of the associated periodic orbit of the Legendre-dual Hamiltonian system (see section 1.1). The Maslov index can actually be defined for periodic orbits of more general Hamiltonian systems, and indeed it can be regarded as a generalization of the Morse index.

We begin by recalling some background from linear symplectic geometry (we refer the reader to [MS98, Chapter 2] for more details). The **standard symplectic structure** on $\mathbb{R}^{2m}$ is given by the skew-symmetric bilinear form $\omega_0 : \mathbb{R}^{2m} \wedge \mathbb{R}^{2m} \to \mathbb{R}$ defined by

$$\omega_0(v, w) = \langle \mathrm{J}_0 v, w \rangle , \qquad \forall v, w \in \mathbb{R}^{2m}.$$

Here, $\langle \cdot, \cdot \rangle$ denotes the Euclidean inner product of $\mathbb{R}^{2m}$, and J_0 denotes the **standard complex structure** on $\mathbb{R}^{2m}$ given in matrix form by

$$\mathrm{J}_0 = \begin{bmatrix} 0 & -\mathrm{I} \\ \mathrm{I} & 0 \end{bmatrix}.$$

The **symplectic group** $\mathrm{Sp}(2m)$ is defined as the subgroup of $\mathrm{GL}(2m)$ given by the matrices A such that $A^*\omega = \omega$, or equivalently

$$\mathrm{Sp}(2m) = \left\{ A \in \mathrm{GL}(2m) \mid A^{\mathrm{T}} \mathrm{J}_0 A = \mathrm{J}_0 \right\},$$

where A^{T} denotes the transpose of A. We denote by $\mathrm{Sp}^0(2m)$ the subset of $\mathrm{Sp}(2m)$ consisting of matrices having 1 as an eigenvalue, and by $\mathrm{Sp}^*(2m)$ the complementary subspace of $\mathrm{Sp}(2m)$, i.e.,

$$\mathrm{Sp}^0(2m) = \{ A \in \mathrm{Sp}(2m) \mid \det(A - \mathrm{I}) = 0 \},$$
$$\mathrm{Sp}^*(2m) = \mathrm{Sp}(2m) \setminus \mathrm{Sp}^0(2m).$$

The space $\mathrm{Sp}^0(2m)$ is a singular hypersurface of $\mathrm{Sp}(2m)$ that separates $\mathrm{Sp}^*(2m)$ in two connected components $\mathrm{Sp}^+(2m)$ and $\mathrm{Sp}^-(2m)$ given by

$$\mathrm{Sp}^{\pm}(2m) = \{ A \in \mathrm{Sp}(2m) \mid \pm \det(A - I) > 0 \}.$$

Every symplectic matrix A admits a unique **polar decomposition**

$$A = \underbrace{(AA^{\mathrm{T}})^{1/2}}_{P} \, \underbrace{(AA^{\mathrm{T}})^{-1/2} A}_{Q},$$

where $P \in \mathrm{Sp}(2m)$ is symmetric and positive-definite, while $Q \in \mathrm{Sp}(2m) \cap \mathrm{O}(2m)$. The map $r(A) = (AA^{\mathrm{T}})^{-1/2} A$ is a retraction

$$r : \mathrm{Sp}(2m) \to \mathrm{Sp}(2m) \cap \mathrm{O}(2m)$$

coming from the deformation retraction $r_t : \mathrm{Sp}(2m) \to \mathrm{Sp}(2m)$ given by

$$r_t(A) = (AA^{\mathrm{T}})^{-t/2}A. \tag{2.8}$$

In particular r is a homotopy equivalence. Now, if we consider $\mathrm{GL}(m,\mathbb{C})$ to be a subgroup of $\mathrm{GL}(2m)$ via the embedding

$$X + iY \longmapsto \begin{bmatrix} X & -Y \\ Y & X \end{bmatrix}, \qquad \forall X + iY \in \mathrm{GL}(m,\mathbb{C}),$$

then $\mathrm{Sp}(2m) \cap \mathrm{O}(2m)$ is identified with the unitary group $\mathrm{U}(m)$. We denote by $\det_{\mathbb{C}}(M) : \mathrm{Sp}(2m) \cap \mathrm{O}(2m) \to S^1 \subset \mathbb{C}$ the complex determinant function, that is

$$\det\nolimits_{\mathbb{C}}\left(\begin{bmatrix} X & -Y \\ Y & X \end{bmatrix}\right) := \det(X + iY).$$

We define the **rotation function** as the composition

$$\rho := \det\nolimits_{\mathbb{C}} \circ r : \mathrm{Sp}(2m) \to S^1.$$

From our discussion above and the properties of the unitary group, it readily follows that this map induces an isomorphism $\pi_1(\rho)$ between the fundamental groups of $\mathrm{Sp}(2m)$ and S^1.

Let $\mathscr{P}$ be the space of continuous paths $\Psi : [0,1] \to \mathrm{Sp}(2m)$ such that $\Psi(0) = \mathrm{I}$. This space is the disjoint union of the subspaces $\mathscr{P}^*$ and $\mathscr{P}^0$ given by those Ψ such that $\Psi(1) \in \mathrm{Sp}^*(2m)$ and $\Psi(1) \in \mathrm{Sp}^0(2m)$ respectively. We fix the $2m \times 2m$ diagonal symplectic matrices

$$W' := \mathrm{diag}(2, 1/2, -1, -1, \dots, -1),$$
$$W'' := -\mathrm{I} = \mathrm{diag}(-1, -1, \dots, -1).$$

Notice that

$$\rho(W') = (-1)^{m-1} = -\rho(W''), \tag{2.9}$$

and therefore W' and W'' belong to different connected components of $\mathrm{Sp}^*(2m)$. Now, consider a path $\Psi \in \mathscr{P}^*$ and fix arbitrarily an auxiliary continuous path $\tilde{\Psi} : [1,2] \to \mathrm{Sp}^*(2m)$ such that $\tilde{\Psi}(1) = \Psi(1)$ and $\tilde{\Psi}(2) \in \{W', W''\}$. We denote by $\Psi \bullet \tilde{\Psi} : [0,2] \to \mathrm{Sp}(2m)$ the concatenation of the paths Ψ and $\tilde{\Psi}$, i.e.,

$$(\Psi \bullet \tilde{\Psi})(t) = \begin{cases} \Psi(t), & t \in [0,1], \\ \tilde{\Psi}(t), & t \in [1,2]. \end{cases}$$

We define the **Maslov index** $\iota(\Psi)$ as

$$\iota(\Psi) := \frac{\vartheta(2) - \vartheta(0)}{\pi}, \tag{2.10}$$

where $\vartheta : [0,2] \to \mathbb{R}$ is a continuous function such that $\rho \circ (\Psi \bullet \tilde{\Psi})(t) = e^{i\vartheta(t)}$. It turns out that this index is well defined (i.e., it does not depend on the choice of the auxiliary path $\tilde{\Psi}$) and, due to (2.9), it is an integer. This notion of Maslov index is due to Conley and Zehnder [CZ84] (indeed, some authors refer to it as the **Conley-Zehnder index**), and it is related to other notions of Maslov index previously defined by Gel'fand and Lidskiĭ [GL58] and Maslov [Mas72]. The original definition of Conley and Zehnder, although equivalent, is different from the one given in (2.10), which is due to Long and Zehnder [LZ90].

The main properties of the Maslov index are summarized in the following statement. We refer the reader to [SZ92, Section 3] for a proof.

Proposition 2.3.1. *The Maslov index satisfies the following properties:*

(Naturality) *For each $\Psi \in \mathscr{P}^*$ and $A \in \mathrm{Sp}(2m)$ we have $\iota(\Psi) = \iota(A^{-1}\Psi A)$.*

(Homotopy) *For any two paths $\Psi, \Psi' \in \mathscr{P}^*$ which are homotopic with fixed end-points we have $\iota(\Psi) = \iota(\Psi')$.* □

Notice that the homotopy property stated above can be rephrased in the following way. Let $\widetilde{\mathrm{Sp}}(2m)$ be the universal cover of the symplectic linear group. Here, we regard an element of $\widetilde{\mathrm{Sp}}(2m)$ covering $A \in \mathrm{Sp}(2m)$ as a homotopy class (with fixed endpoints) of paths $\Psi : [0,1] \to \mathrm{Sp}(2m)$ with $\Psi(0) = \mathrm{I}$ and $\Psi(1) = A$. Then the Maslov index descends to a locally-constant integer-valued function on the space of those $[\Psi] \in \widetilde{\mathrm{Sp}}(2m)$ with $\Psi(1) \in \mathrm{Sp}^*(2m)$. This function does not admit a continuous extension to the whole $\widetilde{\mathrm{Sp}}(2m)$. Following Long [Lon90], we define its extension as the maximal lower semi-continuous one[2]. More precisely, for each path $\Psi \in \mathscr{P}^0$ we define the Maslov index $\iota(\Psi)$ as

$$\iota(\Psi) := \liminf_{\substack{\Psi' \to \Psi \\ \Psi' \in \mathscr{P}^*}} \iota(\Psi'), \tag{2.11}$$

where $\Psi' \to \Psi$ denotes pointwise convergence.

For paths $\Psi \in \mathscr{P}^0$ (which might be regarded as "degenerate" paths), it is useful to consider another index $\nu(\Psi)$ that is equal to the multiplicity of 1 as an eigenvalue of $\Psi(1)$, i.e.,

$$\nu(\Psi) := \dim \ker(\Psi(1) - \mathrm{I}).$$

The homotopy invariance property of the Maslov index in the general case is described by the following statement, which is due to Long. We refer the reader to [Lon90] or [Lon02, page 145] for a proof.

Proposition 2.3.2. *Let $\Psi : [0,1] \times [0,1] \to \mathrm{Sp}(2m)$ be a continuous map such that $\Psi(s,\cdot) \in \mathscr{P}$ and $\nu(\Psi(s,\cdot)) = \nu(\Psi(0,\cdot))$ for each $s \in [0,1]$. Then $\iota(\Psi(s,\cdot)) = \iota(\Psi(0,\cdot))$ for each $s \in [0,1]$.* □

[2]Other extensions of the Maslov index on $\widetilde{\mathrm{Sp}}(2m)$ have been considered in the literature. For instance, in [RS93a], Robbin and Salamon consider the average between the maximal lower semi-continuous extension and the minimal upper semi-continuous extension.

Now, we wish to show how to associate a Maslov index to periodic orbits of Hamiltonian systems. This can be done in very general Hamiltonian systems (for instance for contractible periodic orbits of Hamiltonian systems on any symplectic manifold (W, ω) such that the first Chern class $c_1(\mathrm{T}W)$ vanishes over $\pi_2(W)$, see, e.g., [Sal99]). Here, we restrict to the case of Tonelli Hamiltonian systems on cotangent bundles, and we follow Abbondandolo and Schwarz [AS06, Section 1.2]. Let us fix a 1-periodic Tonelli Hamiltonian $\mathscr{H} : \mathbb{R}/\mathbb{Z} \times \mathrm{T}^*M \to \mathbb{R}$ that is Legendre-dual to a Tonelli Lagrangian $\mathscr{L} : \mathbb{R}/\mathbb{Z} \times \mathrm{T}M \to \mathbb{R}$, and consider a 1-periodic orbit $\Gamma : \mathbb{R}/\mathbb{Z} \to \mathrm{T}^*M$ of the Hamiltonian flow of $\mathscr{H}$. This orbit corresponds to the 1-periodic solution $\gamma = \tau^* \circ \Gamma : \mathbb{R}/\mathbb{Z} \to M$ of the Euler-Lagrange system of $\mathscr{L}$, where $\tau^* : \mathrm{T}^*M \to M$ is the projection onto the base of the cotangent bundle. For simplicity, let us restrict to the case in which the pull-back bundle $\gamma^*\mathrm{T}M$ is trivial. This assumption is always verified if the orbit γ is contractible, or if the manifold M is orientable (in fact, in this latter case, $\gamma^*\mathrm{T}M$ is an oriented vector bundle over the circle and therefore it must be trivial). For the general case, we refer the reader to Weber [Web02].

The Hamiltonian flow of $\mathscr{H}$ defines symplectic transformations of T^*M. In fact, assume that there exists an open set $(t_0, t_1) \times U \subset \mathbb{R} \times \mathrm{T}^*M$ such that $\Phi_{\mathscr{H}}^{t,t'}|_U$ is a well-defined diffeomorphism onto its image for each $t, t' \in (t_0, t_1)$. This diffeomorphism is **symplectic**, meaning

$$(\Phi_{\mathscr{H}}^{t,t'})^*\omega = \omega.$$

In fact, $(\Phi_{\mathscr{H}}^{t',t'})^*\omega = (\mathrm{id}_U)^*\omega = \omega$, and by the Cartan formula

$$\frac{\mathrm{d}}{\mathrm{d}t}(\Phi_{\mathscr{H}}^{t,t'})^*\omega = (\Phi_{\mathscr{H}}^{t,t'})^*\big(\mathrm{d}(\underbrace{X_{\mathscr{H}} \lrcorner \omega}_{=\mathrm{d}\mathscr{H}}) + X_{\mathscr{H}} \lrcorner \underbrace{\mathrm{d}\omega}_{=0}\big) = 0.$$

This implies that the differential $\mathrm{d}\Phi_{\mathscr{H}}^{t,0}(\Gamma(0))$ is a symplectic linear map. Therefore, once we fix a trivialization of $\Gamma^*\mathrm{T}\mathrm{T}^*M$, it defines a path in the symplectic group $\mathrm{Sp}(2m)$ and in particular it has a Maslov index $\iota(\Gamma)$. However, in order to produce an index for periodic orbits we have to make sure that $\iota(\Gamma)$ does not depend on the chosen trivialization, at least if we pick it in a certain family of trivializations which are intrinsically associated to the symplectic manifold T^*M.

We denote by $\mathrm{T}^{\mathrm{ver}}\mathrm{T}^*M$ the vertical subbundle of $\mathrm{T}\mathrm{T}^*M$, i.e., $\mathrm{T}^{\mathrm{ver}}\mathrm{T}^*M = \ker(\mathrm{T}\tau^*)$. It is straightforward to verify that this vector bundle is isomorphic to the pull-back of $\mathrm{T}M$ by the map τ^*. Therefore, the pull-back bundle $\Gamma^*\mathrm{T}^{\mathrm{ver}}\mathrm{T}^*M$ is trivial, being

$$\Gamma^*\mathrm{T}^{\mathrm{ver}}\mathrm{T}^*M \simeq \Gamma^*(\tau^*)^*\mathrm{T}M = \gamma^*\mathrm{T}M.$$

Consider an almost complex structure J on T^*M that is **compatible** with the canonical symplectic structure ω of T^*M. This means precisely that $\omega(\cdot, \mathrm{J}\cdot)$ is a Riemannian metric on T^*M. With respect to this metric, we can fix an orthogonal trivialization

$$\tilde{\phi} : \mathbb{R}/\mathbb{Z} \times \mathbb{R}^m \xrightarrow{\simeq} \Gamma^*\mathrm{T}^{\mathrm{ver}}\mathrm{T}^*M,$$

and then we can extend $\tilde{\phi}$ to a trivialization

$$\phi : \mathbb{R}/\mathbb{Z} \times \mathbb{R}^{2m} \xrightarrow{\simeq} \Gamma^*\mathrm{T}\mathrm{T}^*M,$$

as

$$\phi(t,\cdot) = (-\mathrm{J} \circ \tilde{\phi}(t,\cdot) \circ \mathrm{J}_0) \oplus \tilde{\phi}(t,\cdot), \qquad \forall t \in \mathbb{R}/\mathbb{Z},$$

where J_0 is the standard complex structure on $\mathbb{R}^{2m}$ introduced previously. By construction, the trivialization ϕ sends the vertical Lagrangian subspace

$$\mathbb{V}^m := \{\mathbf{0}\} \times \mathbb{R}^m \subset \mathbb{R}^{2m}$$

diffeomorphically onto the vertical subbundle $\Gamma^*\mathrm{T}^{\mathrm{ver}}\mathrm{T}^*M$, i.e.,

$$\phi|_{\mathbb{R}/\mathbb{Z}\times\mathbb{V}^m} : \mathbb{R}/\mathbb{Z} \times \mathbb{V}^m \xrightarrow{\simeq} \Gamma^*\mathrm{T}^{\mathrm{ver}}\mathrm{T}^*M. \tag{2.12}$$

Let $e_1, \dots, e_m, f_1, \dots, f_m$ be the standard symplectic basis of $\mathbb{R}^{2m}$, which means that $e_1, \dots, e_m$ is an orthonormal basis of $\mathbb{R}^m \times \{0\}$ and $f_1, \dots, f_m$ is an orthonormal basis of $\mathbb{V}^m = \{0\} \times \mathbb{R}^m$ with $f_j = \mathrm{J}_0 e_j$ for each $j \in \{1, \dots, m\}$. For each $t \in \mathbb{R}/\mathbb{Z}$, if we set

$$\begin{aligned} \tilde{f}_j &:= \phi(t, f_j) = \tilde{\phi}(t, f_j), \\ \tilde{e}_j &:= \phi(t, e_j) = -J \circ \tilde{\phi}(t,\cdot) \circ J_0 e_j = -J\tilde{f}_j, \end{aligned}$$

it is straightforward to verify that $\tilde{e}_1, \dots, \tilde{e}_m, \tilde{f}_1, \dots, \tilde{f}_m$ is a symplectic basis of the tangent space $\mathrm{T}_{\Gamma(t)}\mathrm{T}^*M$, which means

$$\begin{aligned} \omega(\tilde{e}_j, \tilde{e}_h) &= \omega(\tilde{f}_j, \tilde{f}_h) = 0, \\ \omega(\tilde{e}_j, \tilde{f}_h) &= \begin{cases} 1 & j = h, \\ 0 & j \neq h. \end{cases} \end{aligned}$$

This shows that the trivialization ϕ is **symplectic**, in the sense that

$$\phi(t,\cdot)^*\omega = \omega_0, \qquad \forall t \in \mathbb{R}/\mathbb{Z}. \tag{2.13}$$

Therefore, the differential of the Hamiltonian flow along Γ defines a continuous path $\Gamma_\phi : [0,1] \to \mathrm{Sp}(2m)$ by

$$\Gamma_\phi(t) := \phi(t,\cdot)^{-1} \circ \mathrm{d}\Phi_{\mathscr{H}}^{t,0}(\Gamma(0)) \circ \phi(0,\cdot), \qquad \forall t \in [0,1].$$

Notice that $\Gamma_\phi(0) = \mathrm{I}$, hence $\Gamma_\phi \in \mathscr{P}$. We wish to define the Maslov index of Γ as the Maslov index of this path. Before doing it, however, we need to verify that this index does not depend on the specific choice of the trivialization ϕ.

We denote by $\mathrm{Sp}(2m, \mathbb{V}^m)$ the subgroup of $\mathrm{Sp}(2m)$ consisting of those matrices that preserve the vertical Lagrangian subspace $\mathbb{V}^m \subset \mathbb{R}^{2m}$, i.e.,

$$\begin{aligned} \mathrm{Sp}(2m, \mathbb{V}^m) &= \{A \in \mathrm{Sp}(2m) \mid A\mathbb{V}^m = \mathbb{V}^m\} \\ &= \left\{ \begin{bmatrix} A_1 & 0 \\ A_2 & A_3 \end{bmatrix} \,\middle|\, A_1^{\mathrm{T}} A_3 = \mathrm{I},\ A_1^{\mathrm{T}} A_2 = A_2^{\mathrm{T}} A_1 \right\}. \end{aligned}$$

Lemma 2.3.3. *Consider the inclusion $j : \mathrm{Sp}(2m, \mathbb{V}^m) \hookrightarrow \mathrm{Sp}(2m)$. Then the induced fundamental group homomorphism $\pi_1(j) : \pi_1(\mathrm{Sp}(2m, \mathbb{V}^m)) \to \pi_1(\mathrm{Sp}(2m))$ is the zero homomorphism.*

Proof. The deformation retraction of equation (2.8) restricts to a deformation retraction of $\mathrm{Sp}(2m, \mathbb{V}^m)$ onto

$$\mathrm{Sp}(2m, \mathbb{V}^m) \cap \mathrm{U}(m) = \left\{ \left[\begin{array}{cc} R & 0 \\ 0 & R \end{array} \right] \,\middle|\, R \in \mathrm{O}(m) \right\}.$$

Therefore, we know that in the following diagram of inclusions

$$\begin{array}{ccc} \mathrm{Sp}(2m, \mathbb{V}^m) & \overset{j}{\hookrightarrow} & \mathrm{Sp}(2m) \\ \sim \big\updownarrow & & \sim \big\updownarrow \\ \mathrm{Sp}(2m, \mathbb{V}^m) \cap \mathrm{U}(m) & \overset{h}{\hookrightarrow} & \mathrm{U}(m) \end{array}$$

the vertical arrows are homotopy equivalences, and in order to conclude we just need to show that h induces the zero homomorphism between fundamental groups. The complex determinant $\det_{\mathbb{C}}$ induces a fundamental group isomorphism

$$\pi_1(\det_{\mathbb{C}}) : \pi_1(U(m)) \xrightarrow{\simeq} \pi_1(S^1) \simeq \mathbb{Z}.$$

Consider an arbitrary $[\vartheta] \in \pi_1(\mathrm{Sp}(2m, \mathbb{V}^m) \cap \mathrm{U}(m))$, i.e.,

$$\vartheta : (\mathbb{R}/\mathbb{Z}, 0) \to (\mathrm{Sp}(2m, \mathbb{V}^m) \cap \mathrm{U}(m), \mathrm{I}).$$

Since $\det_{\mathbb{C}} \circ \vartheta \equiv 1$, we have $\pi_1(\det_{\mathbb{C}}) \circ \pi_1(h)[\vartheta] = [\det_{\mathbb{C}} \circ \vartheta] = 0$, and we conclude $\pi_1(h)[\vartheta] = 0$. □

Lemma 2.3.4. *The Maslov index $\iota(\Gamma_\phi)$ is independent of the trivialization ϕ, provided ϕ satisfies* (2.12) *and* (2.13).

Proof. Let $\psi : \mathbb{R}/\mathbb{Z} \times \mathbb{R}^{2m} \xrightarrow{\simeq} \Gamma^*\mathrm{T}\mathrm{T}^*M$ be another symplectic trivialization that satisfies (2.12). Consider the loop $\vartheta : \mathbb{R}/\mathbb{Z} \to \mathrm{Sp}(2m, \mathbb{V}^m)$ defined by

$$\vartheta(t) = \phi(t, \cdot)^{-1} \circ \psi(t, \cdot), \qquad \forall t \in \mathbb{R}/\mathbb{Z}.$$

Notice that

$$\Gamma_\psi(t) = \vartheta(t)^{-1} \circ \Gamma_\phi(t) \circ \vartheta(0), \qquad \forall t \in \mathbb{R}/\mathbb{Z}.$$

By Lemma 2.3.3 there exists a homotopy $\Theta : [0,1] \times \mathbb{R}/\mathbb{Z} \to \mathrm{Sp}(2m)$ such that $\Theta(0, \cdot) = \vartheta$, $\Theta(s, 0) = \vartheta(0)$ for each $s \in [0,1]$, and $\Theta(1, \cdot) \equiv \vartheta(0)$. Hence we can build a homotopy $\Omega : [0,1] \times [0,1] \to \mathrm{Sp}(2m)$ as

$$\Omega(s,t) = \Theta(s,t)^{-1} \circ \Gamma_\phi(t) \circ \vartheta(0), \qquad \forall (s,t) \in [0,1] \times [0,1],$$

such that $\Omega(0,\cdot) = \Gamma_\psi$, $\Omega(1,\cdot) = \vartheta(0)^{-1} \circ \Gamma_\phi \circ \vartheta(0)$, $\Omega(s,0) = \Gamma_\psi(0)$ and $\Omega(s,1) = \Gamma_\psi(1)$ for each $s \in [0,1]$. Since the homotopy Ω fixes the endpoints of the homotoped path, we have that $\nu(\Omega(s,\cdot)) = \nu(\Gamma_\psi)$ for each $s \in [0,1]$. Hence, by the homotopy invariance of the Maslov index (Proposition 2.3.2) we conclude

$$\iota(\Gamma_\psi) = \iota(\Omega(s,\cdot)), \qquad \forall s \in [0,1],$$

and in particular

$$\iota(\Gamma_\psi) = \iota\left(\vartheta(0)^{-1} \circ \Gamma_\phi \circ \vartheta(0)\right).$$

Finally, by the naturality property of the Maslov index, we conclude

$$\iota(\Gamma_\psi) = \iota\left(\vartheta(0)^{-1} \circ \Gamma_\phi \circ \vartheta(0)\right) = \iota(\Gamma_\phi).$$ □

By the above lemma, we can define the **Maslov index** of the periodic orbit Γ as the integer $\iota(\Gamma) := \iota(\Gamma_\phi)$, where ϕ is any trivialization satisfying (2.12) and (2.13). Similarly we define $\nu(\Gamma) := \nu(\Gamma_\phi)$, that is, the multiplicity of 1 as an eigenvalue of the endomorphism $\mathrm{d}\Phi^{1,0}_{\mathscr{H}}(\Gamma(0))$ of $\mathrm{T}_{\Gamma(0)}\mathrm{T}^*M$. The indices $\iota(\Gamma)$ and $\nu(\Gamma)$ turn out to be equal to the Morse index $\mathrm{ind}(\gamma)$ and the nullity $\mathrm{nul}(\gamma)$, where $\gamma := \tau^* \circ \Gamma$ is the Euler-Lagrange 1-periodic orbit corresponding to Γ. This is an important result first established by Duistermaat [Dui76] in the non-degenerate case $\mathrm{nul}(\gamma) = 0$, and by Viterbo [Vit90] in full generality. Alternative proofs were also given by Long and An [LA98] and Abbondandolo [Abb03]. We refer the reader to these papers or to [Lon02, Section 7.3] for a proof.

Theorem 2.3.5 (Maslov-Morse indices theorem). *Let $\mathscr{H} : \mathbb{R}/\mathbb{Z} \times \mathrm{T}^*M \to \mathbb{R}$ and $\mathscr{L} : \mathbb{R}/\mathbb{Z} \times \mathrm{T}M \to \mathbb{R}$ be Legendre-dual Tonelli Hamiltonians and Lagrangians, and $\Gamma : \mathbb{R}/\mathbb{Z} \to \mathrm{T}^*M$ a Hamiltonian periodic orbit corresponding to the Euler-Lagrange periodic orbit γ. Assume further that $\gamma^*\mathrm{T}M$ is a trivial bundle. Then the Maslov and Morse indices of these orbits are the same, i.e.,*

$$\begin{aligned} \iota(\Gamma) &= \mathrm{ind}(\gamma), \\ \nu(\Gamma) &= \mathrm{nul}(\gamma). \end{aligned}$$ □

An extensive study of the Maslov index has been carried out, starting in the 1990s, by Long and his collaborators, who in particular generalized Bott's iteration theory to the Maslov case and provided many applications to the existence of periodic orbits in Hamiltonian systems more general than the Tonelli ones. We refer the reader to Long's monograph [Lon02] for a detailed account of these developments.

Chapter 3

Functional Setting for the Lagrangian Action

In order to apply the machinery of critical point theory to study the periodic solutions of Tonelli Lagrangian systems on a closed manifold M, one needs to find a nice functional setting for the Lagrangian action: a suitable free loop space on M with a manifold structure, over which the action is regular, say at least C^1, and such that its sublevels satisfy some sort of compactness, such as the Palais-Smale condition. For the special case in which the Lagrangian is a squared Riemannian norm (the "geodesic" case) a suitable free loop space is known to be the Hilbert manifold $W^{1,2}(\mathbb{R}/\mathbb{Z}; M)$. For general Tonelli Lagrangians, a functional setting such that the action functional fulfills the above requirements is not known. However, $W^{1,2}(\mathbb{R}/\mathbb{Z}; M)$ is still a good choice for the subclass of Lagrangians that are fiberwise convex with fiberwise quadratic-growth: the fact that a Lagrangian grows at most quadratically guarantees that the action functional is regular, and the fact that the Lagrangian grows at least quadratically implies the Palais-Smale condition.

In Section 3.1 we recall how to associate Hilbert manifold structures to $W^{1,2}$ path spaces. In Section 3.2 we discuss a few topological properties of the free loop space that will be needed in the forthcoming chapters. In Section 3.3 we introduce the class of convex quadratic-growth Lagrangians. In Section 3.4, we discuss the regularity of the action functional associated to Lagrangians that are fiberwise subquadratic. In Section 3.5 we study the critical points of the action functional associated to convex quadratic-growth Lagrangians. In particular, we prove that, in the $W^{1,2}$ setting, the critical points of the action are always smooth solutions of the Euler-Lagrange system, and we verify that the Palais-Smale condition holds. Finally, in Section 3.6, we define the mean action functional in any integer period.

3.1 Hilbert manifold structures for path spaces

Let M be a smooth closed manifold of dimension $m \geq 1$. A **free path space** of M is, loosely speaking, a set of maps from a compact interval, say $\mathbb{I} := [0,1]$, to the manifold M. It may be the set of all the continuous maps $C^0(\mathbb{I}; M)$, as is common in topology, or the smaller set of all the smooth maps $C^\infty(\mathbb{I}; M)$, as is often considered in differential geometry. Each of these spaces is endowed with the corresponding topology, i.e., the one induced by the uniform convergence for $C^0(\mathbb{I}; M)$ and the one induced by the uniform convergence of all the derivatives for $C^\infty(\mathbb{I}; M)$. In nonlinear analysis it is rather useful to consider an intermediate free path space, such as $W^{1,2}(\mathbb{I}; M)$, that has the structure of an infinite-dimensional Hilbert manifold.

Following Klingenberg [Kli78, Chapter 1], let us give a precise definition of $W^{1,2}(\mathbb{I}; M)$. By the Whitney embedding theorem (see [Lee03, Theorem 10.11]), there exists a proper smooth embedding $M \hookrightarrow \mathbb{R}^{2m+1}$, and we define

$$W^{1,2}(\mathbb{I}; M) := \left\{\gamma \in W^{1,2}(\mathbb{I}; \mathbb{R}^{2m+1}) \mid \gamma(\mathbb{I}) \subset M\right\},$$

where $W^{1,2}(\mathbb{I}; \mathbb{R}^{2m+1})$ is the Sobolev space of absolutely continuous maps $\gamma : \mathbb{I} \to \mathbb{R}^{2m+1}$ whose (weak) derivative is square-integrable. On $W^{1,2}(\mathbb{I}; M)$ we consider the subspace topology induced by $W^{1,2}(\mathbb{I}; \mathbb{R}^{2m+1})$. Now, let us provide M with a Riemannian metric $\langle\cdot,\cdot\rangle_\cdot$, for instance the pull-back of the flat metric on $\mathbb{R}^{2m+1}$ via the embedding $M \hookrightarrow \mathbb{R}^{2m+1}$, with corresponding norm $|\cdot|_\cdot$. For each $\gamma \in W^{1,2}(\mathbb{I}; M)$, we denote by $W^{1,2}(\gamma^*\mathrm{T}M)$ the space of $W^{1,2}$ sections of the pull-back bundle $\gamma^*\mathrm{T}M \to \mathbb{I}$, namely the space of absolutely continuous sections ξ of $\gamma^*\mathrm{T}M \to \mathbb{I}$ that are weakly differentiable and such that

$$\int_0^1 \left[|\xi(t)|^2_{\gamma(t)} + |\nabla_t \xi|^2_{\gamma(t)}\right] \mathrm{d}t < \infty.$$

Here, ∇_t denotes the covariant derivative induced by the Levi-Civita connection of the Riemannian manifold $(M, \langle\cdot,\cdot\rangle_\cdot)$. It turns out that $W^{1,2}(\gamma^*\mathrm{T}M)$ is a separable Hilbert space, with inner product given by

$$\langle\!\langle \xi, \zeta \rangle\!\rangle_\gamma := \int_0^1 \left[\langle \xi(t), \zeta(t)\rangle_{\gamma(t)} + \langle \nabla_t \xi, \nabla_t \zeta\rangle_{\gamma(t)}\right] \mathrm{d}t, \qquad \forall \xi, \zeta \in W^{1,2}(\gamma^*\mathrm{T}M).$$

For each $\varepsilon > 0$ we define

$$U_\varepsilon := \left\{v \in \mathrm{T}M \;\middle|\; |v|_{\tau(v)} < \varepsilon\right\},$$

where $\tau : \mathrm{T}M \to M$ denotes the projection of the tangent bundle onto the base. Let us fix $\gamma \in C^\infty(\mathbb{I}; M)$ and denote by $W^{1,2}(\gamma^* U_\varepsilon)$ the subset of $W^{1,2}(\gamma^*\mathrm{T}M)$ consisting of those sections that take values in U_ε. Notice that, by the Sobolev embedding theorem (see [AF03, page 85]), $W^{1,2}(\gamma^* U_\varepsilon)$ is actually an open subset of $W^{1,2}(\gamma^*\mathrm{T}M)$. If ε is strictly less than the injectivity radius of M, then the exponential map $\exp_q : U_\varepsilon \cap \mathrm{T}_q M \to M$ is a diffeomorphism onto its image and therefore can be used to define an injective map

$$\exp_\gamma : W^{1,2}(\gamma^* U_\varepsilon) \to W^{1,2}(\mathbb{I}; M)$$

by

$$\exp_\gamma(\xi)(t) := \exp_{\gamma(t)}(\xi(t)), \qquad \forall \xi \in W^{1,2}(\gamma^* U_\varepsilon),\ t \in \mathbb{I}.$$

We denote by $\mathscr{U}_\gamma \subset W^{1,2}(\mathbb{I}; M)$ the image of the map $\exp_\gamma$. It is easy to see that $\mathscr{U}_\gamma$ is an open set in $W^{1,2}(\mathbb{I}; M)$, so that $\exp_\gamma$ is a homeomorphism onto it.

Now, notice that each $\gamma \in W^{1,2}(\mathbb{I}; M)$ can be approximated by smooth maps in the topology of the uniform convergence. In fact, consider the Whitney embedding $M \hookrightarrow \mathbb{R}^{2m+1}$ as before and a tubular neighborhood $N \subseteq \mathbb{R}^{2m+1}$ of M with associated retraction $r : N \to M$. Since $C^\infty(\mathbb{I}; \mathbb{R}^{2m+1})$ is dense in $C^0(\mathbb{I}; \mathbb{R}^{2m+1})$, there exists a sequence $\{\gamma_k\} \subset C^\infty(\mathbb{I}; \mathbb{R}^{2m+1})$ that converges uniformly to γ. Up to ignoring a finite number of elements of this sequence, we can assume that each γ_k has support inside N. Hence, the sequence $\{r \circ \gamma_k\} \subset C^\infty(\mathbb{I}; M)$ converges uniformly to γ as well.

This density argument implies that the family $\{\mathscr{U}_\gamma \,|\, \gamma \in C^\infty(\mathbb{I}; M)\}$ is an open cover of the path space $W^{1,2}(\mathbb{I}; M)$. Moreover, for each $\gamma_1, \gamma_2 \in C^\infty(\mathbb{I}; M)$ such that $\mathscr{U}_{\gamma_1} \cap \mathscr{U}_{\gamma_2} \neq \varnothing$, it turns out that the composition

$$\exp_{\gamma_1}^{-1} \circ \exp_{\gamma_2}|_{\exp_{\gamma_2}^{-1}(\mathscr{U}_{\gamma_1} \cap \mathscr{U}_{\gamma_2})} : \exp_{\gamma_2}^{-1}(\mathscr{U}_{\gamma_1} \cap \mathscr{U}_{\gamma_2}) \to \exp_{\gamma_1}^{-1}(\mathscr{U}_{\gamma_1} \cap \mathscr{U}_{\gamma_2})$$

is a diffeomorphism between open subsets of the Hilbert spaces $W^{1,2}(\gamma_2^* \mathrm{T}M)$ and $W^{1,2}(\gamma_1^* \mathrm{T}M)$. Hence we can endow $W^{1,2}(\mathbb{I}; M)$ with a Hilbert manifold structure given by the atlas

$$\left\{ (\mathscr{U}_\gamma, \exp_\gamma^{-1}) \,|\, \gamma \in C^\infty(\mathbb{I}; M) \right\}.$$

The tangent space of $W^{1,2}(\mathbb{I}; M)$ at a point γ is given by

$$\mathrm{T}_\gamma W^{1,2}(\mathbb{I}; M) = W^{1,2}(\gamma^* \mathrm{T}M),$$

and $\langle\!\langle \cdot, \cdot \rangle\!\rangle_\cdot$ is a Hilbert-Riemannian metric on $W^{1,2}(\mathbb{I}; M)$. By means of this metric, the free path space $W^{1,2}(\mathbb{I}; M)$ turns out to be a **complete** Hilbert-Riemannian manifold (see the definition in Section A.3).

Now, following Abbondandolo and Schwarz [AS09], for each smooth closed submanifold $V \subseteq M \times M$, we consider the subspace $W_V^{1,2}(\mathbb{I}; M)$ of $W^{1,2}(\mathbb{I}; M)$ given by the paths whose pair of endpoints lies in V, i.e.,

$$W_V^{1,2}(\mathbb{I}; M) = \left\{ \gamma \in W^{1,2}(\mathbb{I}; M) \,|\, (\gamma(0), \gamma(1)) \in V \right\}.$$

This space turns out to be a smooth submanifold of $W^{1,2}(\mathbb{I}; M)$. In fact, if we consider the smooth submersion $\pi : W^{1,2}(\mathbb{I}; M) \to M \times M$ given by $\pi(\gamma) = (\gamma(0), \gamma(1))$ for each $\gamma \in W^{1,2}(\mathbb{I}; M)$, then $W_V^{1,2}(\mathbb{I}; M)$ is precisely given by the preimage of V by π, i.e.,

$$W_V^{1,2}(\mathbb{I}; M) = \pi^{-1}(V).$$

Let us consider an arbitrary $\gamma \in W^{1,2}_V(\mathbb{I}; M)$, and set

$$\mathbb{V} := \mathrm{T}_{(\gamma(0),\gamma(1))}V.$$

The tangent space of $W^{1,2}_V(\mathbb{I}; M)$ at γ consists of the $W^{1,2}$ sections of $\gamma^*\mathrm{T}M \to \mathbb{I}$ whose pair of endpoints lies in the vector space $\mathbb{V}$, i.e.,

$$\mathrm{T}_\gamma W^{1,2}_V(\mathbb{I}; M) = W^{1,2}_{\mathbb{V}}(\gamma^*\mathrm{T}M) := \left\{\xi \in W^{1,2}(\gamma^*\mathrm{T}M) \mid (\xi(0), \xi(1)) \in \mathbb{V}\right\}.$$

This tangent space is clearly a Hilbert subspace of the tangent space of $W^{1,2}(\mathbb{I}; M)$ at γ. Therefore, the path space $W^{1,2}_V(\mathbb{I}; M)$, together with the metric $\langle\!\langle\cdot,\cdot\rangle\!\rangle$. inherited from $W^{1,2}(\mathbb{I}; M)$, is a complete Hilbert-Riemannian manifold.

If V is a connected 0-dimensional smooth manifold, i.e., $V = \{(q_0, q_1)\}$ for some $q_0, q_1 \in M$, then $W^{1,2}_V(\mathbb{I}; M) = W^{1,2}_{q_0,q_1}(\mathbb{I}; M)$ is precisely the set of $W^{1,2}$ paths joining q_0 and q_1. With this choice of V we have

$$\mathbb{V} = \mathrm{T}_{(\gamma(0),\gamma(1))}V = \{\mathbf{0}\}$$

and therefore

$$\mathrm{T}_\gamma W^{1,2}_{q_0,q_1}(\mathbb{I}; \mathbb{R}^m) = W^{1,2}_0(\gamma^*\mathrm{T}M) := \left\{\xi \in W^{1,2}(\gamma^*\mathrm{T}M) \mid \xi(0) = \xi(1) = \mathbf{0}\right\}.$$

If we rather take V to be the diagonal, i.e., $V = \{(q, q) \mid q \in M\}$, then $W^{1,2}_V(\mathbb{I}; M)$ becomes the **free loop space** of M, namely the space $W^{1,2}(\mathbb{T}; M)$ of $W^{1,2}$ maps from the circle $\mathbb{T} = \mathbb{R}/\mathbb{Z}$ to M. In this case, we have that

$$\mathbb{V} = \mathrm{T}_{(\gamma(0),\gamma(1))}V = \left\{(v, v) \mid v \in \mathrm{T}_{\gamma(0)}M\right\}.$$

Hence, the tangent space of $W^{1,2}(\mathbb{T}; M)$ at γ consists of the $W^{1,2}$ sections of the pull-back bundle $\gamma^*\mathrm{T}M \to \mathbb{I}$ that are 1-periodic, i.e.,

$$\mathrm{T}_\gamma W^{1,2}(\mathbb{T}; M) = \left\{\xi \in W^{1,2}(\gamma^*\mathrm{T}M) \mid \xi(0) = \xi(1)\right\}.$$

Equivalently, the tangent space of $W^{1,2}(\mathbb{T}; M)$ at γ consists in the $W^{1,2}$ sections of the pull-back bundle $\gamma^*\mathrm{T}M \to \mathbb{T}$.

Remark 3.1.1. In order to compute expressions in local coordinates, it is often convenient to work with other charts of the free path space $W^{1,2}(\mathbb{I}; M)$ that are compatible with the ones in the atlas $\left\{(\mathscr{U}_\gamma, \exp_\gamma^{-1}) \mid \gamma \in C^\infty(\mathbb{I}; M)\right\}$ introduced above. These charts are obtained as follows. Let us fix a smooth $\gamma : \mathbb{I} \to M$. Since the interval $\mathbb{I}$ is contractible, the pull-back bundle $\gamma^*\mathrm{T}M \to \mathbb{I}$ admits a smooth trivialization

$$\phi_\gamma : \gamma^*\mathrm{T}M \xrightarrow{\simeq} \mathbb{I} \times \mathbb{R}^m. \tag{3.1}$$

This trivialization induces an isomorphism of Hilbert spaces

$$W^{1,2}(\phi_\gamma) : W^{1,2}(\gamma^*\mathrm{T}M) \xrightarrow{\simeq} W^{1,2}(\mathbb{I}; \mathbb{R}^m)$$

given by

$$W^{1,2}(\phi_\gamma)(\xi)(t) = \pi_2 \circ \phi_\gamma(t, \xi(t)), \qquad \forall \xi \in W^{1,2}(\gamma^*\mathrm{T}M),\ t \in \mathbb{I},$$

where $\pi_2 : \mathbb{I} \times \mathbb{R}^m \to \mathbb{R}^m$ is the projection onto the $\mathbb{R}^m$ factor of $\mathbb{I} \times \mathbb{R}^m$. Therefore, we can provide $W^{1,2}(\mathbb{I}; M)$ with a compatible atlas $\{(\mathscr{U}_\gamma, \Phi_\gamma) \mid \gamma \in C^\infty(\mathbb{I}; M)\}$, whose charts are given by

$$\Phi_\gamma = W^{1,2}(\phi_\gamma) \circ \exp_\gamma^{-1} : \mathscr{U}_\gamma \to W^{1,2}(\mathbb{I}; \mathbb{R}^m). \tag{3.2}$$

This atlas restricts to compatible atlases for the submanifolds $W^{1,2}_{q_0,q_1}(\mathbb{I}; M)$ and $W^{1,2}(\mathbb{T}; M)$. In fact, for each smooth $\gamma : \mathbb{I} \to M$ such that $\gamma(t_0) = q_0$ and $\gamma(t_1) = q_1$, the chart Φ_γ defined in (3.2) restricts to a chart for $W^{1,2}_{q_0,q_1}(\mathbb{I}; M)$ of the form

$$\Phi_\gamma : \mathscr{U}_\gamma \cap W^{1,2}_{q_0,q_1}(\mathbb{I}; M) \to W^{1,2}_0(\mathbb{I}; \mathbb{R}^m),$$

where

$$W^{1,2}_0(\mathbb{I}; \mathbb{R}^m) = \{\zeta \in W^{1,2}(\mathbb{I}; \mathbb{R}^m) \mid \zeta(0) = \zeta(1) = 0\}.$$

As for $W^{1,2}(\mathbb{T}; M)$, for each smooth $\gamma : \mathbb{I} \to M$ with $\gamma(0) = \gamma(1)$ consider the trivialization ϕ_γ as in (3.1) and define the m-dimensional vector subspace $\mathbb{V} \subset \mathbb{R}^m \times \mathbb{R}^m$ by

$$\mathbb{V} = \left\{(\pi_2 \circ \phi_\gamma(0, v), \pi_2 \circ \phi_\gamma(1, v)) \mid v \in \mathrm{T}_{\gamma(0)}M = \mathrm{T}_{\gamma(1)}M\right\}. \tag{3.3}$$

The chart Φ_γ defined in (3.2) restricts to a chart for $W^{1,2}(\mathbb{T}; M)$ of the form

$$\Phi_\gamma : \mathscr{U}_\gamma \cap W^{1,2}(\mathbb{T}; M) \to W^{1,2}_{\mathbb{V}}(\mathbb{I}; \mathbb{R}^m),$$

where

$$W^{1,2}_{\mathbb{V}}(\mathbb{I}; \mathbb{R}^m) = \{\zeta \in W^{1,2}(\mathbb{I}; \mathbb{R}^m) \mid (\zeta(0), \zeta(1)) \in \mathbb{V}\}.$$

Notice that, whenever the bundle $\gamma^*\mathrm{T}M \to \mathbb{T}$ is trivial (e.g., when the manifold M is orientable or if the loop γ is contractible), we can choose ϕ_γ to be a trivialization of the form

$$\phi_\gamma : \gamma^*\mathrm{T}M \xrightarrow{\simeq} \mathbb{T} \times \mathbb{R}^m,$$

so that the vector space $\mathbb{V}$ is simply the diagonal vector subspace of $\mathbb{R}^m \times \mathbb{R}^m$ and therefore $W^{1,2}_{\mathbb{V}}(\mathbb{I}; \mathbb{R}^m) = W^{1,2}(\mathbb{T}; \mathbb{R}^m)$. □

Now, let us consider a nonzero positive integer $n \in \mathbb{N}$ and denote by $\mathbb{T}^{[n]}$ the n-periodic circle $\mathbb{R}/n\mathbb{Z}$. The **n-periodic free loop space** $W^{1,2}(\mathbb{T}^{[n]}; M)$ is diffeomorphic to the 1-periodic one $W^{1,2}(\mathbb{T}; M)$ simply by the map $\gamma \mapsto \tilde\gamma$, where $\tilde\gamma(t) = \gamma(nt)$ for each $t \in \mathbb{I}$. On $W^{1,2}(\mathbb{T}^{[n]}; M)$ we can put the rescaled Hilbert-Riemannian metric given by

$$\langle\!\langle \xi, \zeta \rangle\!\rangle_\gamma := \frac{1}{n} \int_0^n \left[\langle \xi(t), \zeta(t) \rangle_{\gamma(t)} + \langle \nabla_t \xi, \nabla_t \zeta \rangle_{\gamma(t)}\right] \mathrm{d}t,$$
$$\forall \gamma \in W^{1,2}(\mathbb{T}^{[n]}; M),\ \xi, \zeta \in W^{1,2}(\gamma^*\mathrm{T}M).$$

We define the nth **iteration map** $\psi^{[n]} : W^{1,2}(\mathbb{T}; M) \hookrightarrow W^{1,2}(\mathbb{T}^{[n]}; M)$ by

$$\psi^{[n]}(\gamma) = \gamma^{[n]}, \qquad \forall \gamma \in W^{1,2}(\mathbb{T}; M),$$

where, as in Section 2.2, $\gamma^{[n]}$ denotes the composition of γ with the n-fold covering map of the circle $\mathbb{T}^{[n]} \to \mathbb{T}$. We can easily show that the iteration map $\psi^{[n]}$ is smooth. In fact, if we consider two correspondent charts $(\mathscr{U}_\gamma, \exp_\gamma^{-1})$ and $(\mathscr{U}_{\gamma^{[n]}}, \exp_{\gamma^{[n]}}^{-1})$ of $W^{1,2}(\mathbb{T}; M)$ and $W^{1,2}(\mathbb{T}^{[n]}; M)$, the composition

$$\exp_{\gamma^{[n]}}^{-1} \circ \psi^{[n]} \circ \exp_\gamma : \mathscr{U}_\gamma \hookrightarrow \mathscr{U}_{\gamma^{[n]}}$$

is simply the analogous iteration map between the Hilbert spaces $W^{1,2}(\gamma^*\mathrm{T}M)$ and $W^{1,2}(\gamma^{[n]*}\mathrm{T}M)$. This latter map, in turn, is bounded linear and in particular it is smooth. Hence $\psi^{[n]}$ is smooth as well. For each $\gamma \in W^{1,2}(\mathbb{T}; M)$, the differential of the iteration map at γ,

$$\mathrm{d}\psi^{[n]}(\gamma) : W^{1,2}(\gamma^*\mathrm{T}M) \to W^{1,2}(\gamma^{[n]*}\mathrm{T}M),$$

is the iteration map between the above Hilbert spaces, i.e.,

$$\mathrm{d}\psi^{[n]}(\gamma)\xi = \xi^{[n]}, \qquad \forall \xi \in W^{1,2}(\gamma^*\mathrm{T}M).$$

By our choice of the Hilbert-Riemannian metrics on the loop spaces, we have

$$\langle\langle \xi, \xi \rangle\rangle_\gamma = \langle\langle \xi^{[n]}, \xi^{[n]} \rangle\rangle_{\gamma^{[n]}}, \qquad \forall \xi \in W^{1,2}(\gamma^*\mathrm{T}M).$$

This shows that $\psi^{[n]}$ is an isometry, and we can consider $W^{1,2}(\mathbb{T}; M)$ as a submanifold of $W^{1,2}(\mathbb{T}^{[n]}; M)$ with the pulled-back Hilbert-Riemannian metric.

3.2 Topological properties of the free loop space

In terms of homotopy type, the free loop spaces $C^\infty(\mathbb{T}; M)$, $W^{1,2}(\mathbb{T}; M)$ and $C^0(\mathbb{T}; M)$ are indistinguishable. The fact that the inclusion map

$$W^{1,2}(\mathbb{T}; M) \subset C^0(\mathbb{T}; M),$$

which is continuous by the Sobolev embedding theorem, is a homotopy equivalence follows from an abstract theorem of Palais [Pal66, page 5] about dense inclusions of infinite-dimensional vector spaces (see also [Kli78, page 15]). Here, we give a proof that the continuous inclusion

$$C^\infty(\mathbb{T}; M) \subset W^{1,2}(\mathbb{T}; M)$$

is also a homotopy equivalence, by means of a simple convolution argument.

Proposition 3.2.1. *The continuous inclusion $C^\infty(\mathbb{T}; M) \subset W^{1,2}(\mathbb{T}; M)$ is a homotopy equivalence.*

Proof. We will prove the statement by building a continuous homotopy

$$\mathscr{I} : \mathbb{I} \times W^{1,2}(\mathbb{T}; M) \to W^{1,2}(\mathbb{T}; M)$$

such that $\mathscr{I}(0,\cdot)$ is the identity on $W^{1,2}(\mathbb{T}; M)$, and $\mathscr{I}(1,\cdot)$ is a map of the form $\mathscr{I}(1,\cdot) : W^{1,2}(\mathbb{T}; M) \to C^\infty(\mathbb{T}; M)$ that is continuous with respect to the topologies of $W^{1,2}(\mathbb{T}; M)$ and $C^\infty(\mathbb{T}; M)$.

First of all, we define a smooth function $k : \mathbb{R} \to \mathbb{R}$ by

$$k(t) = \begin{cases} C \exp\left(\frac{1}{t^2-1}\right), & \text{if } |t| < 1, \\ 0, & \text{if } |t| \geq 1, \end{cases}$$

where $C > 0$ is a constant such that

$$\int_{\mathbb{R}} k(t)\, \mathrm{d}t = 1.$$

For each $\varepsilon > 0$ we define a smooth function $k_\varepsilon : \mathbb{R} \to \mathbb{R}$ by $k_\varepsilon(t) = \varepsilon^{-1} k(t\varepsilon^{-1})$. Notice that

$$\int_{\mathbb{R}} k_\varepsilon(t)\, \mathrm{d}t = 1, \qquad \forall \varepsilon > 0,$$

and k_ε tends to the Dirac delta as $\varepsilon \to 0$, in the sense of distributions. The functions in the family $\{k_\varepsilon \,|\, \varepsilon > 0\}$ can be used as convolution kernels to regularize $W^{1,2}$ maps (in the convolution literature, this family is sometimes called **approximate identity**, see for instance [Rud73, page 157]). We define

$$\mathscr{K} : (0,\infty) \times W^{1,2}(\mathbb{T}; \mathbb{R}^{2m+1}) \to C^\infty(\mathbb{T}; \mathbb{R}^{2m+1}) \tag{3.4}$$

as $\mathscr{K}(\varepsilon, \zeta) = k_\varepsilon * \zeta$ for each $\varepsilon > 0$ and $\zeta \in W^{1,2}(\mathbb{T}; \mathbb{R}^{2m+1})$, i.e.,

$$\mathscr{K}(\varepsilon, \zeta)(t) = \int_{\mathbb{R}} k_\varepsilon(t-s)\zeta(s)\, \mathrm{d}s \qquad \forall t \in \mathbb{R}.$$

Notice that $\mathscr{K}$ is in fact a map of the form (3.4) (since the convolution of periodic maps is still periodic) and it is a continuous map. Moreover $\mathscr{K}$ can be continuously extended to a continuous map

$$\mathscr{K} : [0,\infty) \times W^{1,2}(\mathbb{T}; \mathbb{R}^{2m+1}) \to W^{1,2}(\mathbb{T}; \mathbb{R}^{2m+1})$$

by setting $\mathscr{K}(0, \zeta) = \zeta$ for each $\zeta \in W^{1,2}(\mathbb{T}; \mathbb{R}^{2m+1})$.

Recall that, by the Whitney embedding theorem, we may assume that M is a closed submanifold of $\mathbb{R}^{2m+1}$. Consider an open tubular neighborhood $N \subseteq \mathbb{R}^{2m+1}$ of M, and a corresponding retraction $r : N \to M$. This retraction induces a smooth map

$$W^{1,2}(r) : W^{1,2}(\mathbb{T}; N) \to W^{1,2}(\mathbb{T}; M)$$

given by $W^{1,2}(r)(\zeta) = r \circ \zeta$, for each $\zeta \in W^{1,2}(\mathbb{T}; N)$. Moreover, $W^{1,2}(r)$ restricts to a smooth map

$$C^\infty(r) : C^\infty(\mathbb{T}; N) \to C^\infty(\mathbb{T}; M).$$

Notice that $W^{1,2}(\mathbb{T}; N)$ and $C^\infty(\mathbb{T}; N)$ are open subsets of $W^{1,2}(\mathbb{T}; \mathbb{R}^{2m+1})$ and $C^\infty(\mathbb{T}; \mathbb{R}^{2m+1})$ respectively.

Now, for each $\gamma \in C^\infty(\mathbb{T}; M)$ we choose a real $\varepsilon_\gamma > 0$ and an open neighborhood $\mathscr{W}_\gamma \subset W^{1,2}(\mathbb{T}; M)$ of γ such that $\mathscr{K}([0, \varepsilon_\gamma] \times \mathscr{W}_\gamma) \subset W^{1,2}(\mathbb{T}; N)$. The family $\mathfrak{W} = \{\mathscr{W}_\gamma \mid \gamma \in C^\infty(\mathbb{T}; M)\}$ is an open cover of $W^{1,2}(\mathbb{T}; M)$, and since this latter is a Hilbert manifold (in particular, it is paracompact) there exists a partition of unity $\{\rho_\gamma \mid \gamma \in C^\infty(\mathbb{T}; M)\}$ subordinated to the open cover $\mathfrak{W}$. We define a smooth function $\varepsilon : W^{1,2}(\mathbb{T}; M) \to (0, \infty)$ by

$$\varepsilon(\zeta) := \sum_{\gamma \in C^\infty(\mathbb{T}; M)} \varepsilon_\gamma \rho_\gamma(\zeta).$$

Notice that $\mathscr{K}(\varepsilon(\zeta), \zeta) \in C^\infty(\mathbb{T}; N)$ for each $\zeta \in W^{1,2}(\mathbb{T}; M)$. A homotopy $\mathscr{J}$ as claimed at the beginning can be built by setting

$$\mathscr{J}(s, \zeta) := W^{1,2}(r) \circ \mathscr{K}(s\varepsilon(\zeta), \zeta). \qquad \square$$

From now on, let us assume that the closed manifold M is connected (for non-connected manifolds, all the forthcoming arguments can be applied separately to each connected component). There is a well-known relationship between the connected components of the free loop space and the fundamental group of the underlying manifold M. We are going to recall it with the next proposition. For each continuous loop $\gamma : \mathbb{T} \to M$ such that $\gamma(0) = q_*$ is a fixed basepoint on the configuration space M, we denote by $C_{[\gamma]}$ the conjugacy class of the element of $\pi_1(M) = \pi_1(M, q_*)$ represented by γ, i.e.,

$$C_{[\gamma]} = \{[\beta][\gamma][\beta]^{-1} \in \pi_1(M) \mid [\beta] \in \pi_1(M)\}.$$

Proposition 3.2.2. *The connected components of the free loop spaces $C^\infty(\mathbb{T}; M)$, $W^{1,2}(\mathbb{T}; M)$ and $C^0(\mathbb{T}; M)$ are in one-to-one correspondence with the conjugacy classes of $\pi_1(M)$. Under this correspondence, the connected component associated to any conjugacy class $C_{[\gamma]} \subseteq \pi_1(M)$ is given by those loops that are homotopic to γ.*

Proof. First of all, notice that there is a one-to-one correspondence among the connected components of $C^\infty(\mathbb{T}; M)$, $W^{1,2}(\mathbb{T}; M)$ and $C^0(\mathbb{T}; M)$, as these spaces are homotopically equivalent. Moreover, since these spaces are all locally path-connected, their connected components are actually path-connected components. Therefore we just need to prove the statement for the path-connected components of $C^0(\mathbb{T}; M)$.

Let us fix an arbitrary basepoint q_* on the manifold M. Notice that any loop $\gamma \in C^0(\mathbb{T}; M)$ is homotopic to a loop $\gamma_* \in C^0(\mathbb{T}; M)$ such that $\gamma_*(0) = q_*$. In

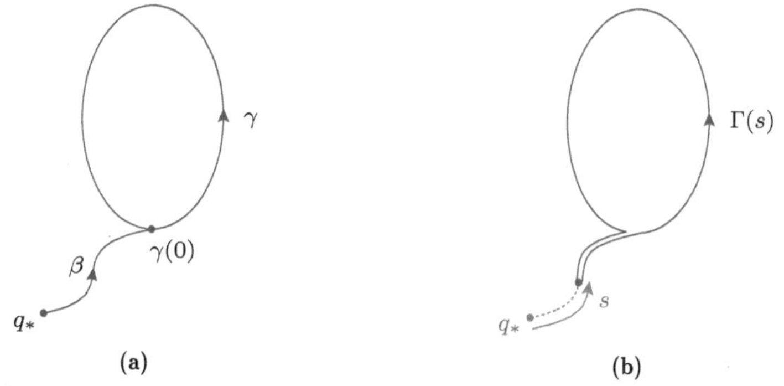

Figure 3.1. **(a)** Example of path β joining the basepoint q_* with the starting point of the loop γ. **(b)** Loop $\Gamma(s)$ of the corresponding homotopy.

fact, let $\beta : \mathbb{I} \to M$ be a continuous path joining the basepoint q_* with $\gamma(0)$, i.e., $\beta(0) = q_*$ and $\beta(1) = \gamma(0)$. We define a homotopy $\Gamma : \mathbb{I} \to C^0(\mathbb{T}; M)$ by

$$\Gamma(s)(t) := \begin{cases} \beta(s + 3t), & t \in \left[0, \frac{1-s}{3}\right], \\ \gamma\left(\frac{3}{1+2s}\left(t - \frac{1-s}{3}\right)\right), & t \in \left[\frac{1-s}{3}, \frac{2+s}{3}\right], \\ \beta(s + 3 - 3t), & t \in \left[\frac{2+s}{3}, 1\right]. \end{cases} \tag{3.5}$$

This homotopy begins at a loop $\gamma_* := \Gamma(0)$ with the desired properties, and ends at $\Gamma(1) = \gamma$ (see Figure 3.1).

Therefore, for any loop γ, its path-connected component in $C^0(\mathbb{T}; M)$ contains a loop γ_* as above that represents an element of the fundamental group $\pi_1(M) = \pi_1(M, q_*)$. In order to conclude the proof, all we have to show is that any given loops $\gamma_0, \gamma_1 \in C^0(\mathbb{T}; M)$ with $\gamma_0(0) = \gamma_1(0) = q_*$ are homotopic if and only if they represent conjugated elements of $\pi_1(M)$, that is, if and only if there exists $[\beta] \in \pi_1(M)$ such that $[\gamma_0] = [\beta][\gamma_1][\beta]^{-1} \in \pi_1(M)$.

If $[\gamma_0] = [\beta][\gamma_1][\beta]^{-1}$ in $\pi_1(M)$, then the loops γ_0 and γ_1 are homotopic (by a homotopy analogous to the one in (3.5)). Conversely, assume that we have a continuous path $\Gamma : \mathbb{I} \to C^0(\mathbb{T}; M)$ joining γ_0 and γ_1, i.e., $\Gamma(0) = \gamma_0$ and $\Gamma(1) = \gamma_1$. We define a loop $\beta \in C^0(\mathbb{T}; M)$ by $\beta(t) := \Gamma(t)(0)$ for each $t \in \mathbb{T}$, and a homotopy $H : \mathbb{I} \times \mathbb{T} \to M$ by

$$H(s,t) := \begin{cases} \beta(3t), & t \in \left[0, \frac{s}{3}\right], \\ \Gamma(s)\left(\frac{3t-s}{3-2s}\right), & t \in \left[\frac{s}{3}, 1 - \frac{s}{3}\right], \\ \beta(3 - 3t), & t \in \left[1 - \frac{s}{3}, 1\right]. \end{cases} \tag{3.6}$$

Notice that the homotopy H preserves the basepoint, namely $H(s, 0) = q_*$ for each $s \in \mathbb{I}$ (see Figure 3.2). Therefore, in $\pi_1(M)$, we have that

$$[\gamma_0] = [H(0, \cdot)] = [H(1, \cdot)] = [\beta][\gamma_1][\beta]^{-1}. \qquad \square$$

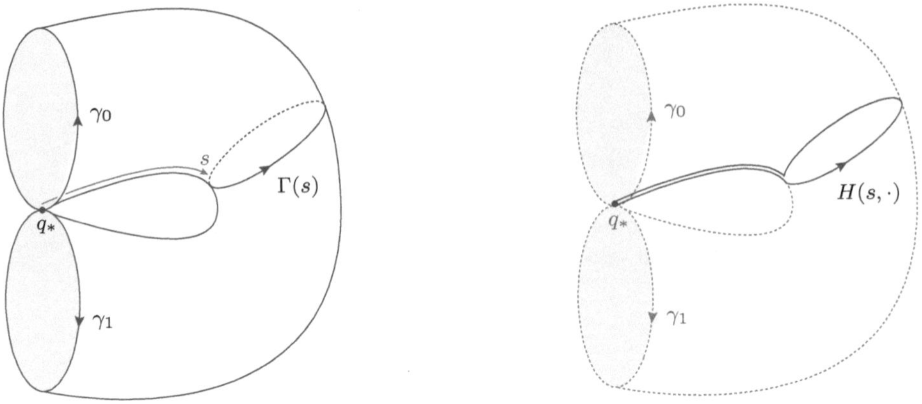

(a) (b)

Figure 3.2. **(a)** An example of a path $\Gamma : \mathbb{I} \to C^0(\mathbb{T}; M)$ joining two loops γ_0 and γ_1 with $q_* = \gamma_0(0) = \gamma_1(0)$. **(b)** The corresponding homotopy H defined by equation (3.6).

Concerning the richness of the homotopy and homology of the free loop space, a first easy remark is that the connected component of contractible loops on M inherits all the homology and homotopy of the manifold M. This is in fact a particular case of the following general property. Let **Top** be the category of topological spaces and continuous maps, $\mathbf{C}$ an arbitrary category and $\mathbf{F} : \mathbf{Top} \to \mathbf{C}$ a functor. We denote by $W^{1,2}_{\mathrm{contr}}(\mathbb{T}; M)$ the connected component of $W^{1,2}(\mathbb{T}; M)$ given by the contractible loops, and by $\iota : M \hookrightarrow W^{1,2}_{\mathrm{contr}}(\mathbb{T}; M)$ the map that sends a point to the constant loop at that point, i.e., $\iota(q)(t) = q$ for each $q \in M$ and $t \in \mathbb{T}$.

Proposition 3.2.3. *The induced morphism* $\mathbf{F}(\iota) : \mathbf{F}(M) \to \mathbf{F}(W^{1,2}_{\mathrm{contr}}(\mathbb{T}; M))$ *is a monomorphism (in the sense of category theory).*

Proof. Let $\mathrm{ev} : W^{1,2}(\mathbb{T}; M) \to M$ be the **evaluation map** defined by

$$\mathrm{ev}(\gamma) = \gamma(0), \qquad \forall \gamma \in W^{1,2}(\mathbb{T}; M).$$

This map is continuous (and even smooth), since it is the restriction of the bounded linear map $\mathrm{ev} : W^{1,2}(\mathbb{T}; \mathbb{R}^{2m+1}) \to \mathbb{R}^{2m+1}$ defined analogously. We obtain the following commutative diagram of continuous maps.

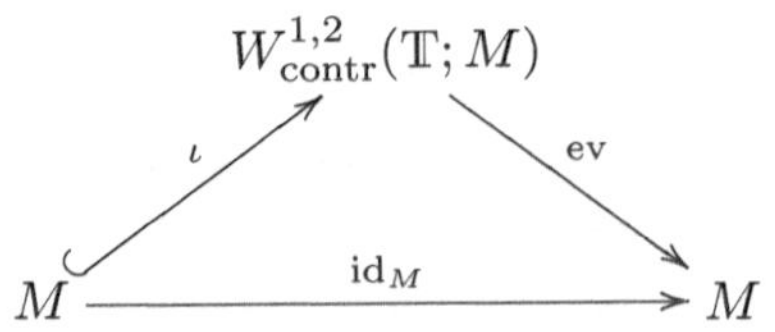

The endomorphism $\mathbf{F}(\mathrm{id}_M) : \mathbf{F}(M) \to \mathbf{F}(M)$ is the identity on $\mathbf{F}(M)$, and in particular it is an isomorphism in the sense of category theory. By functoriality, we have $\mathbf{F}(\mathrm{id}_M) = \mathbf{F}(\mathrm{ev}) \circ \mathbf{F}(\iota)$, and therefore $\mathbf{F}(\iota)$ is a monomorphism in the sense of category theory. □

Notice that, in the category **Grp** of groups and homomorphisms, the notions of monomorphism and isomorphism are the usual ones. Therefore, by choosing $\mathbf{F}$ to be the homotopy functor π_* and the singular homology functor H_* (with arbitrary coefficient group), we obtain the following.

Corollary 3.2.4. *The homotopy homomorphism $\pi_*(\iota)$ and homology homomorphism $\mathrm{H}_*(\iota)$ are injective.* □

An important result due to Vigué-Poirrier and Sullivan implies that the free loop space of M has nontrivial homology with real coefficients in infinitely many degrees, provided the closed manifold M is simply connected. We refer the reader to [VPS76] for a proof of this (and actually of a stronger) result.

Theorem 3.2.5. *Let H_* be the homology functor with real coefficients. If M is a simply connected closed manifold of positive dimension, then $\mathrm{H}_d(W^{1,2}(\mathbb{T};M))$ is nontrivial for infinitely many $d \in \mathbb{N}$.* □

3.3 Convex quadratic-growth Lagrangians

Let M be a closed manifold of dimension $m \geq 1$, over which we fix a Riemannian metric $\langle\cdot,\cdot\rangle$ and a finite atlas (so that every expression in local coordinates will be implicitly understood with respect to some chart of this atlas). We will consider smooth Lagrangian functions $\mathscr{L} : \mathbb{I} \times \mathrm{T}M \to \mathbb{R}$ that satisfy the following two conditions:

(Q1) there is a positive constant ℓ_0 such that

$$\sum_{i,j=1}^{m} \frac{\partial^2 \mathscr{L}}{\partial v^i\,\partial v^j}(t,q,v)\, w^i w^j \geq \ell_0 \, |w|_q^2 ,$$

for all $(t,q,v) \in \mathbb{I} \times \mathrm{T}M$ and all $w \in \mathrm{T}_q M$;

(Q2) there is a positive constant ℓ_1 such that

$$\left| \frac{\partial^2 \mathscr{L}}{\partial v^i\,\partial v^j}(t,q,v) \right| \leq \ell_1,$$

$$\left| \frac{\partial^2 \mathscr{L}}{\partial q^i\,\partial v^j}(t,q,v) \right| \leq \ell_1(1 + |v|_q),$$

$$\left| \frac{\partial^2 \mathscr{L}}{\partial q^i\,\partial q^j}(t,q,v) \right| \leq \ell_1(1 + |v|_q^2),$$

for all $(t,q,v) \in \mathbb{I} \times \mathrm{T}M$ and $i,j = 1,\dots,m$.

If we integrate the inequalities in **(Q2)** along the fibers of $\mathrm{T}M$ we obtain, for some positive constants $\ell_3, \ell_4 > 0$,

$$\left|\frac{\partial \mathscr{L}}{\partial v^i}(t,q,v)\right| \leq \ell_3(1+|v|_q), \tag{3.7}$$

$$\left|\frac{\partial \mathscr{L}}{\partial q^i}(t,q,v)\right| \leq \ell_3(1+|v|_q^2), \tag{3.8}$$

$$|\mathscr{L}(t,q,v)| \leq \ell_4(1+|v|_q^2), \tag{3.9}$$

for all $(t,q,v) \in \mathbb{I} \times \mathrm{T}M$ and $i = 1,\dots,m$. Analogously, integrating the inequality in **(Q1)** along the fibers of $\mathrm{T}M$ we get, for some positive constants $\ell_5, \ell_6 > 0$,

$$\left|\frac{\partial \mathscr{L}}{\partial v^i}(t,q,v)\right| \geq \ell_5(|v|_q - 1),$$

$$\mathscr{L}(t,q,v) \geq \ell_6(|v|_q^2 - 1),$$

for all $(t,q,v) \in \mathbb{I} \times \mathrm{T}M$ and $i = 1,\dots,m$. Notice that the Euler-Lagrange system is unchanged if we add a constant to the Lagrangian $\mathscr{L}$. Therefore, by adding ℓ_6 to $\mathscr{L}$ and setting $\underline{\ell} := \ell_6$ and $\overline{\ell} := \ell_4 + \ell_6$, we infer

$$\underline{\ell}\,|v|_q^2 \leq \mathscr{L}(t,q,v) \leq \overline{\ell}\,(|v|_q^2 + 1), \qquad \forall (t,q,v) \in \mathbb{I} \times \mathrm{T}M. \tag{3.10}$$

From these bounds we readily obtain that $\mathscr{L}$ is a Tonelli Lagrangian. In the following, we will informally refer to the class of smooth Lagrangian functions that satisfy **(Q1)** and **(Q2)** as to the class of **convex quadratic-growth** Lagrangians.

Apparently, conditions **(Q1)** and **(Q2)** depend on the choice of the Riemannian metric and of the finite atlas on M. However, this is not the case, as stated by the following.

Proposition 3.3.1. *Up to changing the constants ℓ_0 and ℓ_1, conditions* **(Q1)** *and* **(Q2)** *are independent of the choice of the Riemannian metric and of the finite atlas on M.*

Proof. Let $\mathfrak{U} = \{\phi_\alpha : U_\alpha \to \mathbb{R}^m \mid \alpha = 0,\dots,u\}$ be the previously fixed atlas on M, and consider another finite atlas $\mathfrak{U}' = \{\phi_\beta : U'_\beta \to \mathbb{R}^m \mid \beta = 0,\dots,u'\}$. As in section 1.1, we denote by $\mathrm{T}\mathfrak{U} = \{\mathrm{T}\phi_\alpha : \mathrm{T}U_\alpha \to \mathbb{R}^m \times \mathbb{R}^m \mid \alpha = 0,\dots,u\}$ and $\mathrm{T}\mathfrak{U}' = \{\mathrm{T}\phi_\beta : \mathrm{T}U'_\beta \to \mathbb{R}^m \times \mathbb{R}^m \mid \beta = 0,\dots,u'\}$ the associated atlases on the tangent bundle of M. We denote the components of the introduced charts by

$$\phi_\alpha = (q_\alpha^1,\dots,q_\alpha^m), \qquad \phi_\beta = (q_\beta^1,\dots,q_\beta^m),$$

$$\mathrm{T}\phi_\alpha = (q_\alpha^1,\dots,q_\alpha^m,v_\alpha^1,\dots,v_\alpha^m), \qquad \mathrm{T}\phi_\beta = (q_\beta^1,\dots,q_\beta^m,v_\beta^1,\dots,v_\beta^m).$$

Notice that, whenever the domains of two charts $\phi_\alpha \in \mathfrak{U}$ and $\phi_\beta \in \mathfrak{U}'$ intersect, on the intersection we have

$$
\begin{aligned}
v_\alpha^i &= \sum_{k=1}^m v_\beta^k \frac{\partial q_\alpha^i}{\partial q_\beta^k}, & \frac{\partial v_\alpha^i}{\partial v_\beta^j} &= \frac{\partial q_\alpha^i}{\partial q_\beta^j},\\
\frac{\partial v_\alpha^i}{\partial q_\beta^j} &= \sum_{k=1}^m v_\beta^k \frac{\partial^2 q_\alpha^i}{\partial q_\beta^j\,\partial q_\beta^k}, & \frac{\partial^2 v_\alpha^i}{\partial q_\beta^j\,\partial q_\beta^h} &= \sum_{k=1}^m v_\beta^k \frac{\partial^3 q_\alpha^i}{\partial q_\beta^j\,\partial q_\beta^h\,\partial q_\beta^k},\\
&&&\forall i,j,h \in \{1,\dots,m\}.
\end{aligned}
$$

By the compactness of M, there exists a constant $c > 0$ such that, for each $\phi_\alpha \in \mathfrak{U}$ and $\phi_\beta \in \mathfrak{U}'$ with $U_\alpha \cap U'_\beta \neq \varnothing$ and for each $q \in U_\alpha \cap U'_\beta$, $v \in \mathrm{T}_qM$ and $i, j \in \{1,\dots,m\}$, we have

$$
\begin{aligned}
\left|\frac{\partial q_\alpha^i}{\partial q_\beta^j}(q)\right| &\leq c, & \left|\frac{\partial^2 q_\alpha^i}{\partial q_\beta^j\,\partial q_\beta^k}(q)\right| &\leq c,\\
\left|\frac{\partial v_\alpha^i}{\partial q_\beta^j}(q,v)\right| &\leq c\,|v|_q, & \left|\frac{\partial^2 v_\alpha^i}{\partial q_\beta^j\,\partial q_\beta^h}(q,v)\right| &\leq c\,|v|_q.
\end{aligned}
$$

This, together with **(Q2)**, implies that

$$
\left|\frac{\partial^2 \mathscr{L}}{\partial v_\beta^i\,\partial v_\beta^j}(t,q,v)\right| = \left|\sum_{h,k=1}^m \frac{\partial^2 \mathscr{L}}{\partial v_\alpha^h\,\partial v_\alpha^k}(t,q,v)\,\frac{\partial q_\alpha^h}{\partial q_\beta^i}(q)\,\frac{\partial q_\alpha^k}{\partial q_\beta^j}(q)\right| \leq \ell_1 c^2 m^2.
$$

By (3.7) and (3.8), we have

$$
\begin{aligned}
\left|\frac{\partial^2 \mathscr{L}}{\partial q_\beta^i\,\partial v_\beta^j}(t,q,v)\right| &= \left|\frac{\partial}{\partial q_\beta^i}\left(\sum_{h=1}^m \frac{\partial \mathscr{L}}{\partial v_\alpha^h}(t,q,v)\,\frac{\partial q_\alpha^h}{\partial q_\beta^j}(q)\right)\right|\\
&= \left|\sum_{h=1}^m\left(\frac{\partial \mathscr{L}}{\partial v_\alpha^h}(t,q,v)\,\frac{\partial^2 q_\alpha^h}{\partial q_\beta^i\,\partial q_\beta^j}(q)\right.\right.\\
&\qquad + \sum_{k=1}^m\left(\frac{\partial^2 \mathscr{L}}{\partial q_\alpha^k\,\partial v_\alpha^h}(t,q,v)\,\frac{\partial q_\alpha^k}{\partial q_\beta^i}(q)\,\frac{\partial q_\alpha^h}{\partial q_\beta^j}(q)\right.\\
&\qquad \left.\left.\left. + \frac{\partial^2 \mathscr{L}}{\partial v_\alpha^k\,\partial v_\alpha^h}(t,q,v)\,\frac{\partial v_\alpha^k}{\partial q_\beta^i}(q)\,\frac{\partial q_\alpha^h}{\partial q_\beta^j}(q)\right)\right)\right|\\
&\leq \sum_{h=1}^m\left(\ell_3(1+|v|_q)c + \sum_{k=1}^m\left(\ell_1(1+|v|_q)c^2 + \ell_1 c^2|v|_q\right)\right)\\
&\leq (m\ell_3 c + 2m^2\ell_1 c^2)(1+|v|_q).
\end{aligned}
$$

Moreover

$$
\begin{aligned}
\left|\frac{\partial^2 \mathscr{L}}{\partial q^i_\beta\,\partial q^j_\beta}(t,q,v)\right| &= \left|\frac{\partial}{\partial q^i_\beta}\left(\sum_{h=1}^m\left(\frac{\partial \mathscr{L}}{\partial q^h_\alpha}(t,q,v)\,\frac{\partial q^h_\alpha}{\partial q^j_\beta}(q) + \frac{\partial \mathscr{L}}{\partial v^h_\alpha}(t,q,v)\,\frac{\partial v^h_\alpha}{\partial q^j_\beta}(q)\right)\right)\right| \\
&= \left|\sum_{h=1}^m\left(\frac{\partial \mathscr{L}}{\partial q^h_\alpha}(t,q,v)\,\frac{\partial^2 q^h_\alpha}{\partial q^i_\beta\,\partial q^j_\beta}(q) + \frac{\partial \mathscr{L}}{\partial v^h_\alpha}(t,q,v)\,\frac{\partial^2 v^h_\alpha}{\partial q^i_\beta\,\partial q^j_\beta}(q)\right.\right. \\
&\qquad + \sum_{k=1}^m\left(\frac{\partial^2 \mathscr{L}}{\partial q^k_\alpha\,\partial q^h_\alpha}(t,q,v)\,\frac{\partial q^k_\alpha}{\partial q^i_\beta}(q)\,\frac{\partial q^h_\alpha}{\partial q^j_\beta}(q)\right. \\
&\qquad + \frac{\partial^2 \mathscr{L}}{\partial v^k_\alpha\,\partial q^h_\alpha}(t,q,v)\,\frac{\partial v^k_\alpha}{\partial q^i_\beta}(q)\,\frac{\partial q^h_\alpha}{\partial q^j_\beta}(q) \\
&\qquad + \frac{\partial^2 \mathscr{L}}{\partial q^k_\alpha\,\partial v^h_\alpha}(t,q,v)\,\frac{\partial q^k_\alpha}{\partial q^i_\beta}(q)\,\frac{\partial v^h_\alpha}{\partial q^j_\beta}(q) \\
&\qquad \left.\left.\left. + \frac{\partial^2 \mathscr{L}}{\partial v^k_\alpha\,\partial v^h_\alpha}(t,q,v)\,\frac{\partial v^k_\alpha}{\partial q^i_\beta}(q)\,\frac{\partial v^h_\alpha}{\partial q^j_\beta}(q)\right)\right)\right| \\
&\leq \sum_{h=1}^m\Bigg(\ell_3(1+|v|_q^2)c + \ell_3(1+|v|_q)c|v|_q \\
&\qquad + \sum_{k=1}^m\big(\ell_1(1+|v|_q^2)c^2 + \ell_1(1+|v|_q)c^2|v|_q \\
&\qquad + \ell_1(1+|v|_q)c^2|v|_q + \ell_1 c^2|v|_q^2\big)\Bigg) \\
&\leq (3m\ell_3 c + 6m^2\ell_1 c^2)(1+|v|_q^2).
\end{aligned}
$$

These estimates prove that condition **(Q2)** still holds with respect to the atlas $\mathfrak{U}'$, up to replacing the constant ℓ_1 with $(3m\ell_3 c + 6m^2\ell_1 c^2)$. Condition **(Q1)** holds with respect to the atlas $\mathfrak{U}'$, even without changing the constant ℓ_0. In fact, for each $(t,q,v) \in \mathbb{R}\times \mathrm{T}M$ and $w \in \mathrm{T}_q M$, we have

$$
\begin{aligned}
\sum_{i,j=1}^m \frac{\partial^2 \mathscr{L}}{\partial v^i_\alpha\,\partial v^j_\alpha}(t,q,v)\,w^i_\alpha w^j_\alpha &= \left.\frac{\mathrm{d}^2}{\mathrm{d}s^2}\right|_{s=0} \mathscr{L}(t,q,v+sw) \\
&= \sum_{i,j=1}^m \frac{\partial^2 \mathscr{L}}{\partial v^i_\beta\,\partial v^j_\beta}(t,q,v)\,w^i_\beta w^j_\beta,
\end{aligned}
$$

which implies that condition **(Q1)** is coordinate independent. Finally, the fact that conditions **(Q1)** and **(Q2)** still hold with respect to a different Riemannian metric follows from the compactness of M, together with the fact that all the Riemannian metrics on it are locally equivalent. Namely, if $\langle\!\langle\cdot,\cdot\rangle\!\rangle_\cdot$ is another Riemannian metric on M with correspondent norm $\|\cdot\|_\cdot$, there exists a constant $d \geq 1$ such that

$$d^{-1}|v|_q \leq \|v\|_q \leq d\,|v|_q, \qquad \forall (q,v) \in \mathrm{T}M. \qquad \square$$

Classical examples of convex quadratic-growth Lagrangians are given by functions $\mathscr{L} : \mathbb{I} \times \mathrm{T}M \to \mathbb{R}$ of the form

$$\mathscr{L}(t,q,v) = \alpha(t,q)[v,v] + \beta(t,q)v - V(t,q),$$

where α is a time-dependent symmetric and positive-definite $(0,2)$-tensor field on M, β is a time-dependent one-form on M and $V : \mathbb{I} \times M \to \mathbb{R}$ is a function. In physics, Lagrangians of this form are called **electro-magnetic**: α is the kinetic tensor of the system, β is the magnetic form and V is the potential energy.

3.4 Regularity of the action functional

Let $\mathscr{L} : \mathbb{I} \times \mathrm{T}M \to \mathbb{R}$ be a smooth Lagrangian with associated action functional

$$\mathscr{A}(\gamma) = \int_0^1 \mathscr{L}(t,\gamma(t),\dot\gamma(t))\,\mathrm{d}t, \qquad \forall \gamma \in W^{1,2}(\mathbb{I};M).$$

We require $\mathscr{L}$ to satisfy **(Q2)**, and in particular, by (3.9), to have subquadratic-growth on each fiber of $\mathrm{T}M$. This implies that every $W^{1,2}$ curve has finite action. Namely, $\mathscr{A}$ is a well-defined functional of the form $\mathscr{A} : W^{1,2}(\mathbb{I};M) \to \mathbb{R}$. Its regularity is described by the following statement, essentially due to Benci [Ben86], and by Proposition 3.4.3, which is due to Abbondandolo and Schwarz [AS09].

Proposition 3.4.1. *Let $\mathscr{L} : \mathbb{I} \times \mathrm{T}M \to \mathbb{R}$ be a smooth Lagrangian satisfying* **(Q2)**. *Then its action functional $\mathscr{A} : W^{1,2}(\mathbb{I};M) \to \mathbb{R}$ is C^1, and the differential $\mathrm{d}\mathscr{A} : \mathrm{T}W^{1,2}(\mathbb{I};M) \to \mathbb{R}$ is Gâteaux-differentiable and locally Lipschitz-continuous.*

Remark 3.4.1 (Localization). In order to do analysis with the Lagrangian action functional locally at a curve $\gamma \in C^\infty(\mathbb{I};M)$, it is often worthwhile to employ the chart $\Phi_\gamma : \mathscr{U}_\gamma \to W^{1,2}(\mathbb{I};\mathbb{R}^m)$ introduced in Remark 3.1.1. More precisely, let U be a sufficiently small open neighborhood of the origin in $\mathbb{R}^m$, so that the closed set $W^{1,2}(\mathbb{I};\overline{U}) \subset W^{1,2}(\mathbb{I};\mathbb{R}^m)$ is contained in the image of the chart Φ_γ, and let ϕ_γ be the trivialization (3.1) that enters the definition of Φ_γ. We write

$$\Pi_\gamma := \Phi_\gamma^{-1}|_{W^{1,2}(\mathbb{I};U)} : W^{1,2}(\mathbb{I};U) \to \mathscr{U}_\gamma$$

and we define a smooth embedding $\pi_\gamma : \mathbb{I} \times \overline{U} \times \mathbb{R}^m \hookrightarrow \mathbb{I} \times \mathrm{T}M$ in the following way: for each $(t,q,v) \in \mathbb{I} \times \overline{U} \times \mathbb{R}^m$ we set

$$\pi_\gamma(t,q,v) := (t,q',v'),$$

where

$$q' = \exp_{\gamma(t)} \circ \phi_\gamma^{-1}(t,q),$$
$$v' = \mathrm{d}(\exp_{\gamma(t)} \circ \phi_\gamma^{-1})(t,q)(0,v) + \frac{\mathrm{d}}{\mathrm{d}t}(\exp_{\gamma(t)} \circ \phi_\gamma^{-1}(t,q)).$$

Notice that, if $\mathscr{L}$ satisfies **(Q1)** [resp. **(Q2)**], then the pull-back Lagrangian

$$\mathscr{L} \circ \pi_\gamma : \mathbb{I} \times \overline{U} \times \mathbb{R}^m \to \mathbb{R}$$

satisfies **(Q1)** [resp. **(Q2)**] as well, up to changing the constant ℓ_0 [resp. ℓ_1]. Moreover, the pulled-back functional

$$\mathscr{A} \circ \Pi_\gamma : W^{1,2}(\mathbb{I}; U) \to \mathbb{R}$$

turns out to be the action functional associated to the Lagrangian $\mathscr{L} \circ \pi_\gamma$, i.e.,

$$\mathscr{A} \circ \Pi_\gamma(\zeta) = \int_0^1 \mathscr{L} \circ \pi_\gamma(t, \zeta(t), \dot\zeta(t))\, \mathrm{d}t, \qquad \forall \zeta \in W^{1,2}(\mathbb{I}; U).$$

This localization procedure can be also employed when we consider the action functional $\mathscr{A}$ restricted to either $W^{1,2}_{q_0,q_1}(\mathbb{I}; M)$ (for any $q_0, q_1 \in M$) or $W^{1,2}(\mathbb{T}; M)$. In fact, in the first case, for each smooth $\gamma : \mathbb{I} \to M$ with $\gamma(0) = q_0$ and $\gamma(1) = q_1$, we localize $\mathscr{A} : W^{1,2}_{q_0,q_1}(\mathbb{I}; M) \to \mathbb{R}$ by considering the restriction of $\mathscr{A} \circ \Pi_\gamma$ to the open set

$$W^{1,2}_0(\mathbb{I}; U) := W^{1,2}(\mathbb{I}; U) \cap W^{1,2}_0(\mathbb{I}; \mathbb{R}^m).$$

In the second case, for each smooth $\gamma : \mathbb{I} \to M$ with $\gamma(0) = \gamma(1)$, we localize $\mathscr{A} : W^{1,2}(\mathbb{T}; M) \to \mathbb{R}$ by considering the restriction of $\mathscr{A} \circ \Pi_\gamma$ to the open set

$$W^{1,2}_{\mathbb{V}}(\mathbb{I}; U) := W^{1,2}(\mathbb{I}; U) \cap W^{1,2}_{\mathbb{V}}(\mathbb{I}; \mathbb{R}^m),$$

where the vector space $\mathbb{V} \subset \mathbb{R}^m \times \mathbb{R}^m$ is defined in (3.3). Moreover, if the pull-back bundle $\gamma^* \mathrm{T}M \to \mathbb{T}$ is trivial, with a suitable trivialization the vector space $\mathbb{V}$ turns out to be the diagonal in $\mathbb{R}^m \times \mathbb{R}^m$, and therefore

$$W^{1,2}_{\mathbb{V}}(\mathbb{I}; U) = W^{1,2}(\mathbb{T}; U).$$ □

The proof of Proposition 3.4.1 makes use of the following elementary fact about convergences in metric spaces.

Lemma 3.4.2. *In a metric space, a sequence $\{x_j\}$ converges to x as $j \to \infty$ if and only if every subsequence $\{x_{j(k)}\}$ has a further subsequence $\{x_{j(k(h))}\}$ that converges to x as $h \to \infty$.*

Proof. The "only if" part is trivial. For the other implication, assume that $\{x_j\}$ does not converge to x as $j \to \infty$. This means that there exists $\varepsilon > 0$ such that, for each $k \in \mathbb{N}$, there exists $j(k) > k$ such that $\mathrm{dist}(x_{j(k)}, x) > \varepsilon$. Hence the subsequence $\{x_{j(k)}\}$ has no subsequence converging to x. □

Proof of Proposition 3.4.1. Since the statement is of a local nature, by adopting the localization argument of Remark 3.4.1 we can assume that our Lagrangian has the form $\mathscr{L} : \mathbb{I} \times \overline{U} \times \mathbb{R}^m \to \mathbb{R}$, where U is a bounded open subset of $\mathbb{R}^m$, so that the action functional $\mathscr{A}$ is defined on the open subset $W^{1,2}(\mathbb{I}; U)$ of the Sobolev space $W^{1,2}(\mathbb{I}; \mathbb{R}^m)$. On this latter space we can consider either the usual norm $\|\cdot\|_{W^{1,2}}$ or the equivalent norm $|||\cdot|||_{W^{1,2}}$ given by

$$|||\xi|||_{W^{1,2}} = \big(|\xi(0)|^2 + \|\dot{\xi}\|_{L^2}^2\big)^{1/2}, \qquad \forall \xi \in W^{1,2}(\mathbb{I}; \mathbb{R}^m).$$

It will be useful for the reader to keep in mind the following inequalities:

$$\|\xi\|_{L^1} \leq \|\xi\|_{L^2} \leq \|\xi\|_{L^\infty} \leq |\xi(0)| + \|\dot{\xi}\|_{L^1} \leq \sqrt{2}\big(|\xi(0)|^2 + \|\dot{\xi}\|_{L^1}^2\big)^{1/2} \leq \sqrt{2}|||\xi|||_{W^{1,2}}, \qquad \forall \xi \in W^{1,2}(\mathbb{I}; \mathbb{R}^m).$$

Consider $\zeta \in W^{1,2}(\mathbb{I}; U)$ and $\xi \in W^{1,2}(\mathbb{I}; \mathbb{R}^m)$. For each $\varepsilon \in \mathbb{R}$ with sufficiently small absolute value we have that $\zeta + \varepsilon\xi$ belongs to $W^{1,2}(\mathbb{I}; U)$. Then

$$\begin{aligned} &\frac{\mathscr{A}(\zeta + \varepsilon\xi) - \mathscr{A}(\zeta)}{\varepsilon} \\ &\quad = \int_0^1\!\!\int_0^1 \Big[\langle \partial_q \mathscr{L}(t, \zeta + s\varepsilon\xi, \dot{\zeta} + s\varepsilon\dot{\xi}), \xi\rangle + \langle \partial_v \mathscr{L}(t, \zeta + s\varepsilon\xi, \dot{\zeta} + s\varepsilon\dot{\xi}), \dot{\xi}\rangle\Big]\, dt\, ds, \end{aligned}$$

where we denoted by $\partial_q \mathscr{L}$ and $\partial_v \mathscr{L}$ the gradients of $\mathscr{L}$ with respect to the q and v variables respectively. By (3.7) and (3.8) we can apply the dominated convergence theorem to assert that, for $\varepsilon \to 0$, the above quantity tends to

$$\begin{aligned} \mathrm{d}\mathscr{A}(\zeta)\,\xi &:= \lim_{\varepsilon \to 0} \frac{\mathscr{A}(\zeta + \varepsilon\xi) - \mathscr{A}(\zeta)}{\varepsilon} \\ &= \int_0^1 \Big[\langle \partial_q \mathscr{L}(t, \zeta, \dot{\zeta}), \xi\rangle + \langle \partial_v \mathscr{L}(t, \zeta, \dot{\zeta}), \dot{\xi}\rangle\Big]\, dt. \end{aligned}$$

The functional $\mathrm{d}\mathscr{A}(\zeta) : W^{1,2}(\mathbb{I}; \mathbb{R}^m) \to \mathbb{R}$ is the Gâteaux-differential of the action $\mathscr{A}$ at ζ, being a bounded linear functional on $W^{1,2}(\mathbb{I}; \mathbb{R}^m)$. Now, we want to prove that the map $\mathrm{d}\mathscr{A} : W^{1,2}(\mathbb{I}; U) \to W^{1,2}(\mathbb{I}; \mathbb{R}^m)^*$ is continuous, namely we want to prove that, for an arbitrary sequence $\{\zeta_j\} \subset W^{1,2}(\mathbb{I}; U)$ converging to some $\zeta \in W^{1,2}(\mathbb{I}; U)$, we have

$$\lim_{j \to \infty} \sup \big\{ \mathrm{d}\mathscr{A}(\zeta_j)\,\xi - \mathrm{d}\mathscr{A}(\zeta)\,\xi \;\big|\; |||\xi|||_{W^{1,2}} = 1 \big\} = 0.$$

First of all, for each $\zeta, \lambda \in W^{1,2}(\mathbb{I}; U)$ and $\xi \in W^{1,2}(\mathbb{I}; \mathbb{R}^m)$, we have the estimate

$$\begin{aligned} |(\mathrm{d}\mathscr{A}(\zeta) - \mathrm{d}\mathscr{A}(\lambda))\,\xi| &\leq \|\partial_q \mathscr{L}(\cdot, \zeta, \dot{\zeta}) - \partial_q \mathscr{L}(\cdot, \lambda, \dot{\lambda})\|_{L^1} \|\xi\|_{L^\infty} \\ &\quad + \|\partial_v \mathscr{L}(\cdot, \zeta, \dot{\zeta}) - \partial_v \mathscr{L}(\cdot, \lambda, \dot{\lambda})\|_{L^2} \|\dot{\xi}\|_{L^2} \\ &\leq \big(\sqrt{2}\|\partial_q \mathscr{L}(\cdot, \zeta, \dot{\zeta}) - \partial_q \mathscr{L}(\cdot, \lambda, \dot{\lambda})\|_{L^1} \\ &\quad + \|\partial_v \mathscr{L}(\cdot, \zeta, \dot{\zeta}) - \partial_v \mathscr{L}(\cdot, \lambda, \dot{\lambda})\|_{L^2}\big) |||\xi|||_{W^{1,2}}. \end{aligned} \tag{3.11}$$

Let us assume that the sequence $\{\zeta_j\} \subset W^{1,2}(\mathbb{I};U)$ converges to ζ in $W^{1,2}$. This implies that $\{\zeta_j\}$ converges to ζ uniformly and $\{\dot\zeta_j\}$ converges to $\dot\zeta$ in L^2. In particular there exists a function $f \in L^2(\mathbb{I};\mathbb{R})$ such that $|\dot\zeta_j| < f$ almost everywhere for all j. The L^2 convergence implies that, for every subsequence $\{\dot\zeta_{j(k)}\}$, there exists a further subsequence $\{\dot\zeta_{j(k(h))}\}$ converging to $\dot\zeta$ almost everywhere. Therefore, by (3.7) and (3.8), we can apply the dominated convergence theorem as before to get

$$\partial_q\mathscr{L}(\cdot,\zeta_{j(k(h))},\dot\zeta_{j(k(h))}) \underset{h\to\infty}{\longrightarrow} \partial_q\mathscr{L}(\cdot,\zeta,\dot\zeta) \text{ in } L^1(\mathbb{I};\mathbb{R}^m),$$
$$\partial_v\mathscr{L}(\cdot,\zeta_{j(k(h))},\dot\zeta_{j(k(h))}) \underset{h\to\infty}{\longrightarrow} \partial_v\mathscr{L}(\cdot,\zeta,\dot\zeta) \text{ in } L^2(\mathbb{I};\mathbb{R}^m).$$

These convergences, together with the estimate in (3.11), imply that the subsequence $\{\mathrm{d}\mathscr{A}(\zeta_{j(k(h))})\}$ converges to $\mathrm{d}\mathscr{A}(\zeta)$ in $W^{1,2}(\mathbb{I};\mathbb{R}^m)^*$ as $h\to\infty$. By Lemma 3.4.2 we conclude that the whole sequence $\{\mathrm{d}\mathscr{A}(\zeta_j)\}$ converges to $\mathrm{d}\mathscr{A}(\zeta)$ in $W^{1,2}(\mathbb{I};\mathbb{R}^m)^*$ as $j\to\infty$. By the total differential theorem, the functional $\mathscr{A}$ is C^1 with Fréchet differential $\mathrm{d}\mathscr{A}$.

Now, consider $\zeta \in W^{1,2}(\mathbb{I};U)$, $\xi,\sigma \in W^{1,2}(\mathbb{I};\mathbb{R}^m)$ and $\varepsilon\in\mathbb{R}$ with sufficiently small absolute value so that $\zeta+\varepsilon\sigma$ belongs to $W^{1,2}(\mathbb{I};U)$. We have

$$\begin{aligned}
&\frac{\mathrm{d}\mathscr{A}(\zeta+\varepsilon\sigma)\,\xi - \mathrm{d}\mathscr{A}(\zeta)\,\xi}{\varepsilon}\\
&= \int_0^1\!\!\int_0^1 \Big[\langle\partial^2_{vv}\mathscr{L}(t,\zeta+s\varepsilon\sigma,\dot\zeta+s\varepsilon\dot\sigma)\dot\xi,\dot\sigma\rangle + \langle\partial^2_{vq}\mathscr{L}(t,\zeta+s\varepsilon\sigma,\dot\zeta+s\varepsilon\dot\sigma)\xi,\dot\sigma\rangle\\
&\quad + \langle\partial^2_{qv}\mathscr{L}(t,\zeta+s\varepsilon\sigma,\dot\zeta+s\varepsilon\dot\sigma)\dot\xi,\sigma\rangle + \langle\partial^2_{qq}\mathscr{L}(t,\zeta+s\varepsilon\sigma,\dot\zeta+s\varepsilon\dot\sigma)\xi,\sigma\rangle\Big]\,\mathrm{d}t\,\mathrm{d}s.
\end{aligned}$$

By **(Q2)**, we can apply the dominated convergence theorem to assert that the above quantity converges, as $\varepsilon\to 0$, to

$$\begin{aligned}
\mathrm{Hess}\mathscr{A}(\zeta)[\sigma,\xi] &:= \lim_{\varepsilon\to 0}\frac{\mathrm{d}\mathscr{A}(\zeta+\varepsilon\sigma)\,\xi - \mathrm{d}\mathscr{A}(\zeta)\,\xi}{\varepsilon}\\
&= \int_0^1 \Big[\langle\partial^2_{vv}\mathscr{L}(t,\zeta,\dot\zeta)\dot\xi,\dot\sigma\rangle + \langle\partial^2_{vq}\mathscr{L}(t,\zeta,\dot\zeta)\xi,\dot\sigma\rangle\\
&\quad + \langle\partial^2_{qv}\mathscr{L}(t,\zeta,\dot\zeta)\dot\xi,\sigma\rangle + \langle\partial^2_{qq}\mathscr{L}(t,\zeta,\dot\zeta)\xi,\sigma\rangle\Big]\,\mathrm{d}t.
\end{aligned}$$

Since $\mathrm{Hess}\mathscr{A}(\zeta) : W^{1,2}(\mathbb{I};\mathbb{R}^m)\times W^{1,2}(\mathbb{I};\mathbb{R}^m) \to \mathbb{R}$ is a bounded symmetric bilinear form on $W^{1,2}(\mathbb{I};\mathbb{R}^m)$, the bounded linear map

$$\mathrm{d}^2\mathscr{A}(\zeta) : W^{1,2}(\mathbb{I};\mathbb{R}^m) \to W^{1,2}(\mathbb{I};\mathbb{R}^m)^*$$

given by

$$(\mathrm{d}^2\mathscr{A}(\zeta)\sigma)\xi := \mathrm{Hess}\mathscr{A}(\zeta)[\sigma,\xi], \qquad \forall \xi,\sigma\in W^{1,2}(\mathbb{I};\mathbb{R}^m),$$

is the Gâteaux-differential of $\mathrm{d}\mathscr{A}$ at ζ. By **(Q2)**, we obtain the estimate

$$\begin{aligned}|(\mathrm{d}^2\mathscr{A}(\zeta)\sigma)\xi| \leq &\ell_1\|\dot\xi\|_{L^2}\|\dot\sigma\|_{L^2} + \ell_1\big(\|\xi\|_{L^\infty}\|\dot\sigma\|_{L^1} + \|\dot\zeta\|_{L^2}\|\xi\|_{L^\infty}\|\dot\sigma\|_{L^2}\big)\\ &+ \ell_1\big(\|\dot\xi\|_{L^1}\|\sigma\|_{L^\infty} + \|\dot\zeta\|_{L^2}\|\dot\xi\|_{L^2}\|\sigma\|_{L^\infty}\big)\\ &+ \ell_1\big(\|\xi\|_{L^\infty}\|\sigma\|_{L^\infty} + \|\xi\|_{L^\infty}\|\sigma\|_{L^\infty}\|\dot\zeta\|_{L^2}^2\big)\\ \leq &\underbrace{\ell_1\big((2\sqrt{2}+3) + 4\|\zeta\|_{W^{1,2}} + 2\|\zeta\|_{W^{1,2}}^2\big)}_{=:\omega(\|\zeta\|_{W^{1,2}})}\|\xi\|_{W^{1,2}}\|\sigma\|_{W^{1,2}}.\end{aligned}$$

Notice that the function $\omega : [0,\infty) \to [0,\infty)$ is continuous and monotone increasing. For each $R > 0$ and, for each $\zeta, \lambda \in W^{1,2}(\mathbb{I}; U)$ with $\|\zeta\|_{W^{1,2}} < R$ and $\|\lambda\|_{W^{1,2}} < R$, we have

$$\begin{aligned}\|\mathrm{d}\mathscr{A}(\lambda) - \mathrm{d}\mathscr{A}(\zeta)\|_{(W^{1,2})^*} &= \left\| \int_0^1 \frac{\mathrm{d}}{\mathrm{d}r}\mathrm{d}\mathscr{A}((1-r)\zeta + r\lambda)\,\mathrm{d}r \right\|_{(W^{1,2})^*}\\ &= \left\| \int_0^1 \mathrm{d}^2\mathscr{A}((1-r)\zeta + r\lambda)(\lambda-\zeta)\,\mathrm{d}r \right\|_{(W^{1,2})^*}\\ &\leq \int_0^1 \omega\big(\|(1-r)\zeta + r\lambda\|_{W^{1,2}}\big)\,\|\lambda-\zeta\|_{W^{1,2}}\,\mathrm{d}r\\ &\leq \omega(\max\{\|\zeta\|_{W^{1,2}}, \|\lambda\|_{W^{1,2}}\})\,\|\lambda-\zeta\|_{W^{1,2}}\\ &< \omega(R)\,\|\lambda-\zeta\|_{W^{1,2}},\end{aligned}$$

which proves that $\mathrm{d}\mathscr{A}$ is locally Lipschitz-continuous. □

Proposition 3.4.3. *Let $\mathscr{L} : \mathbb{I} \times \mathrm{T}M \to \mathbb{R}$ be a smooth Lagrangian satisfying* **(Q2)**. *Then its action functional $\mathscr{A} : W^{1,2}(\mathbb{I}; M) \to \mathbb{R}$ is C^2 if and only if, for each $(t,q) \in \mathbb{I} \times M$, the function $\mathscr{L}(t,q,\cdot) : \mathrm{T}_qM \to \mathbb{R}$ is a polynomial of degree at most two. Moreover, if $\mathscr{A}$ is C^2, then it is C^∞.*

Proof. Adopting the localization argument of Remark 3.4.1, we will assume that our Lagrangian has the form $\mathscr{L} : \mathbb{I} \times \overline{U} \times \mathbb{R}^m \to \mathbb{R}$, where U is a bounded open subset of $\mathbb{R}^m$. Throughout this proof, on the Sobolev space $W^{1,2}(\mathbb{I}; \mathbb{R}^m)$ we will consider the equivalent norm $\|\cdot\|_{W^{1,2}}$ given by

$$\|\xi\|_{W^{1,2}} = \big(|\xi(0)|^2 + \|\dot\xi\|_{L^2}^2\big)^{1/2}, \qquad \forall \xi \in W^{1,2}(\mathbb{I}; \mathbb{R}^m).$$

If $\mathscr{L}$ is fiberwise a polynomial of degree at most 2, then after the localization it has the form

$$\mathscr{L}(t,q,v) = \langle a(t,q)v, v\rangle + \langle b(t,q), v\rangle - c(t,q), \qquad \forall (t,q,v) \in \mathbb{I} \times \overline{U} \times \mathbb{R}^m,$$

where $a : \mathbb{I} \times \overline{U} \to \mathbb{R}^{m\times m}$, $b : \mathbb{I} \times \overline{U} \to \mathbb{R}^m$ and $c : \mathbb{I} \times \overline{U} \to \mathbb{R}$. We denote by $\mathrm{Bil}(W^{1,2}(\mathbb{I}; \mathbb{R}^m))$ the Banach space of bounded bilinear forms on $W^{1,2}(\mathbb{I}; \mathbb{R}^m)$.

We have already proved in Proposition 3.4.1 that the action functional admits a second Gâteaux differential $\mathrm{d}^2\mathscr{A}$. For each $\zeta \in W^{1,2}(\mathbb{I}; U)$ and $\xi, \lambda \in W^{1,2}(\mathbb{I}; \mathbb{R}^m)$ we have

$$(\mathrm{d}^2\mathscr{A}(\zeta)\sigma)\xi = \mathscr{P}(\zeta)[\sigma,\xi] + \mathscr{Q}(\zeta)[\sigma,\xi] + \mathscr{Q}(\zeta)[\xi,\sigma] + \mathscr{R}(\zeta)[\sigma,\xi],$$

where $\mathscr{P}, \mathscr{Q}, \mathscr{R} : W^{1,2}(\mathbb{I}; \mathbb{R}^m) \to \mathrm{Bil}(W^{1,2}(\mathbb{I}; \mathbb{R}^m))$ are given by

$$\mathscr{P}(\zeta)[\sigma,\xi] = \int_0^1 \langle \partial^2_{vv}\mathscr{L}(t,\zeta,\dot\zeta)\dot\xi, \dot\sigma\rangle \mathrm{d}t = \int_0^1 \langle (a(t,\zeta) + a(t,\zeta)^T)\dot\xi, \dot\sigma\rangle \mathrm{d}t,$$
$$\mathscr{Q}(\zeta)[\sigma,\xi] = \int_0^1 \langle \partial^2_{vq}\mathscr{L}(t,\zeta,\dot\zeta)\xi, \dot\sigma\rangle \mathrm{d}t,$$
$$\mathscr{R}(\zeta)[\sigma,\xi] = \int_0^1 \langle \partial^2_{qq}\mathscr{L}(t,\zeta,\dot\zeta)\xi, \sigma\rangle \mathrm{d}t.$$

For each $\zeta, \lambda \in W^{1,2}(\mathbb{I}; U)$ and $\xi, \sigma \in W^{1,2}(\mathbb{I}; \mathbb{R}^m)$, we have the estimates

$$\begin{aligned}
|(\mathscr{P}(\zeta) - \mathscr{P}(\lambda))[\sigma,\xi]| &\le 2\|a(\cdot,\zeta) - a(\cdot,\lambda)\|_{L^\infty}\|\dot\xi\|_{L^2}\|\dot\sigma\|_{L^2}\\
&\le 2\|a(\cdot,\zeta) - a(\cdot,\lambda)\|_{L^\infty}\|\xi\|_{W^{1,2}}\|\sigma\|_{W^{1,2}},\\
|(\mathscr{Q}(\zeta) - \mathscr{Q}(\lambda))[\sigma,\xi]| &\le \|\partial^2_{vq}\mathscr{L}(\cdot,\zeta,\dot\zeta) - \partial^2_{vq}\mathscr{L}(\cdot,\lambda,\dot\lambda)\|_{L^2}\|\xi\|_{L^\infty}\|\dot\sigma\|_{L^2}\\
&\le \sqrt{2}\|\partial^2_{vq}\mathscr{L}(\cdot,\zeta,\dot\zeta) - \partial^2_{vq}\mathscr{L}(\cdot,\lambda,\dot\lambda)\|_{L^2}\|\xi\|_{W^{1,2}}\|\sigma\|_{W^{1,2}},\\
|(\mathscr{R}(\zeta) - \mathscr{R}(\lambda))[\sigma,\xi]| &\le \|\partial^2_{qq}\mathscr{L}(\cdot,\zeta,\dot\zeta) - \partial^2_{qq}\mathscr{L}(\cdot,\lambda,\dot\lambda)\|_{L^1}\|\xi\|_{L^\infty}\|\sigma\|_{L^\infty}\\
&\le 2\|\partial^2_{qq}\mathscr{L}(\cdot,\zeta,\dot\zeta) - \partial^2_{qq}\mathscr{L}(\cdot,\lambda,\dot\lambda)\|_{L^1}\|\xi\|_{W^{1,2}}\|\sigma\|_{W^{1,2}}.
\end{aligned}$$

Now, let us consider an arbitrary sequence $\{\zeta_j\} \subset W^{1,2}(\mathbb{I}; U)$ converging to ζ in $W^{1,2}$. This implies that $\{\zeta_j\}$ converges to ζ uniformly and $\{\dot\zeta_j\}$ converges to $\dot\zeta$ in L^2. In particular there exists a function $f \in L^2(\mathbb{I}; \mathbb{R})$ such that $|\dot\zeta_j| < f$ almost everywhere for all j. The L^2 convergence implies that, for every subsequence $\{\dot\zeta_{j(k)}\}$, there exists a further subsequence $\{\dot\zeta_{j(k(h))}\}$ converging to $\dot\zeta$ almost everywhere. Therefore, by **(Q2)**, we can apply the dominated convergence theorem to get

$$a(\cdot,\zeta_{j(k(h))}) \underset{h\to\infty}{\longrightarrow} a(\cdot,\zeta) \text{ in } L^\infty(\mathbb{I}; \mathbb{R}^m),$$
$$\partial^2_{vq}\mathscr{L}(\cdot,\zeta_{j(k(h))},\dot\zeta_{j(k(h))}) \underset{h\to\infty}{\longrightarrow} \partial^2_{vq}\mathscr{L}(\cdot,\zeta,\dot\zeta) \text{ in } L^2(\mathbb{I}; \mathbb{R}^m),$$
$$\partial^2_{qq}\mathscr{L}(\cdot,\zeta_{j(k(h))},\dot\zeta_{j(k(h))}) \underset{h\to\infty}{\longrightarrow} \partial^2_{qq}\mathscr{L}(\cdot,\zeta,\dot\zeta) \text{ in } L^1(\mathbb{I}; \mathbb{R}^m).$$

These convergences, together with the above estimates on $\mathscr{P}$, $\mathscr{Q}$ and $\mathscr{R}$, imply that $\{\mathrm{d}^2\mathscr{A}(\zeta_{j(k(h))})\}$ converges to $\mathrm{d}^2\mathscr{A}(\zeta)$ in $\mathrm{Bil}(W^{1,2}(\mathbb{I}; \mathbb{R}^m))$ as $h \to \infty$. By Lemma 3.4.2 we conclude that the full sequence $\{\mathrm{d}^2\mathscr{A}(\zeta_j)\}$ converges to $\mathrm{d}^2\mathscr{A}(\zeta)$ in $\mathrm{Bil}(W^{1,2}(\mathbb{I}; \mathbb{R}^m))$ as $h \to \infty$. Therefore, by the total differential theorem, the functional $\mathscr{A}$ is C^2 with second Fréchet differential $\mathrm{d}^2\mathscr{A}$. By an inductive argument, one can easily prove that $\mathscr{A}$ is actually C^∞. In fact, knowing that $\mathscr{A}$ is

C^{p-1} for some integer $p > 2$, one easily shows that the pth Gâteaux differential of $\mathscr{A}$, seen as a symmetric bounded multilinear form on $W^{1,2}(\mathbb{I};\mathbb{R}^m)$, is given by

$$\mathrm{d}^p\mathscr{A}(\zeta)[\sigma_1,\sigma_2,\ldots,\sigma_p] = \sum_{j_1,\ldots,j_p=1}^{2m} \int_0^1 [\partial_{j_1}\ldots\partial_{j_p}\mathscr{L}(t,\zeta(t),\dot\zeta(t))]\Sigma_1^{j_1}(t)\ldots\Sigma_p^{j_p}(t)\,\mathrm{d}t,$$

$$\forall\zeta\in W^{1,2}(\mathbb{I};U),\ \Sigma_1,\ldots,\Sigma_p\in W^{1,2}(\mathbb{I};\mathbb{R}^m),$$

where, for each $j = 1,\ldots,p$, we have set $\Sigma_j = (\sigma_j,\dot\sigma_j) \in L^2(\mathbb{I};\mathbb{R}^m\times\mathbb{R}^m)$ and we have adopted the shorthand notation

$$\partial_j = \begin{cases} \partial_{q^j}, & \text{if } j\in\{1,\ldots,m\},\\ \partial_{v^{j-m}}, & \text{if } j\in\{m+1,\ldots,2m\}. \end{cases}$$

Beside the calligraphic complications, with a reasoning analogous to the case $p = 2$ one can prove that the pth Gâteaux differential $\mathrm{d}^p\mathscr{A}$ is continuous, and then conclude that $\mathscr{A}$ is C^p by the total differential theorem. We leave the details to the reader.

In order to conclude the proof, it only remains to establish the "only if" part of the first statement of the proposition. Let us assume that, for some $(t_*,q_*)\in\mathbb{I}\times U$, the function $\mathscr{L}(t_*,q_*,\cdot):\mathbb{R}^m\to\mathbb{R}$ is not a polynomial of degree at most 2. Equivalently, we have that the matrix-valued map $\partial^2_{vv}\mathscr{L}(t_*,q_*,\cdot):\mathbb{R}^m\to\mathbb{R}^{m\times m}$ is not constant. By continuity, there exist an open neighborhood $J\times V\subseteq\mathbb{I}\times U$ of (t_*,q_*), three nonzero vectors $v_*,w_*,z_*\in\mathbb{R}^m$ and $c_*>0$ such that

$$\langle\big[\partial^2_{vv}\mathscr{L}(t,q_*,0)-\partial^2_{vv}\mathscr{L}(t,q,v_*)\big]w_*,z_*\rangle \geq c_*, \qquad \forall(t,q)\in J\times V.$$

Let $\zeta:\mathbb{I}\to U$ be the stationary curve $\zeta\equiv q_*$. For each $\varepsilon>0$ sufficiently small, there exists a measurable subset $J_\varepsilon\subseteq J$ having Lebesgue measure $\mu_{\mathrm{Leb}}(J_\varepsilon)=\varepsilon$, and we can define a curve $\zeta_\varepsilon\in W^{1,2}(\mathbb{I};V)$ by

$$\zeta_\varepsilon(t) := q_* + \left(\int_0^t \chi_{J_\varepsilon}(s)\,\mathrm{d}s\right)v_*, \qquad \forall t\in\mathbb{I}.$$

Here χ_{J_ε} is the characteristic functions of set J_ε. The sequence $\{\zeta_\varepsilon\}$ clearly converges to ζ in $W^{1,2}$ as $\varepsilon\to 0$. Now, notice that the proof of the continuity of $\mathscr{Q}$ and $\mathscr{R}$ given above did not need the assumption that $\mathscr{L}$ is a fiberwise polynomial of degree at most 2, but only the **(Q1)** assumption. Therefore, we have that $\{\mathscr{Q}(\zeta_\varepsilon)\}$ and $\{\mathscr{R}(\zeta_\varepsilon)\}$ converge respectively to $\mathscr{Q}(\zeta)$ and $\mathscr{R}(\zeta)$ in $\mathrm{Bil}(W^{1,2}(\mathbb{I};\mathbb{R}^m))$ as $\varepsilon\to 0$. If we show that $\{\mathscr{P}(\zeta_\varepsilon)\}$ does not converge to $\mathscr{P}(\zeta)$ in $\mathrm{Bil}(W^{1,2}(\mathbb{I};\mathbb{R}^m))$ as $\varepsilon\to 0$, we will immediately obtain that $\mathrm{d}^2\mathscr{A}$ is not continuous at ζ. For each $\varepsilon>0$ as before, we define the vector fields $\xi_\varepsilon,\sigma_\varepsilon\in W^{1,2}(\mathbb{I};\mathbb{R}^m)$ by

$$\xi_\varepsilon(t) := \left(\frac{1}{\sqrt{\varepsilon}\,|w_*|}\int_0^t \chi_{J_\varepsilon}(s)\,\mathrm{d}s\right)w_*, \qquad \sigma_\varepsilon(t) := \left(\frac{1}{\sqrt{\varepsilon}\,|z_*|}\int_0^t \chi_{J_\varepsilon}(s)\,\mathrm{d}s\right)z_*,$$

$$\forall t\in\mathbb{I}.$$

Notice that $\|\xi_\varepsilon\|_{W^{1,2}} = \|\sigma_\varepsilon\|_{W^{1,2}} = 1$. Therefore, we conclude

$$
\begin{aligned}
&\|\mathscr{P}(\zeta) - \mathscr{P}(\zeta_\varepsilon)\|_{\mathrm{Bil}(W^{1,2})} \\
&\qquad = \max\left\{ |(\mathscr{P}(\zeta) - \mathscr{P}(\zeta_\varepsilon))[\xi,\sigma]| \;\middle|\; \|\xi\|_{W^{1,2}} = \|\sigma\|_{W^{1,2}} = 1 \right\} \\
&\qquad \geq |(\mathscr{P}(\zeta) - \mathscr{P}(\zeta_\varepsilon))[\xi_\varepsilon,\sigma_\varepsilon]| \\
&\qquad = \left| \int_0^1 \langle [\partial^2_{vv}\mathscr{L}(t,\zeta,\dot\zeta) - \partial^2_{vv}\mathscr{L}(t,\zeta_\varepsilon,\dot\zeta_\varepsilon)]\dot\xi_\varepsilon(t), \dot\sigma_\varepsilon(t)\rangle \,\mathrm{d}t \right| \\
&\qquad = \left| \frac{1}{\varepsilon\,|w_*|\,|z_*|} \int_{J_\varepsilon} \underbrace{\langle [\partial^2_{vv}\mathscr{L}(t,q_*,0) - \partial^2_{vv}\mathscr{L}(t,\zeta_\varepsilon,v_*)] w_*, z_* \rangle}_{\geq c_*} \,\mathrm{d}t \right| \\
&\qquad \geq \frac{c_*}{|w_*|\,|z_*|}.
\end{aligned}
$$

□

Remark 3.4.2. In [AS09, Proposition 2.3], Abbondandolo and Schwarz actually proved a stronger version of Proposition 3.4.3. Under the same assumptions, they proved that $\mathscr{A}$ is twice Fréchet-differentiable if and only if, for each $(t,q) \in \mathbb{I} \times M$, the function $\mathscr{L}(t,q,\cdot) : \mathrm{T}_qM \to \mathbb{R}$ is a polynomial of degree at most 2, and in this case $\mathscr{A}$ is even C^∞. □

Let $V \subset M \times M$ be either a 0-dimensional manifold $\{(q_0,q_1)\}$ or the diagonal $\{(q,q)\,|\,q \in M\}$. With the notation of Section 3.1 we consider the Hilbert submanifold $W^{1,2}_V(\mathbb{I};M) \subset W^{1,2}(\mathbb{I};M)$, i.e.,

$$
W^{1,2}_V(\mathbb{I};M) = \begin{cases} W^{1,2}_{q_0,q_1}(\mathbb{I};M) & \text{if } V = \{(q_0,q_1)\}, \\ W^{1,2}(\mathbb{T};M) & \text{if } V = \{(q,q)\,|\,q \in M\}. \end{cases}
$$

If the action functional $\mathscr{A} : W^{1,2}(\mathbb{I};M) \to \mathbb{R}$ is C^p, for some positive integer p, then its restriction to $W^{1,2}_V(\mathbb{I};M)$ must be C^p as well. Furthermore, it is easy to check that, after minor modifications, the proof of Theorem 3.4.3 goes through even for the restricted action functional, and therefore we have the following.

Proposition 3.4.4. *Let $\mathscr{L} : \mathbb{I} \times \mathrm{T}M \to \mathbb{R}$ be a smooth Lagrangian satisfying* **(Q2)**, *and $V \subset M \times M$ be either $\{(q_0,q_1)\}$ or $\{(q,q)\,|\,q \in M\}$. Then the associated action functional $\mathscr{A} : W^{1,2}_V(\mathbb{I};M) \to \mathbb{R}$ is C^2 if and only if, for each $(t,q) \in \mathbb{I} \times M$, the function $\mathscr{L}(t,q,\cdot) : \mathrm{T}_qM \to \mathbb{R}$ is a polynomial of degree at most 2.* □

3.5 Critical points of the action functional

Let $\mathscr{L} : \mathbb{I} \times \mathrm{T}M \to \mathbb{R}$ be a convex quadratic-growth Lagrangian, and V a smooth submanifold of $M \times M$. We consider the Euler-Lagrange system of $\mathscr{L}$, which in local coordinates can be written as

$$
\frac{\mathrm{d}}{\mathrm{d}t}\frac{\partial \mathscr{L}}{\partial v^j}(t,\zeta,\dot\zeta) - \frac{\partial \mathscr{L}}{\partial q^j}(t,\zeta,\dot\zeta) = 0, \qquad \forall j = 1,\ldots,m, \tag{3.12}
$$

together with the nonlocal boundary conditions

$$(\zeta(0), \zeta(1)) \in V, \tag{3.13}$$
$$\partial_v \mathscr{L}(0, \zeta(0), \dot{\zeta}(0))\, v_0 = \partial_v \mathscr{L}(1, \zeta(1), \dot{\zeta}(1))\, v_1, \qquad \forall (v_0, v_1) \in \mathrm{T}_{(\zeta(0),\zeta(1))} V, \tag{3.14}$$

where $\partial_v \mathscr{L}$ denotes the fiberwise derivative of $\mathscr{L}$ (see Section 1.1).

Remark 3.5.1. If V is the 0-dimensional manifold $\{(q_0, q_1)\}$, then the boundary condition (3.14) is always satisfied, while the boundary condition (3.13) becomes

$$\zeta(0) = q_0, \qquad \zeta(1) = q_1.$$

If V is the diagonal $\{(q,q) \,|\, q \in M\}$ and the Lagrangian $\mathscr{L}$ is 1-periodic in time (i.e., if it has the form $\mathscr{L} : \mathbb{T} \times \mathrm{T}M \to \mathbb{R}$), then the boundary conditions (3.13) and (3.14) can be rewritten as

$$(\zeta(0), \dot{\zeta}(0)) = (\zeta(1), \dot{\zeta}(1)).$$

Namely, in this case, the solutions of the Euler-Lagrange system subject to the imposed boundary conditions are precisely the 1-periodic orbits. □

Proposition 3.5.1. *Let $\mathscr{L} : \mathbb{I} \times \mathrm{T}M \to \mathbb{R}$ be a convex quadratic-growth Lagrangian, and $V \subset M \times M$ be either $\{(q_0, q_1)\}$ or $\{(q,q) \,|\, q \in M\}$. Then the critical points of the associated action functional $\mathscr{A} : W^{1,2}_V(\mathbb{I}; M) \to \mathbb{R}$ are precisely the (smooth) solutions $\zeta : \mathbb{I} \to M$ of the Euler-Lagrange system* (3.12) *subject to the boundary conditions* (3.13) *and* (3.14).

Proof. Since the statement is of a local nature, we can adopt the localization argument of Remark 3.4.1. More precisely, with the notation employed there, fix a smooth curve $\gamma : \mathbb{I} \to M$ such that $(\gamma(0), \gamma(1)) \in V$ and consider the associated maps

$$\pi_\gamma : \mathbb{I} \times \overline{U} \times \mathbb{R}^m \hookrightarrow \mathbb{I} \times \mathrm{T}M,$$
$$\Pi_\gamma = \Phi_\gamma^{-1} : W^{1,2}_{\mathbb{V}}(\mathbb{I}; U) \to \mathscr{U}_\gamma \cap W^{1,2}_V(\mathbb{I}; M).$$

Here, $\mathbb{V}$ is the trivial vector space $\{\mathbf{0}\}$ if $V = \{(q_0, q_1)\}$, and is defined by (3.3) if $V = \{(q,q) \,|\, q \in M\}$. Note that a curve $\zeta \in \mathscr{U}_\gamma$ is a (smooth) solution of the Euler-Lagrange system of $\mathscr{L}$ subject to the boundary conditions (3.13) and (3.14) if and only if the curve $\zeta_\gamma := \Phi_\gamma(\zeta)$ is a smooth solution of the Euler-Lagrange system of $\mathscr{L}_\gamma := \mathscr{L} \circ \pi_\gamma$ subject to the boundary conditions

$$(\zeta_\gamma(0), \zeta_\gamma(1)) \in \mathbb{V}, \tag{3.15}$$
$$\langle \partial_v \mathscr{L}_\gamma(0, \zeta_\gamma(0), \dot{\zeta}_\gamma(0)), v_0 \rangle = \langle \partial_v \mathscr{L}_\gamma(1, \zeta_\gamma(1), \dot{\zeta}_\gamma(1)), v_1 \rangle, \quad \forall (v_0, v_1) \in \mathbb{V}. \tag{3.16}$$

From now on, we will simply write $\mathscr{L}$ and $\mathscr{A}$ for $\mathscr{L}_\gamma$ and $\mathscr{A} \circ \Pi_\gamma$ respectively.

Let $\zeta \in W^{1,2}_{\mathbb{V}}(\mathbb{I};U)$ be a critical point of $\mathscr{A}$, which means that, for each $\xi \in W^{1,2}_{\mathbb{V}}(\mathbb{I};\mathbb{R}^m)$, we have

$$\mathrm{d}\mathscr{A}(\zeta)\,\xi = \int_0^1 \Big[\langle \partial_q \mathscr{L}(t,\zeta,\dot\zeta),\xi\rangle + \langle \partial_v \mathscr{L}(t,\zeta,\dot\zeta),\dot\xi\rangle\Big]\,\mathrm{d}t = 0. \tag{3.17}$$

If we take any smooth ξ such that $\xi(0)=\xi(1)=\mathbf{0}$, integrating the above equation by parts we obtain

$$\int_0^1 \Big\langle -\int_0^t \partial_q \mathscr{L}(s,\zeta(s),\dot\zeta(s))\,\mathrm{d}s + \partial_v \mathscr{L}(t,\zeta(t),\dot\zeta(t)),\dot\xi(t)\Big\rangle\,\mathrm{d}t = 0.$$

By the Du Bois-Reymond lemma, there exists a vector $w_\zeta \in \mathbb{R}^m$ such that, for almost every $t \in \mathbb{I}$, we have

$$-\int_0^t \partial_q \mathscr{L}(s,\zeta(s),\dot\zeta(s))\,\mathrm{d}s + \partial_v \mathscr{L}(t,\zeta(t),\dot\zeta(t)) = w_\zeta. \tag{3.18}$$

Now, consider the Hamiltonian[1] $\mathscr{H}:\mathbb{I}\times\overline{U}\times\mathbb{R}^m \to \mathbb{R}$ which is Legendre-dual to $\mathscr{L}$. Notice that $\mathscr{L}$ is a Tonelli Lagrangian, since it satisfies **(Q1)**, and therefore $\mathscr{H}$ is a Tonelli Hamiltonian. By Proposition 1.2.2(ii), we have

$$\partial_p \mathscr{H}(t,q,\cdot) = \partial_v \mathscr{L}(t,q,\cdot)^{-1} : \mathbb{R}^m \xrightarrow{\simeq} \mathbb{R}^m, \qquad \forall (t,q) \in \mathbb{I}\times\overline{U},$$

which, together with (3.18), implies

$$\dot\zeta(t) = \partial_p \mathscr{H}\left(t,\zeta(t),w_\zeta + \int_0^t \partial_q \mathscr{L}(s,\zeta(s),\dot\zeta(s))\,\mathrm{d}s\right) \tag{3.19}$$

for almost every $t \in \mathbb{I}$. We can conclude that ζ is smooth by means of a standard **boot-strap** argument: we know that ζ is continuous, since it is in $W^{1,2}_{\mathbb{V}}(\mathbb{I};U)$; if we assume that it is C^p, for some integer $p \geq 0$, then the right-hand side of equation (3.19) is C^p in the variable t, which forces $\dot\zeta$ to be C^p and therefore ζ to be C^{p+1}. Now, by applying a different integration by parts in (3.17), we obtain that

$$\begin{aligned} 0 = &\int_0^1 \langle \partial_q \mathscr{L}(t,\zeta,\dot\zeta) - \frac{\mathrm{d}}{\mathrm{d}t}\partial_v \mathscr{L}(t,\zeta,\dot\zeta),\xi\rangle\,\mathrm{d}t \\ &+ \langle \partial_v \mathscr{L}(0,\zeta(0),\dot\zeta(0)),\xi(0)\rangle - \langle \partial_v \mathscr{L}(1,\zeta(1),\dot\zeta(1)),\xi(1)\rangle \end{aligned} \tag{3.20}$$

for each $\xi \in W^{1,2}_{\mathbb{V}}(\mathbb{I};\mathbb{R}^m)$. Since this equality holds in particular for all the smooth ξ such that $\xi(0)=\xi(1)=\mathbf{0}$, the fundamental lemma of the calculus of variations

[1] In order to simplify the notation, we are making the identification $\mathbb{R}^m \equiv (\mathbb{R}^m)^*$ via the Euclidean inner product.

implies that the curve ζ satisfies the Euler-Lagrange system of $\mathscr{L}$. Therefore, equation (3.20) reduces to

$$\langle \partial_v \mathscr{L}(0,\zeta(0),\dot\zeta(0)),\xi(0)\rangle - \langle \partial_v \mathscr{L}(1,\zeta(1),\dot\zeta(1)),\xi(1)\rangle = 0, \qquad \forall \xi \in W^{1,2}_{\mathrm{V}}(\mathbb{I};\mathbb{R}^m),$$

which is equivalent to the boundary condition (3.16).

Conversely, let $\zeta : \mathbb{I} \to U$ be a solution of the Euler-Lagrange system of $\mathscr{L}$ subject to the boundary conditions (3.15) and (3.16). Since, by condition **(Q1)**, the Lagrangian $\mathscr{L}$ is non-degenerate (see Section 1.1), the orbit ζ must be smooth. Therefore, an integration by parts as in equation (3.20) shows that $\mathrm{d}\mathscr{A}(\zeta)$ vanishes. □

In order to study the critical points of the Lagrangian action functional by means of critical point theory, we have to make sure that its sublevels satisfy some sort of compactness. A sufficient requirement, as discussed in the Appendix, is given by the **Palais-Smale condition.** In [Ben86], Benci proved that this condition holds provided the involved Lagrangian is convex quadratic-growth. Here, we give a proof of this statement following [AF07].

Proposition 3.5.2. *Let $\mathscr{L} : \mathbb{I}\times\mathrm{T}M \to \mathbb{R}$ be a convex quadratic-growth Lagrangian, and $V \subset M \times M$ be either $\{(q_0, q_1)\}$ or $\{(q,q) \mid q \in M\}$. Then the associated action functional $\mathscr{A} : W^{1,2}_V(\mathbb{I};M) \to \mathbb{R}$ satisfies the Palais-Smale condition.*

Proof. Consider an arbitrary sequence $\{\zeta_j \mid j \in \mathbb{N}\} \subset W^{1,2}_V(\mathbb{I};M)$ such that

$$c := \sup\{\mathscr{A}(\zeta_j) \mid j \in \mathbb{N}\} < +\infty, \tag{3.21}$$

$$\lim_{j\to\infty} \|\mathrm{d}\mathscr{A}(\zeta_j)\|_{\zeta_j} = 0, \tag{3.22}$$

where

$$\|\mathrm{d}\mathscr{A}(\zeta_j)\|_{\zeta_j} = \sup\left\{ |\mathrm{d}\mathscr{A}(\zeta_j)\xi| \;\middle|\; \xi \in \mathrm{T}_{\zeta_j} W^{1,2}_V(\mathbb{I};M),\ \|\xi\|_{\zeta_j} = 1 \right\}.$$

In order to conclude we have to prove that there exists a subsequence that converges to some $\zeta \in W^{1,2}_V(\mathbb{I};M)$ in the $W^{1,2}$ topology.

First of all, without loss of generality, we can assume that the Lagrangian $\mathscr{L}$ satisfies (3.10). The sequence of nonnegative real numbers

$$\left\{ \int_0^1 |\dot\zeta_j(t)|^2_{\zeta_j(t)}\mathrm{d}t \;\middle|\; j \in \mathbb{N} \right\}$$

is bounded, for

$$\int_0^1 |\dot\zeta_j(t)|^2_{\zeta_j(t)}\mathrm{d}t \le \underline{\ell}^{-1} \int_0^1 \mathscr{L}(t,\zeta_j(t),\dot\zeta_j(t))\,\mathrm{d}t = \underline{\ell}^{-1}\mathscr{A}(\zeta_j) \le \underline{\ell}^{-1}c.$$

For every $t_0, t_1 \in \mathbb{I}$ with $t_0 \leq t_1$, we have

$$\begin{aligned}\operatorname{dist}(\zeta_j(t_0), \zeta_j(t_1)) &\leq \int_{t_0}^{t_1} |\dot{\zeta}_j(t)|_{\zeta_j(t)} \mathrm{d}t \\ &\leq |t_1 - t_0|^{1/2} \left(\int_0^1 |\dot{\zeta}_j(t)|^2_{\zeta_j(t)} \mathrm{d}t \right)^{1/2} \\ &\leq |t_1 - t_0|^{1/2} (\underline{\ell}^{-1} c)^{1/2}.\end{aligned}$$

This proves that the sequence $\{\zeta_j \,|\, j \in \mathbb{N}\}$ is equi-1/2-Hölder continuous. By the Arzelà-Ascoli theorem, up to passing to a subsequence, $\{\zeta_j \,|\, j \in \mathbb{N}\}$ converges uniformly to some continuous curve $\zeta : \mathbb{I} \to M$ with $(\zeta(0), \zeta(1)) \in V$. Up to further ignoring a finite number of entries of the sequence, we can assume that $\{\zeta_j \,|\, j \in \mathbb{N}\}$ and ζ are contained in some coordinate open set $\mathscr{U}_\gamma \subseteq W^{1,2}_V(\mathbb{I}; M)$ (see Section 3.1 for the notation), where $\gamma : \mathbb{I} \to M$ is a smooth curve which is arbitrarily C^0-close to ζ. Adopting the localization argument of Remark 3.4.1, we consider the maps

$$\pi_\gamma : \mathbb{I} \times \overline{U} \times \mathbb{R}^m \hookrightarrow \mathbb{I} \times \mathrm{T}M,$$
$$\Pi_\gamma = \Phi_\gamma^{-1} : W^{1,2}_{\mathbb{V}}(\mathbb{I}; U) \to \mathscr{U}_\gamma \cap W^{1,2}_V(\mathbb{I}; M),$$

where $\mathbb{V}$ is the trivial vector space $\{\mathbf{0}\}$ if $V = \{(q_0, q_1)\}$ and is defined by (3.3) if $V = \{(q, q) \,|\, q \in M\}$. From now on, we can consider $\mathscr{L}$ to be a Lagrangian of the form $\mathscr{L} : \mathbb{I} \times \overline{U} \times \mathbb{R}^m \to \mathbb{R}$, so that $\{\zeta_j \,|\, j \in \mathbb{N}\}$ and ζ are contained in $W^{1,2}_{\mathbb{V}}(\mathbb{I}; U)$. Therefore, equation (3.22) can be written as

$$\lim_{j \to \infty} \|\mathrm{d}\mathscr{A}(\zeta_j)\|_{(W^{1,2}_{\mathbb{V}})^*} = 0. \tag{3.23}$$

In order to conclude, we must show that $\{\zeta_j \,|\, j \in \mathbb{N}\}$ admits a subsequence that converges to ζ in the Hilbert space $W^{1,2}_{\mathbb{V}}(\mathbb{I}; \mathbb{R}^m)$.

The sequence $\{\zeta_j \,|\, j \in \mathbb{N}\}$ is bounded in $W^{1,2}_{\mathbb{V}}(\mathbb{I}; \mathbb{R}^m)$. Thus, up to a choice of a subsequence, we can assume that it converges to some ζ_* in the weak $W^{1,2}$ topology and uniformly. This forces $\zeta_* = \zeta \in W^{1,2}_{\mathbb{V}}(\mathbb{I}; U)$. Now, it only remains to establish the convergence in the (strong) $W^{1,2}$ topology. By (3.23), we have

$$\begin{aligned}0 &= \lim_{j \to \infty} \mathrm{d}\mathscr{A}(\zeta_j)(\zeta_j - \zeta) \\ &= \underbrace{\int_0^1 \langle \partial_q \mathscr{L}(t, \zeta_j, \dot{\zeta}_j), \zeta_j - \zeta \rangle \,\mathrm{d}t}_{=: I_j} + \underbrace{\int_0^1 \langle \partial_v \mathscr{L}(t, \zeta_j, \dot{\zeta}_j), \dot{\zeta}_j - \dot{\zeta} \rangle \,\mathrm{d}t}_{=: II_j}.\end{aligned}$$

By (3.8), the sequence $\{\partial_q \mathscr{L}(\cdot, \zeta_j, \dot{\zeta}_j) \,|\, j \in \mathbb{N}\}$ is bounded in $L^1(\mathbb{I}; \mathbb{R}^m)$ and, since the sequence $\{\zeta_j - \zeta \,|\, j \in \mathbb{N}\}$ converges to zero uniformly, we obtain that

$$\lim_{j \to \infty} I_j = 0,$$

which also implies

$$\lim_{j\to\infty} \mathit{II}_j = 0. \tag{3.24}$$

Now, condition **(Q1)** implies that, for almost every $t \in \mathbb{I}$,

$$\begin{aligned}
&\langle \partial_v \mathscr{L}(t, \zeta_j, \dot\zeta_j), \dot\zeta_j - \dot\zeta\rangle - \langle \partial_v \mathscr{L}(t, \zeta_j, \dot\zeta), \dot\zeta_j - \dot\zeta\rangle \\
&\qquad = \int_0^1 \langle \partial_{vv} \mathscr{L}(t, \zeta_j, \dot\zeta + s(\dot\zeta_j - \dot\zeta))\,(\dot\zeta_j - \dot\zeta), \dot\zeta_j - \dot\zeta\rangle \mathrm{d}s \\
&\qquad \geq \ell_0 |\dot\zeta_j(t) - \dot\zeta(t)|^2.
\end{aligned}$$

Integrating this inequality, we get

$$\ell_0 \int_0^1 |\dot\zeta_j - \dot\zeta|^2 \,\mathrm{d}t \leq \underbrace{\int_0^1 \langle \partial_v \mathscr{L}(t, \zeta_j, \dot\zeta_j), \dot\zeta_j - \dot\zeta\rangle \,\mathrm{d}t}_{=\mathit{II}_j} - \underbrace{\int_0^1 \langle \partial_v \mathscr{L}(t, \zeta_j, \dot\zeta), \dot\zeta_j - \dot\zeta\rangle \,\mathrm{d}t}_{=:\mathit{III}_j}.$$

By (3.7), we have that

$$\partial_v \mathscr{L}(\cdot, \zeta_j, \dot\zeta) \underset{j\to\infty}{\longrightarrow} \partial_v \mathscr{L}(\cdot, \zeta, \dot\zeta) \text{ in } L^2(\mathbb{I}; \mathbb{R}^m),$$

and since the sequence $\{\dot\zeta_j \,|\, j \in \mathbb{N}\}$ converges to $\dot\zeta$ weakly in the L^2 topology, we conclude that

$$\lim_{j\to\infty} \mathit{III}_j = 0. \tag{3.25}$$

Equations (3.24) and (3.25) imply

$$\lim_{j\to\infty} \int_0^1 |\dot\zeta_j - \dot\zeta|^2 \,\mathrm{d}t \leq \lim_{j\to\infty} \frac{1}{\ell_0}(\mathit{II}_j - \mathit{III}_j) = 0,$$

and therefore that $\{\zeta_j \,|\, j \in \mathbb{N}\}$ converges to ζ in the (strong) $W^{1,2}$ topology. □

A first easy consequence of the Palais-Smale condition is the existence of minima for the action functional.

Proposition 3.5.3. *Let $\mathscr{L} : \mathbb{I}\times \mathrm{T}M \to \mathbb{R}$ be a convex quadratic-growth Lagrangian, $V \subset M \times M$ be either $\{(q_0, q_1)\}$ or $\{(q, q) \,|\, q \in M\}$, $\mathscr{A} : W^{1,2}_V(\mathbb{I}; M) \to \mathbb{R}$ be the action functional associated to $\mathscr{L}$, and $\mathscr{C}$ be either $W^{1,2}_V(\mathbb{I}; M)$ or a connected component of it. Then $\mathscr{A}|_{\mathscr{C}}$ admits a global minimum which is a (smooth) solution of the Euler-Lagrange system of $\mathscr{L}$ subject to the boundary conditions (3.13) and (3.14).*

Proof. The action functional $\mathscr{A}$ is bounded from below by the infimum of $\mathscr{L}$, which is a real number since $\mathscr{L}$ satisfies **(Q1)**. We set $\mathfrak{U} := \{\{\zeta\} \mid \zeta \in \mathscr{C}\}$. By applying the Minimax theorem as in Remark A.6.1, we obtain that

$$\inf_{\zeta \in \mathscr{C}} \{\mathscr{A}(\zeta)\} = \operatorname*{minimax}_{\mathfrak{U}} \mathscr{A}$$

is a critical value of $\mathscr{A}$. Therefore, there exists a critical point $\zeta_{\mathscr{C}}$ of $\mathscr{A}$ such that

$$\mathscr{A}(\zeta_{\mathscr{C}}) = \inf_{\zeta \in \mathscr{C}} \{\mathscr{A}(\zeta)\}.$$

By Proposition 3.5.1, $\zeta_{\mathscr{C}}$ is a (smooth) solution of the Euler-Lagrange system of $\mathscr{L}$ subject to the boundary conditions (3.13) and (3.14). □

As a particular case of the above proposition, we reobtain the Tonelli theorem (cf. Theorem 1.3.1) for convex quadratic-growth Lagrangians. Notice that, in this case, the regularity of the action minimizer (cf. Theorem 1.3.7) is always guaranteed.

Theorem 3.5.4 (Tonelli theorem for convex quadratic-growth Lagrangians). *Let $\mathscr{L} : \mathbb{I} \times \mathrm{T}M \to \mathbb{R}$ be a convex quadratic-growth Lagrangian. For each interval $[t_0, t_1] \subseteq \mathbb{I}$ and for all $q_0, q_1 \in M$ there exists an action minimizer[2] (with respect to $\mathscr{L}$) γ with $\gamma(t_0) = q_0$ and $\gamma(t_1) = q_1$. Furthermore, γ is a smooth solution of the Euler-Lagrange system of $\mathscr{L}$.*

Proof. We denote by $\mathscr{A}^{t_0,t_1}$ the action functional of $\mathscr{L}$ defined on the space $W^{1,2}_{q_0,q_1}([t_0,t_1]; M)$, i.e.,

$$\mathscr{A}^{t_0,t_1}(\zeta) = \int_{t_0}^{t_1} \mathscr{L}(t, \zeta(t), \dot\zeta(t))\, \mathrm{d}t, \qquad \forall \zeta \in W^{1,2}([t_0,t_1]; M).$$

Notice that, for what concerns the functional properties of the action, dealing with the interval $[t_0, t_1]$ is equivalent to dealing with the interval $\mathbb{I}$. By Proposition 3.5.3, $\mathscr{A}^{t_0,t_1}$ admits a global minimum ζ, which is a smooth solution of the Euler-Lagrange system of $\mathscr{L}$. Now, let us assume without loss of generality that the Lagrangian $\mathscr{L}$ satisfies (3.10). For each absolutely continuous curve $\lambda : \mathbb{I} \to \mathbb{R}$, the quantity

$$\int_{t_0}^{t_1} \mathscr{L}(t, \lambda(t), \dot\lambda(t))\, \mathrm{d}t$$

is finite if and only $\lambda \in W^{1,2}([t_0,t_1]; M)$, for

$$\underline{\ell} \int_{t_0}^{t_1} |\dot\lambda(t)|^2_{\lambda(t)}\, \mathrm{d}t \le \int_{t_0}^{t_1} \mathscr{L}(t, \lambda(t), \dot\lambda(t))\, \mathrm{d}t \le \overline{\ell} \left((t_1 - t_0) + \int_{t_0}^{t_1} |\dot\lambda(t)|^2_{\lambda(t)}\, \mathrm{d}t \right).$$

This shows that the minima of $\mathscr{A}^{t_0,t_1}$ are also action minimizers in the sense of Section 1.3. □

[2]Here we mean action minimizer in the sense of Section 1.3.

In the time-periodic case, Proposition 3.5.3 implies the following elementary multiplicity result for periodic orbits.

Theorem 3.5.5. *Let* $\mathscr{L} : \mathbb{T} \times \mathrm{T}M \to \mathbb{R}$ *be a* 1*-periodic convex quadratic-growth Lagrangian. For each conjugacy class* C *of* $\pi_1(M)$*, the Euler-Lagrange system of* $\mathscr{L}$ *admits a* 1*-periodic orbit* γ *that is homotopic to some (and therefore all)* $\zeta : \mathbb{T} \to M$ *representing* C*. Moreover,* γ *minimizes the action among all the absolutely continuous* 1*-periodic curves that are homotopic to* ζ*.*

Proof. Consider an arbitrary ζ that represents C. By Proposition 3.2.2, the space of those $W^{1,2}$ loops that are homotopic to ζ is a connected component $\mathscr{C}$ of $W^{1,2}(\mathbb{T}; M)$. Let $\mathscr{A} : W^{1,2}(\mathbb{T}; M) \to \mathbb{R}$ be the action functional associated to $\mathscr{L}$. By Proposition 3.5.3, $\mathscr{A}|_{\mathscr{C}}$ admits a global minimum γ, which is a (smooth) 1-periodic solution of the Euler-Lagrange system of $\mathscr{L}$. Now, as we showed in the proof of Theorem 3.5.4, for each absolutely continuous curve $\lambda : \mathbb{T} \to M$ the quantity

$$\int_0^1 \mathscr{L}(t, \lambda(t), \dot{\lambda}(t)) \, \mathrm{d}t$$

is finite if and only if $\lambda \in W^{1,2}(\mathbb{T}; M)$. Therefore, γ minimizes the action among all the absolutely continuous 1-periodic curves that are homotopic to ζ. □

3.6 The mean action functional in higher periods

Let $\mathscr{L} : \mathbb{T} \times \mathrm{T}M \to \mathbb{R}$ be a 1-periodic convex quadratic-growth Lagrangian. For each period $n \in \mathbb{N}$, consider the nth iteration map

$$\psi^{[n]} : W^{1,2}(\mathbb{T}; M) \hookrightarrow W^{1,2}(\mathbb{T}^{[n]}; M)$$

introduced in section 3.1. By this map we interpret $W^{1,2}(\mathbb{T}; M)$ as a Hilbert submanifold of $W^{1,2}(\mathbb{T}^{[n]}; M)$. A natural extension of the action functional $\mathscr{A}$ to $W^{1,2}(\mathbb{T}^{[n]}; M)$ is given by the **mean action functional** $\mathscr{A}^{[n]} : W^{1,2}(\mathbb{T}^{[n]}; M) \to \mathbb{R}$, defined as in section 2.2 by

$$\mathscr{A}^{[n]}(\zeta) = \frac{1}{n} \int_0^n \mathscr{L}(t, \zeta(t), \dot{\zeta}(t)) \, \mathrm{d}t, \qquad \forall \zeta \in W^{1,2}(\mathbb{T}^{[n]}; M).$$

Notice that $\mathscr{A}^{[n]} \circ \psi^{[n]} = \mathscr{A}$. Of course, $\mathscr{A}^{[n]}$ has the same functional properties of $\mathscr{A}$. In particular, it is $C^{1,1}$ and its critical points are the n-periodic solutions of the Euler-Lagrange system of $\mathscr{L}$.

Chapter 4

Discretizations

The $W^{1,2}$ functional setting for the action functional $\mathscr{A}$, introduced in Chapter 3, presents several drawbacks. First of all, the regularity that we can expect for $\mathscr{A}$ is only $C^{1,1}$, at least if we work with a general convex quadratic-growth Lagrangian. This prevents the applicability of all those abstract results that require more smoothness, for instance the Morse lemma from critical point theory. Moreover, the $W^{1,2}$ topology is sometimes uncomfortable to work with. In fact, on several occasions it is useful to deal with a topology that is as strong as the C^1 topology, or at least the $W^{1,\infty}$ topology. This would guarantee that the restriction of the action functional $\mathscr{A}$ to a small neighborhood of a loop γ only depends on the values that the Lagrangian assumes on a small neighborhood of the support of the lifted loop $(\gamma, \dot\gamma)$ in the tangent bundle of the configuration space. In order to overcome these difficulties, in this chapter we develop a discretization technique that provides a suitable finite-dimensional setting for the Lagrangian action functional. This approach was pioneered by Morse (see [Mil63, Section 16] or [Kli78, Section A.1]) with his broken geodesics approximation of the path space, and then further investigated in the Finsler case by Rademacher [Rad92] and by Bangert and Long [BL10]. In the Hamiltonian formulation, an analogous approach based on generating functions was followed by Chaperon [Cha84] and later by Laudenbach and Sikorav [LS85] and by Robbin and Salamon [RS93b], see also the book of McDuff and Salamon [MS98, Section 9.2] and the bibliography therein.

In Section 4.1 we prove a stronger version of the Weierstrass theorem (Theorem 1.3.4) asserting the uniqueness of action minimizing curves joining two sufficiently close given points. This result, as well as the forthcoming arguments in the chapter, are valid for the class of convex quadratic-growth Lagrangian functions defined in Chapter 3. In Section 4.2 we introduce the discretization technique, that basically consists in reducing our analysis to the spaces Λ_k (for each integer k sufficiently large) of continuous loops that are k-broken solutions of the Euler-Lagrange system. These loop spaces are finite-dimensional submanifolds of $W^{1,2}(\mathbb{T};M)$, and in particular all the reasonable topologies coincide on them. In Section 4.3 we study the discrete action functional, that is, the restriction of the action functional to the broken Euler-Lagrange loop spaces. We prove that it is smooth and it has compact sublevels, and hence that it is suitable for a Morse-theoretic analysis. In Section 4.4 we study its critical points, proving that they correspond to critical points of the full action functional. Moreover, we prove that the Morse index and nullity of corresponding critical points of the action and of the discrete action functionals are the same. In Section 4.5 we show that the discretization technique can be used to build finite-dimensional homotopic approximations of the action sublevels and, more importantly, that the action and the discrete action functionals have the same local homology groups. In Section 4.6, we apply minimax techniques (see Section A.6) to the discrete action functional in order to prove an important multiplicity result for its critical points, originally due to Benci [Ben86]. This result implies that the Euler-Lagrange system of a convex quadratic-growth Lagrangian has infinitely many 1-periodic solutions, provided the configuration space has finite or abelian fundamental group. Finally, in Section 4.7, we introduce the notation concerning discretizations in an arbitrary period, and we define the iteration map in the discrete setting of broken Euler-Lagrange loops.

4.1 Uniqueness of the action minimizers

Throughout this chapter, M will be a smooth closed manifold of dimension $m \geq 1$, equipped with a Riemannian metric $\langle\cdot,\cdot\rangle$. that turns it into a complete metric space with respect to the induced Riemannian distance $\mathrm{dist} : M \times M \to [0,\infty)$. Let $\mathscr{L} : \mathbb{T} \times \mathrm{T}M \to \mathbb{R}$ be a 1-periodic convex quadratic-growth Lagrangian. As we have already shown in Section 3.3, without loss of generality we can assume that there exist two positive constants $\underline{\ell} \leq \overline{\ell}$ such that

$$\underline{\ell}\,|v|_q^2 \leq \mathscr{L}(t,q,v) \leq \overline{\ell}(|v|_q^2+1), \qquad \forall (t,q,v) \in \mathbb{T} \times \mathrm{T}M. \tag{4.1}$$

For every compact interval $[t_0,t_1] \subset \mathbb{R}$ and for every absolutely continuous curve $\zeta : [t_0,t_1] \to M$, we denote by $\mathscr{A}^{t_0,t_1}(\zeta)$ the usual Lagrangian action of the curve, i.e.,

$$\mathscr{A}^{t_0,t_1}(\zeta) = \int_{t_0}^{t_1} \mathscr{L}(t,\zeta(t),\dot\zeta(t))\,\mathrm{d}t \in \mathbb{R} \cup \{+\infty\}\,.$$

Notice that, by (4.1), $\mathscr{A}^{t_0,t_1}(\zeta)$ is finite if and only if the derivative of ζ is an L^2 curve. In other words, $\mathscr{A}^{t_0,t_1}$ is a well-defined functional of the form

$$\mathscr{A}^{t_0,t_1} : W^{1,2}([t_0,t_1];M) \to \mathbb{R},$$

and we know, by Proposition 3.4.1, that it is $C^{1,1}$ and twice Gâteaux-differentiable. The Weierstrass theorem (Theorem 1.3.4) asserts that, for each $C_0 > 0$, the restriction of the action functional $\mathscr{A}^{t_0,t_1}$ to the space

$$W^{1,2}_{q_0,q_1}([t_0,t_1];M) = \{\zeta \in W^{1,2}([t_0,t_1];M) \mid \zeta(t_0) = q_0,\ \zeta(t_1) = q_1\}$$

admits a unique global minimum provided that $t_1 - t_0$ is small enough and $\mathrm{dist}(q_0,q_1) \leq C_0(t_1 - t_0)$. For our purposes, we will need the following stronger version of this result, that requires the convex quadratic-growth assumptions on the involved Lagrangian. Under slightly stronger assumptions this result can be found in Mañé [Mn91, page 53], while the current version is taken from [Maz11].

Theorem 4.1.1 (Uniqueness of the action minimizers). *Let $\mathscr{L} : \mathbb{T} \times \mathrm{T}M \to \mathbb{R}$ be a 1-periodic convex quadratic-growth Lagrangian. Then there exist $\varepsilon_0 = \varepsilon_0(\mathscr{L}) > 0$ and $\rho_0 = \rho_0(\mathscr{L}) > 0$ such that, for each interval $[t_0,t_1] \subset \mathbb{R}$ with $0 < t_1 - t_0 \leq \varepsilon_0$ and for all $q_0, q_1 \in M$ with $\mathrm{dist}(q_0,q_1) < \rho_0$, there is a unique action minimizer (with respect to $\mathscr{L}$) $\gamma_{q_0,q_1} : [t_0,t_1] \to M$ with $\gamma_{q_0,q_1}(t_0) = q_0$ and $\gamma_{q_0,q_1}(t_1) = q_1$. Moreover, γ_{q_0,q_1} is a smooth solution of the Euler-Lagrange system of $\mathscr{L}$.*

Proof. Let us consider two arbitrary points $q_0, q_1 \in M$ and two arbitrary real numbers $t_0 < t_1$. We set

$$\rho := \mathrm{dist}(q_0,q_1), \qquad \varepsilon := t_1 - t_0.$$

In order to prove the statement, we have to show that, for ρ and ε sufficiently small, the functional $\mathscr{A}^{t_0,t_1} : W^{1,2}_{q_0,q_1}([t_0,t_1];M) \to \mathbb{R}$ admits a unique global minimum. Let us fix a real constant $\mu > 1$. By compactness, the manifold M admits a finite atlas $\mathfrak{U} = \{(U_\alpha, \phi_\alpha) \mid \alpha = 1,\dots,u\}$ such that for all $\alpha \in \{1,\dots,u\}$, $q, q' \in U_\alpha$ and $v \in \mathrm{T}_q M$ we have

$$\mu^{-1} \left|\phi_\alpha(q) - \phi_\alpha(q')\right| \leq \mathrm{dist}(q,q') \leq \mu \left|\phi_\alpha(q) - \phi_\alpha(q')\right|, \tag{4.2}$$

$$\mu^{-1} \left|\mathrm{d}\phi_\alpha(q)v\right| \leq |v|_q \leq \mu \left|\mathrm{d}\phi_\alpha(q)v\right|. \tag{4.3}$$

Here we have denoted by $|\cdot|$ the Euclidean norm in $\mathbb{R}^m$ and by $|\cdot|_q$ the Riemannian norm in $\mathrm{T}_q M$ as usual. Without loss of generality, we can further assume that the image $\phi_\alpha(U_\alpha)$ of every chart is a convex ball of $\mathbb{R}^m$. Let $\mathrm{Leb}(\mathfrak{U})$ denote the Lebesgue number of the atlas $\mathfrak{U}$ and consider the two points $q_0, q_1 \in M$ chosen at the beginning with $\mathrm{dist}(q_0,q_1) = \rho$. By definition of the Lebesgue number, the Riemannian closed ball

$$\overline{B(q_0, \mathrm{Leb}(\mathfrak{U})/2)} = \{q \in M \mid \mathrm{dist}(q,q_0) \leq \mathrm{Leb}(\mathfrak{U})/2\}$$

is contained in a coordinate open set U_α for some $\alpha \in \{1,\dots,u\}$. Therefore if we require that $\rho \leq \mathrm{Leb}(\mathfrak{U})/2$ the points q_0 and q_1 lie in the same open set U_α.

Let $r : [t_0,t_1] \to U_\alpha$ be the segment from q_0 to q_1 given by

$$r(t) = \phi_\alpha^{-1}\left(\frac{t_1 - t}{\varepsilon}\,\phi_\alpha(q_0) + \frac{t - t_0}{\varepsilon}\,\phi_\alpha(q_1)\right), \qquad \forall t \in [t_0,t_1].$$

Assuming without loss of generality that $\mathscr{L}$ satisfies (4.1), by (4.2) and (4.3) we obtain

$$\begin{aligned}\mathscr{A}^{t_0,t_1}(r) &\leq \overline{\ell}\left(\int_{t_0}^{t_1}|\dot r(t)|^2_{r(t)}\mathrm{d}t+\varepsilon\right)\leq \overline{\ell}\left(\varepsilon\max_{t\in[t_0,t_1]}\left\{|\dot r(t)|^2_{r(t)}\right\}+\varepsilon\right)\\ &\leq \overline{\ell}\left(\mu^2\frac{|\phi_\alpha(q_1)-\phi_\alpha(q_0)|^2}{\varepsilon}+\varepsilon\right)\leq \overline{\ell}\left(\mu^4\frac{\mathrm{dist}(q_0,q_1)^2}{\varepsilon}+\varepsilon\right)\\ &\leq \overline{\ell}\mu^4\left(\frac{\rho^2}{\varepsilon}+\varepsilon\right)=C\left(\frac{\rho^2}{\varepsilon}+\varepsilon\right).\end{aligned}$$

Notice that the positive constant $C:=\overline{\ell}\mu^4$ does not depend on q_0,q_1 and $[t_0,t_1]$. This estimate, in turn, furnishes an upper bound for the action of the minima, i.e.,

$$\min\left\{\mathscr{A}^{t_0,t_1}(\zeta)\;\middle|\;\zeta\in W^{1,2}_{q_0,q_1}([t_0,t_1];M)\right\}\leq C\left(\frac{\rho^2}{\varepsilon}+\varepsilon\right).$$

In particular, the action sublevel

$$\mathscr{U}^{t_0,t_1}_{q_0,q_1}(\rho,\varepsilon)=\left\{\zeta\in W^{1,2}_{q_0,q_1}([t_0,t_1];M)\;\middle|\;\mathscr{A}^{t_0,t_1}(\zeta)\leq C\left(\frac{\rho^2}{\varepsilon}+\varepsilon\right)\right\}\tag{4.4}$$

is not empty. By Theorem 3.5.4, this action sublevel must contain a global minimum γ_{q_0,q_1} of $\mathscr{A}^{t_0,t_1}$, which is also a smooth solution of the Euler-Lagrange system of $\mathscr{L}$. All we have to do in order to conclude is to show that, for ρ and ε sufficiently small, the sublevel $\mathscr{U}^{t_0,t_1}_{q_0,q_1}=\mathscr{U}^{t_0,t_1}_{q_0,q_1}(\rho,\varepsilon)$ cannot contain other minima of the action.

By the first inequality in (4.1) we have

$$\int_{t_0}^{t_1}|\dot\zeta(t)|^2_{\zeta(t)}\mathrm{d}t\leq\underline{\ell}^{-1}\mathscr{A}^{t_0,t_1}(\zeta),\qquad \forall\zeta\in W^{1,2}_{q_0,q_1}([t_0,t_1];M),$$

and this, in turn, gives the following bound for all $\zeta\in\mathscr{U}^{t_0,t_1}_{q_0,q_1}$:

$$\begin{aligned}\max_{t\in[t_0,t_1]}\mathrm{dist}(\zeta(t_0),\zeta(t))^2&\leq\left(\int_{t_0}^{t_1}|\dot\zeta(t)|_{\zeta(t)}\mathrm{d}t\right)^2\leq\varepsilon\int_{t_0}^{t_1}|\dot\zeta(t)|^2_{\zeta(t)}\mathrm{d}t\\ &\leq\varepsilon\underline{\ell}^{-1}\mathscr{A}^{t_0,t_1}(\zeta)\leq C\underline{\ell}^{-1}(\rho^2+\varepsilon^2).\end{aligned}$$

Therefore, all the curves $\zeta\in\mathscr{U}^{t_0,t_1}_{q_0,q_1}(\rho,\varepsilon)$ have image inside the coordinate open set $U_\alpha\subseteq M$ provided ρ and ε satisfy

$$\rho^2+\varepsilon^2\leq\frac{\underline{\ell}}{4C}\mathrm{Leb}(\mathfrak{U})^2.\tag{4.5}$$

This allows us to restrict our attention to the open set U_α. From now on we will identify U_α with $\phi_\alpha(U_\alpha)\subseteq\mathbb{R}^m$, so that

$$q_0\equiv\phi_\alpha(q_0)\in\mathbb{R}^m,\qquad q_1\equiv\phi_\alpha(q_1)\in\mathbb{R}^m.$$

Without loss of generality we can also assume that $q_0 \equiv \phi_\alpha(q_0) = \mathbf{0} \in \mathbb{R}^m$. We will also consider $\mathscr{L}$ to be a convex quadratic-growth Lagrangian of the form

$$\mathscr{L} : \mathbb{T} \times \phi_\alpha(U_\alpha) \times \mathbb{R}^m \to \mathbb{R}$$

by means of the identification

$$\mathscr{L}(t, q, v) \equiv \mathscr{L}(t, \phi_\alpha^{-1}(q), \mathrm{d}\phi_\alpha^{-1}(\phi_\alpha(q))v).$$

Now, consider the following closed convex subset of $W^{1,2}([t_0, t_1]; \mathbb{R}^m)$:

$$\mathscr{C}_{q_0,q_1}^{t_0,t_1} = \mathscr{C}_{q_0,q_1}^{t_0,t_1}(\rho, \varepsilon) = \Big\{ \zeta \in W^{1,2}([t_0, t_1]; \mathbb{R}^m) \,\Big|\, , \\ \zeta(t_0) = q_0 = \mathbf{0},\ \zeta(t_1) = q_1,\ \|\dot\zeta\|_{L^2}^2 \le \mu C \underline{\ell}^{-1} \left(\frac{\rho^2}{\varepsilon} + \varepsilon \right) \Big\}. \quad (4.6)$$

Since $\|\zeta\|_{L^\infty}^2 \le \varepsilon \|\dot\zeta\|_{L^2}^2$, for ρ and ε sufficiently small all the curves $\zeta \in \mathscr{C}_{q_0,q_1}^{t_0,t_1}$ have support inside the open set U_α. Moreover, by (4.1), (4.3) and (4.4) we have

$$\|\dot\zeta\|_{L^2}^2 \le \mu \int_{t_0}^{t_1} |\dot\zeta(t)|_{\zeta(t)}^2 \mathrm{d}t \le \mu \underline{\ell}^{-1} \mathscr{A}^{t_0,t_1}(\zeta) \le \mu C \underline{\ell}^{-1} \left(\frac{\rho^2}{\varepsilon} + \varepsilon \right), \qquad \forall \zeta \in \mathscr{U}_{q_0,q_1}^{t_0,t_1},$$

which implies

$$\mathscr{U}_{q_0,q_1}^{t_0,t_1} \subseteq \mathscr{C}_{q_0,q_1}^{t_0,t_1}.$$

Now, since all the minima of $\mathscr{A}^{t_0,t_1}$ lie in the closed convex set $\mathscr{C}_{q_0,q_1}^{t_0,t_1}$, in order to conclude that γ_{q_0,q_1} is the unique minimum we only need to show that the Hessian of the action is positive-definite on $\mathscr{C}_{q_0,q_1}^{t_0,t_1}$ provided ρ and ε are sufficiently small. Namely, we need to show that there exist $\rho_0 > 0$ and $\varepsilon_0 > 0$ such that, for all $\rho \in (0, \rho_0)$ and $\varepsilon \in (0, \varepsilon_0]$, we have

$$\mathrm{Hess}\mathscr{A}^{t_0,t_1}(\zeta)[\sigma, \sigma] > 0, \quad \forall \zeta \in \mathscr{C}_{q_0,q_1}^{t_0,t_1}(\rho, \varepsilon),\ \sigma \in W_0^{1,2}([t_0, t_1]; \mathbb{R}^m). \quad (4.7)$$

As usual, we have denoted by $W_0^{1,2}([t_0, t_1]; \mathbb{R}^m)$ the tangent space to $\mathscr{C}_{q_0,q_1}^{t_0,t_1}$ at ζ, i.e.,

$$W_0^{1,2}([t_0, t_1]; \mathbb{R}^m) = \left\{ \sigma \in W^{1,2}([t_0, t_1]; \mathbb{R}^m) \,|\, \sigma(t_0) = \sigma(t_1) = \mathbf{0} \right\}.$$

For every $\zeta \in \mathscr{C}_{q_0,q_1}^{t_0,t_1}$ and $\sigma \in W_0^{1,2}([t_0, t_1]; \mathbb{R}^m)$, we have

$$\begin{aligned} &\mathrm{Hess}\mathscr{A}^{t_0,t_1}(\zeta)[\sigma, \sigma] \\ &= \int_{t_0}^{t_1} \Big(\langle \partial_{vv}^2 \mathscr{L}(t, \zeta, \dot\zeta)\dot\sigma, \dot\sigma \rangle + 2\langle \partial_{vq}^2 \mathscr{L}(t, \zeta, \dot\zeta)\sigma, \dot\sigma \rangle + \langle \partial_{qq}^2 \mathscr{L}(t, \zeta, \dot\zeta)\sigma, \sigma \rangle \Big) \mathrm{d}t \\ &\ge \int_{t_0}^{t_1} \ell_0 |\dot\sigma|^2 \mathrm{d}t - \underbrace{\int_{t_0}^{t_1} 2\ell_1(1 + \mu|\dot\zeta|)\, |\sigma|\, |\dot\sigma| \,\mathrm{d}t}_{=: I_1} - \underbrace{\int_{t_0}^{t_1} \ell_1(1 + \mu^2|\dot\zeta|^2)\, |\sigma|^2 \,\mathrm{d}t}_{=: I_2}, \end{aligned}$$

where ℓ_0 and ℓ_1 are the positive constants that appear in **(Q1)** and **(Q2)** with respect to the atlas $\mathfrak{U}$. Now, the quantities I_1 and I_2 can be estimated from above as follows:

$$\begin{aligned} I_1 &\leq 2\ell_1\mu\|\sigma\|_{L^\infty}\left(\|\dot\sigma\|_{L^1} + \left\||\dot\zeta|\cdot|\dot\sigma|\right\|_{L^1}\right)\\ &\leq 2\ell_1\mu\sqrt{\varepsilon}\|\dot\sigma\|_{L^2}\left(\sqrt{\varepsilon}\|\dot\sigma\|_{L^2} + \|\dot\zeta\|_{L^2}\|\dot\sigma\|_{L^2}\right)\\ &= 2\ell_1\mu\|\dot\sigma\|_{L^2}^2\left(\varepsilon + \sqrt{\varepsilon}\|\dot\zeta\|_{L^2}\right),\\ I_2 &\leq \ell_1\mu^2\left(\|\sigma\|_{L^2}^2 + \|\sigma\|_{L^\infty}^2\|\dot\zeta\|_{L^2}^2\right) \leq \ell_1\mu^2\|\dot\sigma\|_{L^2}^2\left(\varepsilon^2 + \varepsilon\|\dot\zeta\|_{L^2}^2\right), \end{aligned}$$

and since by (4.6) we have

$$\|\dot\zeta\|_{L^2}^2 \leq \mu C\underline{\ell}^{-1}\left(\frac{\rho^2}{\varepsilon} + \varepsilon\right),$$

we conclude that

$$\begin{aligned} &\mathrm{Hess}\mathscr{A}^{t_0,t_1}(\zeta)[\sigma,\sigma]\\ &\quad\geq \ell_0\|\dot\sigma\|_{L^2}^2 - I_1 - I_2\\ &\quad\geq \|\dot\sigma\|_{L^2}^2\underbrace{\left(\ell_0 - 2\ell_1\mu\left(\sqrt{\mu C\underline{\ell}^{-1}} + 1\right)(\rho+\varepsilon) - \ell_1\mu^2\left(\mu C\underline{\ell}^{-1} + 1\right)(\rho^2 + \varepsilon^2)\right)}_{=:\,F(\rho,\varepsilon)}. \end{aligned}$$

Notice that the quantity $F(\rho,\varepsilon)$ is independent of the specific choice of the points q_0, q_1 and of the interval $[t_0,t_1]$, but depends only on $\rho = \mathrm{dist}(q_0,q_1)$ and $\varepsilon = t_1 - t_0$. Moreover, there exist $\rho_0 > 0$ and $\varepsilon_0 > 0$ small enough so that for all $\rho \in (0,\rho_0)$ and $\varepsilon \in (0,\varepsilon_0]$ the quantity $F(\rho,\varepsilon)$ is positive. This proves (4.7). □

Now, we wish to prove that the unique action minimizers γ_{q_0,q_1}, whose existence is asserted by the previous theorem, depend smoothly on their endpoints q_0 and q_1. If ρ_0 is the constant given by Theorem 4.1.1, we denote by $\Delta(\rho_0)$ the open neighborhood of the diagonal submanifold of $M \times M$ given by

$$\Delta(\rho_0) = \{(q_0,q_1) \in M \times M \mid \mathrm{dist}(q_0,q_1) < \rho_0\}.$$

Theorem 4.1.2 (Smooth dependence on endpoints). *With the notation of Theorem 4.1.1, for each real interval $[t_0,t_1] \subset \mathbb{R}$ with $0 < t_1 - t_0 \leq \varepsilon_0$ the assignment*

$$(q_0,q_1) \mapsto \gamma_{q_0,q_1} : [t_0,t_1] \to M \tag{4.8}$$

defines a smooth map $\Delta(\rho_0) \to C^\infty([t_0,t_1];M)$.

Proof. By Theorem 4.1.1, the map given by (4.8) is well defined and has the form $\Delta(\rho_0) \to C^\infty([t_0,t_1];M)$. Thus, we just need to show that the dependence of γ_{q_0,q_1}

on (q_0, q_1) is smooth. For $\varepsilon := t_1 - t_0 \in (0, \varepsilon_0]$ and $(q_0, q_1) \in \Delta(\rho_0)$, in the proof of Theorem 4.1.1 we have already shown that $\gamma_{q_0,q_1} : [t_0, t_1] \to M$ has image contained in a coordinate neighborhood $U_\alpha \subseteq M$ that we can identify with an open set of $\mathbb{R}^m$. Since γ_{q_0,q_1} is a smooth solution of the Euler-Lagrange system of $\mathscr{L}$, we have

$$\Phi_{\mathscr{L}}^{t,t_0}(q_0, v_0) = (\gamma_{q_0,q_1}(t), \dot{\gamma}_{q_0,q_1}(t)), \qquad \forall t \in [t_0, t_1],$$

where $v_0 = \dot{\gamma}_{q_0,q_1}(t_0)$ and $\Phi_{\mathscr{L}}^{t,t_0}$ is the Euler-Lagrange flow associated to $\mathscr{L}$ (see section 1.1). We define

$$Q_{\mathscr{L}}^{t,t_0} := \tau \circ \Phi_{\mathscr{L}}^{t,t_0} : U'_\alpha \times V'_\alpha \to U_\alpha, \qquad \forall t \in [t_0, t_1],$$

where $U'_\alpha \subset U_\alpha$ is a small neighborhood of q_0, $V'_\alpha \subset \mathbb{R}^m$ is a small neighborhood of v_0, and $\tau : \mathbb{R}^m \times \mathbb{R}^m \to \mathbb{R}^m$ is the projection onto the first m components, i.e., $\tau(q, v) = q$ for all $(q, v) \in \mathbb{R}^m \times \mathbb{R}^m$. We claim that

$$\mathrm{d}Q_{\mathscr{L}}^{t_1,t_0}(q_0, v_0)(\{\mathbf{0}\} \times \mathbb{R}^m) = \mathbb{R}^m. \tag{4.9}$$

In fact, assume by contradiction that (4.9) does not hold. Then there exists a nonzero vector $v \in \mathbb{R}^m$ such that

$$\left.\frac{\mathrm{d}}{\mathrm{d}s}\right|_{s=0} Q_{\mathscr{L}}^{t_1,t_0}(q_0, v_0 + sv) = \mathbf{0}.$$

Let us define the curve $\sigma : [t_0, t_1] \to \mathbb{R}^m$ by

$$\sigma(t) := \left.\frac{\mathrm{d}}{\mathrm{d}s}\right|_{s=0} Q_{\mathscr{L}}^{t,t_0}(q_0, v_0 + sv), \qquad \forall t \in [t_0, t_1].$$

Thus $\sigma(t_0) = \sigma(t_1) = \mathbf{0}$. Now notice that the curve $t \mapsto Q_{\mathscr{L}}^{t,t_0}(q_0, v_0 + sv)$ is a solution of the Euler-Lagrange system of $\mathscr{L}$. By differentiating this system in s at $s = 0$, we obtain that σ is a solution of the linearized Euler-Lagrange system

$$\frac{\mathrm{d}}{\mathrm{d}t}(a\,\dot{\sigma} + b\,\sigma) - b^T\,\dot{\sigma} - c\,\sigma = \mathbf{0},$$

where, for each $t \in [t_0, t_1]$, we have put

$$\begin{aligned} a(t) &= \partial^2_{vv}\mathscr{L}(t, \gamma_{q_0,q_1}, \dot{\gamma}_{q_0,q_1}), \\ b(t) &= \partial^2_{vq}\mathscr{L}(t, \gamma_{q_0,q_1}, \dot{\gamma}_{q_0,q_1}), \\ c(t) &= \partial^2_{qq}\mathscr{L}(t, \gamma_{q_0,q_1}, \dot{\gamma}_{q_0,q_1}). \end{aligned}$$

This implies that

$$\begin{aligned} \mathrm{Hess}\mathscr{A}^{t_0,t_1}(\gamma_{q_0,q_1})[\sigma, \sigma] &= \int_{t_0}^{t_1} \big(\langle a\,\dot{\sigma}, \dot{\sigma}\rangle + 2\langle b\,\sigma, \dot{\sigma}\rangle + \langle c\,\sigma, \sigma\rangle\big)\,\mathrm{d}t \\ &= \int_{t_0}^{t_1} \langle -\tfrac{\mathrm{d}}{\mathrm{d}t}(a\,\dot{\sigma} + b\,\sigma) + b^T\,\dot{\sigma} + c\,\sigma, \sigma\rangle\,\mathrm{d}t \\ &= 0, \end{aligned}$$

which contradicts the positive definitiveness of $\mathrm{Hess}\mathscr{A}^{t_0,t_1}(\gamma_{q_0,q_1})$ (see (4.7) in the proof of Theorem 4.1.1). Therefore, (4.9) must hold.

By the implicit function theorem we obtain a neighborhood $U_{q_0,q_1} \subset \mathbb{R}^m \times \mathbb{R}^m$ of (q_0, q_1), a neighborhood $U_{v_0} \subset \mathbb{R}^m$ of v_0, and a smooth map $w_0 : U_{q_0,q_1} \to U_{v_0}$ such that, for each $(q_0', q_1', v_0') \in U_{q_0,q_1} \times U_{v_0}$, we have $Q_{\mathscr{L}}^{t_1,t_0}(q_0', v_0') = q_1'$ if and only if $v_0' = w_0(q_0', q_1')$. Thus we can define a smooth map from U_{q_0,q_1} to $C^\infty([t_0, t_1]; U_\alpha)$ by

$$(q_0', q_1') \mapsto \zeta_{q_0',q_1'}, \tag{4.10}$$

where for each $t \in [t_0, t_1]$ we have

$$\zeta_{q_0',q_1'}(t) := Q_{\mathscr{L}}^{t,t_0}(q_0', w_0(q_0', q_1')).$$

In order to conclude we only have to show that the map in (4.10) coincides with the one in (4.8) on U_{q_0,q_1} provided this latter neighborhood is small enough, i.e., after possibly shrinking U_{q_0,q_1} we have to show that $\zeta_{q_0',q_1'}$ is a unique action minimizer for all $(q_0', q_1') \in U_{q_0,q_1}$. This is easily seen as follows. By construction, the curves $\zeta_{q_0',q_1'}$ are critical points of the action

$$\mathscr{A}^{t_0,t_1} : W^{1,2}_{q_0',q_1'}([t_0, t_1]; U_\alpha) \to \mathbb{R},$$

being solutions of the Euler-Lagrange system of $\mathscr{L}$. By the arguments in the proof of Theorem 4.1.1, each of these curves $\zeta_{q_0',q_1'}$ is a unique action minimizer if and only if it lies in the convex set $\mathscr{C}^{t_0,t_1}_{q_0',q_1'}$ defined in (4.6). We already know that

$$\zeta_{q_0,q_1} = \gamma_{q_0,q_1} \in \mathscr{C}^{t_0,t_1}_{q_0,q_1}.$$

Since the map in (4.10) is smooth, for (q_0', q_1') close to (q_0, q_1) we obtain that the curve $\zeta_{q_0',q_1'}$ is C^1-close to $\zeta_{q_0,q_1} = \gamma_{q_0,q_1}$, and therefore

$$\zeta_{q_0',q_1'} \in \mathscr{C}^{t_0,t_1}_{q_0',q_1'}. \qquad \square$$

4.2 The broken Euler-Lagrange loop spaces

For each $k \in \mathbb{N}$, let $\mathbb{Z}_k := \mathbb{Z}/k\mathbb{Z}$ be the cyclic group of order k. We consider the space $C^\infty_k(\mathbb{T}; M)$ of **continuous and k-broken smooth loops**, which consists of those continuous loops $\zeta : \mathbb{T} \to M$ such that, for each $j \in \mathbb{Z}_k$, the restriction $\zeta|_{[j/k,(j+1)/k]}$ is smooth. We can endow $C^\infty_k(\mathbb{T}; M)$ with a topology that turns it into a Fréchet manifold (see for instance [Ham82, Section I.4]) in the following way. Let us denote by $\mathscr{F}_k$ the product of Fréchet manifolds

$$C^\infty([0, \tfrac{1}{k}]; M) \times C^\infty([\tfrac{1}{k}, \tfrac{2}{k}]; M) \times \cdots \times C^\infty([\tfrac{k-1}{k}, 1]; M).$$

Consider the smooth submersion

$$\pi_k : \mathscr{F}_k \to \underbrace{M \times \cdots \times M}_{2k \text{ times}}$$

given by

$$\pi_k(\zeta_0, \dots, \zeta_{k-1}) = (\zeta_0(0), \zeta_0(\tfrac{1}{k}), \zeta_1(\tfrac{1}{k}), \zeta_1(\tfrac{2}{k}), \dots, \zeta_{k-1}(\tfrac{k-1}{k}), \zeta_{k-1}(1)),$$
$$\forall (\zeta_0, \dots, \zeta_{k-1}) \in \mathscr{F}_k.$$

Moreover, consider the smooth submanifold of the $2k$-fold product of M given by

$$\Omega_k := \left\{ (q_0, q_0', q_1, q_1', \dots, q_{k-1}, q_{k-1}') \in M \times \cdots \times M \,|\, q_j' = q_{j+1} \ \forall j \in \mathbb{Z}_k \right\}.$$

The space $C_k^\infty(\mathbb{T}; M)$ is precisely the preimage of Ω_k by the submersion π_k, i.e.,

$$C_k^\infty(\mathbb{T}; M) = \pi_k^{-1}(\Omega_k).$$

Therefore $\mathscr{F}_k$ induces a Fréchet manifold structure on $C_k^\infty(\mathbb{T}; M)$ that makes it a closed Fréchet submanifold[1]. It can be easily shown that $C_k^\infty(\mathbb{T}; M)$, with the topology induced by its Fréchet structure, smoothly embeds into the Hilbert manifold $W^{1,2}(\mathbb{T}; M)$.

Let $\mathscr{L} : \mathbb{T} \times TM \to \mathbb{R}$ be a 1-periodic convex quadratic-growth Lagrangian. We consider the positive constants $\varepsilon_0 = \varepsilon_0(\mathscr{L})$ and $\rho_0 = \rho_0(\mathscr{L})$ given by Theorem 4.1.1 and, for each $k \in \mathbb{N}$, we denote by $\Delta_k = \Delta_k(\rho_0(\mathscr{L}))$ the neighborhood of the diagonal submanifold of the k-fold product $M \times \cdots \times M$ given by

$$\Delta_k := \left\{ (q_0, \dots, q_{k-1}) \in M \times \cdots \times M \,|\, \mathrm{dist}(q_j, q_{j+1}) < \rho_0 \ \forall j \in \mathbb{Z}_k \right\}.$$

By Theorem 4.1.2, for each integer $k \geq 1/\varepsilon_0(\mathscr{L})$, we can define a smooth embedding

$$\lambda_k = \lambda_{k,\mathscr{L}} : \Delta_k \hookrightarrow C_k^\infty(\mathbb{T}; M) \tag{4.11}$$

in the following way: for each $\boldsymbol{q} = (q_0, \dots, q_{k-1}) \in \Delta_k$ we put $\lambda_k(\boldsymbol{q}) := \gamma_{\boldsymbol{q}}$, where $\gamma_{\boldsymbol{q}}$ is the loop whose restrictions $\gamma_{\boldsymbol{q}}|_{[j/k,(j+1)/k]}$ are the unique action minimizers (with respect to $\mathscr{L}$) with endpoints q_j and q_{j+1}, for each $j \in \mathbb{Z}_k$. We define the **k-broken Euler-Lagrange loop space** (with respect to $\mathscr{L}$) as

$$\Lambda_k = \Lambda_{k,\mathscr{L}} := \lambda_k(\Delta_k).$$

Notice that Λ_k is a smooth submanifold of $C_k^\infty(\mathbb{T}; M)$ (and of $W^{1,2}(\mathbb{T}; M)$) with finite dimension km, where m is the dimension of M.

Remark 4.2.1 (Localization). The differentiable atlas of $W^{1,2}(\mathbb{T}; M)$ introduced in Remark 3.1.1 induces a nice atlas on the open manifold Δ_k in the following way. Fix a point $\boldsymbol{q}' \in \Delta_k$ and assume, for simplicity, that the associated loop

[1] This construction is an example of **fiber product** of Fréchet manifolds (see [Ham82, page 93]).

$\gamma_{\boldsymbol{q}'} = \lambda_k(\boldsymbol{q}')$ is contractible. Consider any smooth contractible loop $\gamma : \mathbb{T} \to M$. Since the vector bundle $\gamma^*\mathrm{T}M \to \mathbb{T}$ is trivial, we will implicitly identify its total space $\gamma^*\mathrm{T}M$ with $\mathbb{T} \times \mathbb{R}^m$. There exists a bounded open neighborhood $U \subset \mathbb{R}^m$ of the origin such that, for each $t \in \mathbb{T}$, the exponential map $\exp_{\gamma(t)}$ is well defined on $\overline{U}$ and is a diffeomorphism onto its image. By requiring γ to be sufficiently C^0-close to $\gamma_{\boldsymbol{q}'}$, we can assume that $\gamma_{\boldsymbol{q}'}(t)$ is contained in $\exp_{\gamma(t)}(U)$, for each $t \in \mathbb{T}$. Therefore, we can define a diffeomorphism

$$\Pi : W^{1,2}(\mathbb{T}; U) \xrightarrow{\simeq} \mathscr{U} \subset W^{1,2}(\mathbb{T}; M)$$

by

$$\Pi(\zeta)(t) := \exp_{\gamma(t)}(\zeta(t)), \qquad \forall \zeta \in W^{1,2}(\mathbb{T}; U),\ t \in \mathbb{T},$$

so that $\Phi := \Pi^{-1} : \mathscr{U} \longrightarrow W^{1,2}(\mathbb{T}; U)$ is a chart of $W^{1,2}(\mathbb{T}; M)$ whose domain contains $\gamma_{\boldsymbol{q}'}$. Now, we define $U_k := \lambda_k^{-1}(\mathscr{U})$, which is an open neighborhood of $\boldsymbol{q}'$ in Δ_k. On this open set we can build a chart

$$\phi_k : U_k \to \underbrace{U \times \cdots \times U}_{k \text{ times}}$$

for Δ_k by

$$\phi_k(\boldsymbol{q}) = \left(\exp_{\gamma(0)}^{-1}(q_0), \exp_{\gamma(1/k)}^{-1}(q_1), \ldots, \exp_{\gamma((k-1)/k)}^{-1}(q_{k-1})\right), \qquad \forall \boldsymbol{q} = (q_0, \ldots, q_{k-1}) \in U_k.$$

If we set $W_k := \phi_k(U_k)$, we obtain an embedding $\tilde{\lambda}_k$ such that the following diagram commutes.

$$\begin{array}{ccc} U_k & \xrightarrow[\simeq]{\phi_k} & W_k \\ {\scriptstyle \lambda_k} \big\downarrow & & \big\downarrow {\scriptstyle \tilde{\lambda}_k} \\ \mathscr{U} & \xrightarrow[\simeq]{\Phi} & W^{1,2}(\mathbb{T}; U) \end{array} \tag{4.12}$$

Now, we define the smooth embedding $\pi : \mathbb{T} \times U \times \mathbb{R}^m \hookrightarrow \mathbb{T} \times \mathrm{T}M$ by

$$\pi(t, q, v) = \left(t, \exp_{\gamma(t)}(q), \mathrm{d}(\exp_{\gamma(t)})(q)v + \frac{\mathrm{d}}{\mathrm{d}t}(\exp_{\gamma(t)}(q))\right), \qquad \forall (t, q, v) \in \mathbb{T} \times U \times \mathbb{R}^m,$$

so that $\mathscr{A} \circ \Pi : W^{1,2}(\mathbb{T}; U) \to \mathbb{R}$ is the action functional associated to the convex quadratic-growth Lagrangian $\mathscr{L} \circ \pi : \mathbb{T} \times \overline{U} \times \mathbb{R}^m \to \mathbb{R}$, i.e.,

$$\mathscr{A} \circ \Pi(\sigma) = \int_0^1 \mathscr{L} \circ \pi(t, \zeta(t), \dot{\zeta}(t))\, \mathrm{d}t, \qquad \forall \zeta \in W^{1,2}(\mathbb{T}; U).$$

With the notation of (4.11), the embedding $\tilde{\lambda}_k$ in diagram (4.12) is given by $\lambda_{k,\mathscr{L}\circ\pi}$. Namely, for each $\boldsymbol{q} = (q_0, \ldots, q_{k-1}) \in W_k$, $\gamma_{\boldsymbol{q}} = \tilde{\lambda}_k(\boldsymbol{q})$ is the unique loop such that $\gamma_{\boldsymbol{q}}|_{[j/k,(j+1)/k]}$ is a unique action minimizer (with respect to $\mathscr{L}\circ\pi$) with endpoints q_j and q_{j+1}, for each $j \in \mathbb{Z}_k$.

The general case in which $\gamma^*_{\boldsymbol{q}'}\mathrm{T}M$ is possibly nontrivial can be handled in a similar way by means of the charts of $W^{1,2}(\mathbb{T};M)$ introduced in Remark 3.1.1. We leave the details to the reader. From now on and until the end of section 4.4, for notational convenience, we will restrict ourselves to the connected components of Δ_k given by the $\boldsymbol{q}$'s such that $\gamma^*_{\boldsymbol{q}}\mathrm{T}M$ is trivial (but all the results will hold in the general case). □

4.3 The discrete action functional

Let $\mathscr{A} : W^{1,2}(\mathbb{T};M) \to \mathbb{R}$ be the action functional associated to the Lagrangian $\mathscr{L}$ of the previous section, i.e.,

$$\mathscr{A}(\zeta) = \int_0^1 \mathscr{L}(t, \zeta(t), \dot{\zeta}(t))\,\mathrm{d}t, \qquad \forall \zeta \in W^{1,2}(\mathbb{T};M).$$

We define the **discrete action functional** $\mathscr{A}_k : \Delta_k \to \mathbb{R}$ as the composition $\mathscr{A}\circ\lambda_k$, i.e.,

$$\mathscr{A}_k(\boldsymbol{q}) = \int_0^1 \mathscr{L}(t, \gamma_{\boldsymbol{q}}(t), \dot{\gamma}_{\boldsymbol{q}}(t))\,\mathrm{d}t, \qquad \forall \boldsymbol{q} \in \Delta_k,$$

where $\gamma_{\boldsymbol{q}} = \lambda_k(\boldsymbol{q})$. Despite the lack of C^2 regularity of $\mathscr{A}$ (see Proposition 3.4.4), the discrete action functional is always smooth, as asserted by the following statement.

Proposition 4.3.1. *The discrete action functional $\mathscr{A}_k : \Delta_k \to \mathbb{R}$ is C^∞.*

Proof. Since the statement is of a local nature, by adopting the localization argument of Remark 4.2.1 we can assume that our Lagrangian function has the form $\mathscr{L} : \mathbb{T} \times \overline{U} \times \mathbb{R}^m \to \mathbb{R}$, where U is a bounded open subset of $\mathbb{R}^m$, so that the discrete action functional has the form $\mathscr{A}_k : W_k \to \mathbb{R}$, where W_k is a suitable bounded open subset of the k-fold product of U. The result can be easily proved by induction. By Proposition 3.4.1 we know that $\mathscr{A}$ is $C^{1,1}$ and twice Gâteaux differentiable, and so is $\mathscr{A}_k$. Now, let us assume that $\mathscr{A}_k$ is C^{p-1}, for some integer $p \geq 2$. For each $(\boldsymbol{q}, \boldsymbol{v}) \in W_k \times \mathbb{R}^{km}$ we define the vector field

$$\sigma_{\boldsymbol{q},\boldsymbol{v}} := \mathrm{d}\lambda_k(\boldsymbol{q})\boldsymbol{v} \in C^\infty_k(\mathbb{T};\mathbb{R}^m).$$

We also define the smooth vector field

$$\Sigma_{\boldsymbol{q},\boldsymbol{v}} := (\sigma_{\boldsymbol{q},\boldsymbol{v}}, \dot{\sigma}_{\boldsymbol{q},\boldsymbol{v}}) : \mathbb{T} \setminus \{\tfrac{j}{k} \,|\, j \in \mathbb{Z}_k\} \to \mathrm{T}M.$$

A straightforward computation shows that the pth Gâteaux differential of $\mathscr{A}_k$, seen as a symmetric multilinear form, is given by

$$\begin{aligned}
&\mathrm{d}^p\mathscr{A}_k(\boldsymbol{q})\,[\boldsymbol{v}',\boldsymbol{v}'',\dots,\boldsymbol{v}^{(p)}]\\
&\quad= \sum_{i_1,\dots,i_p=1}^{2m} \int_0^1 \left[\partial_{i_1}\dots\partial_{i_p}\mathscr{L}(t,\gamma_{\boldsymbol{q}}(t),\dot\gamma_{\boldsymbol{q}}(t))\right] \Sigma^{i_1}_{\boldsymbol{q},\boldsymbol{v}'}(t)\dots\Sigma^{i_p}_{\boldsymbol{q},\boldsymbol{v}^{(p)}}(t)\,\mathrm{d}t,\\
&\qquad\qquad\qquad\qquad\forall \boldsymbol{q}\in W_k,\ \boldsymbol{v}',\boldsymbol{v}'',\dots,\boldsymbol{v}^{(p)}\in\mathbb{R}^{km},
\end{aligned}$$

where we have set

$$\partial_i = \begin{cases} \partial_{q^i}, & \text{if } i\in\{1,\dots,m\},\\ \partial_{v^{i-m}}, & \text{if } i\in\{m+1,\dots,2m\}. \end{cases}$$

Since $\lambda_k : W_k \hookrightarrow C^\infty_k(\mathbb{T};M)$ is smooth, if $\boldsymbol{q}_n\to\boldsymbol{q}$ as $n\to\infty$ then we have

$$\gamma_{\boldsymbol{q}_n} \underset{n\to\infty}{\longrightarrow} \gamma_{\boldsymbol{q}} \text{ in } C^\infty_k(\mathbb{T};\mathbb{R}^m)$$

and, for each $\boldsymbol{v}\in\mathbb{R}^m$,

$$\sigma_{\boldsymbol{q}_n,\boldsymbol{v}} \underset{n\to\infty}{\longrightarrow} \sigma_{\boldsymbol{q},\boldsymbol{v}} \text{ in } C^\infty_k(\mathbb{T};\mathbb{R}^m).$$

Hence, for each $\boldsymbol{v}',\boldsymbol{v}'',\dots,\boldsymbol{v}^{(p)}\in\mathbb{R}^{km}$, we have

$$[\partial_{i_1}\dots\partial_{i_p}\mathscr{L}(\cdot,\gamma_{\boldsymbol{q}_n},\dot\gamma_{\boldsymbol{q}_n})]\Sigma^{i_1}_{\boldsymbol{q}_n,\boldsymbol{v}'}\dots\Sigma^{i_p}_{\boldsymbol{q}_n,\boldsymbol{v}^{(p)}} \underset{n\to\infty}{\longrightarrow} [\partial_{i_1}\dots\partial_{i_p}\mathscr{L}(\cdot,\gamma_{\boldsymbol{q}},\dot\gamma_{\boldsymbol{q}})]\Sigma^{i_1}_{\boldsymbol{q},\boldsymbol{v}'}\dots\Sigma^{i_p}_{\boldsymbol{q},\boldsymbol{v}^{(p)}}$$

uniformly in $t\in\mathbb{T}\setminus\left\{\frac{j}{k}\,\middle|\,j\in\mathbb{Z}_k\right\}$, which implies

$$\mathrm{d}^p\mathscr{A}_k(\boldsymbol{q}_n)[\boldsymbol{v}',\dots,\boldsymbol{v}^{(p)}] \to \mathrm{d}^p\mathscr{A}_k(\boldsymbol{q})[\boldsymbol{v}',\dots,\boldsymbol{v}^{(p)}].$$

By the total differential theorem we conclude that $\mathscr{A}_k$ is C^p, and by induction we obtain the claim. □

Now, we want to show that the discrete action functional $\mathscr{A}_k$ is suitable for Morse theory. In fact, every closed sublevel of $\mathscr{A}_k$ is a compact subset of Δ_k, provided the discretization pass $k\in\mathbb{N}$ is large enough.

Proposition 4.3.2. *For each $c\in\mathbb{R}$ there exists $\bar k=\bar k(c)\in\mathbb{N}$ such that, for each $k\geq\bar k$, the closed sublevel $\mathscr{A}_k^{-1}(-\infty,c]$ is compact.*

Proof. Consider the compact subset of Δ_k defined by

$$C_k := \{(q_0,\dots,q_{k-1})\in\Delta_k \,|\, \mathrm{dist}(q_j,q_{j+1})\leq\rho_0/2\ \forall j\in\mathbb{Z}_k\}.$$

In order to prove the statement, we just need to show that

$$\lim_{k\to\infty}\min\{\mathscr{A}_k(\boldsymbol{q})\,|\,\boldsymbol{q}\in\partial C_k\}=\infty.$$

As explained in Section 3.3, without loss of generality we can assume that there exists a constant $\underline{\ell} > 0$ such that

$$\mathscr{L}(t,q,v) \geq \underline{\ell}\,|v|_q^2, \qquad \forall (t,q,v) \in \mathbb{T} \times \mathrm{T}M.$$

For each $\boldsymbol{q} = (q_1, \ldots, q_k)$ that belongs to the boundary of C_k, we have that $\mathrm{dist}(q_j, q_{j+1}) = \rho_0/2$ for some $j \in \mathbb{Z}_k$ and therefore we obtain the desired estimate

$$\begin{aligned}
\mathscr{A}_k(\boldsymbol{q}) &\geq \int_{j/k}^{(j+1)/k} \mathscr{L}(t, \gamma_{\boldsymbol{q}}(t), \dot\gamma_{\boldsymbol{q}}(t))\,\mathrm{d}t \geq \int_{j/k}^{(j+1)/k} \underline{\ell}\,|\dot\gamma_{\boldsymbol{q}}(t)|^2_{\gamma_{\boldsymbol{q}}(t)}\,\mathrm{d}t \\
&\geq k\,\underline{\ell}\left(\int_{j/k}^{(j+1)/k} |\dot\gamma_{\boldsymbol{q}}(t)|_{\gamma_{\boldsymbol{q}}(t)}\,\mathrm{d}t\right)^2 \geq k\,\underline{\ell}\,\mathrm{dist}(q_j, q_{j+1})^2 \\
&\geq k\,\underline{\ell}\,(\rho_0/2)^2.
\end{aligned}$$

□

4.4 Critical points of the discrete action

We know from Proposition 3.5.1 that the critical points of the action functional $\mathscr{A}$ are precisely the smooth 1-periodic solution of the Euler-Lagrange system of $\mathscr{L}$. This implies that, for each $k \in \mathbb{N}$ sufficiently large, the loop γ belongs to the k-broken Euler-Lagrange loop space Λ_k, and therefore the corresponding point $\boldsymbol{q} = \lambda_k^{-1}(\gamma) \in \Delta_k$ is a critical point of the discrete action $\mathscr{A}_k = \mathscr{A} \circ \lambda_k$. Now, we want to prove the converse implication, namely that the critical points of $\mathscr{A}_k$ correspond to critical points of $\mathscr{A}$.

Since our arguments will have a local nature, we will implicitly adopt the localization argument of Remark 4.2.1. Hence, we will assume that our Lagrangian function has the form $\mathscr{L} : \mathbb{T} \times \overline{U} \times \mathbb{R}^m \to \mathbb{R}$, where U is a bounded open subset of $\mathbb{R}^m$. Consequently, the associated action functional and discrete action functional will have the form $\mathscr{A} : W^{1,2}(\mathbb{T}; U) \to \mathbb{R}$ and $\mathscr{A}_k : W_k \to \mathbb{R}$, where W_k is a suitable bounded open subset of the k-fold product of U.

Proposition 4.4.1. *For each $\boldsymbol{q} \in W_k$ and $\boldsymbol{v} = (v_1, \ldots, v_k) \in \mathbb{R}^m \times \cdots \times \mathbb{R}^m$ we have*

$$\mathrm{d}\mathscr{A}_k(\boldsymbol{q})\,\boldsymbol{v} = \sum_{j=0}^{k-1} \langle \partial_v \mathscr{L}(\tfrac{j}{k}, \gamma_{\boldsymbol{q}}(\tfrac{j}{k}), \dot\gamma_{\boldsymbol{q}}(\tfrac{j}{k}^-)) - \partial_v \mathscr{L}(\tfrac{j}{k}, \gamma_{\boldsymbol{q}}(\tfrac{j}{k}), \dot\gamma_{\boldsymbol{q}}(\tfrac{j}{k}^+)), v_j \rangle,$$

where $\gamma_{\boldsymbol{q}} = \lambda_k(\boldsymbol{q})$.

Proof. For each $\boldsymbol{v} \in \mathbb{R}^m \times \cdots \times \mathbb{R}^m$, let $\sigma_{\boldsymbol{q},\boldsymbol{v}} := \mathrm{d}\lambda_k(\boldsymbol{q})\boldsymbol{v} \in C_k^\infty(\mathbb{T}; \mathbb{R}^m)$. Since $\gamma_{\boldsymbol{q}}$ is a solution of the Euler-Lagrange system of $\mathscr{L}$ on each interval $[\frac{j}{k}, \frac{j+1}{k}]$, integration by parts gives

$$
\begin{aligned}
\mathrm{d}\mathscr{A}_k(\boldsymbol{q})\,\boldsymbol{v} &= \sum_{j=0}^{k-1} \int_{j/k}^{(j+1)/k} \big(\langle \partial_v \mathscr{L}(t,\gamma_{\boldsymbol{q}},\dot\gamma_{\boldsymbol{q}}), \dot\sigma_{\boldsymbol{q},\boldsymbol{v}}\rangle + \langle \partial_q \mathscr{L}(t,\gamma_{\boldsymbol{q}},\dot\gamma_{\boldsymbol{q}}), \sigma_{\boldsymbol{q},\boldsymbol{v}}\rangle\big)\mathrm{d}t \\
&= \sum_{j=0}^{k-1} \Big[\langle \partial_v \mathscr{L}(\tfrac{j+1}{k}, \gamma_{\boldsymbol{q}}(\tfrac{j+1}{k}), \dot\gamma_{\boldsymbol{q}}(\tfrac{j+1^-}{k})), \sigma_{\boldsymbol{q},\boldsymbol{v}}(\tfrac{j+1}{k})\rangle \\
&\qquad - \langle \partial_v \mathscr{L}(\tfrac{j}{k}, \gamma_{\boldsymbol{q}}(\tfrac{j}{k}), \dot\gamma_{\boldsymbol{q}}(\tfrac{j^+}{k})), \sigma_{\boldsymbol{q},\boldsymbol{v}}(\tfrac{j}{k})\rangle \\
&\qquad + \int_{j/k}^{(j+1)/k} \langle \underbrace{-\tfrac{\mathrm{d}}{\mathrm{d}t}\partial_v \mathscr{L}(t,\gamma_{\boldsymbol{q}},\dot\gamma_{\boldsymbol{q}}) + \partial_q \mathscr{L}(t,\gamma_{\boldsymbol{q}},\dot\gamma_{\boldsymbol{q}})}_{=0}, \sigma_{\boldsymbol{q},\boldsymbol{v}}\rangle \mathrm{d}t \Big] \\
&= \sum_{j=0}^{k-1} \langle \partial_v \mathscr{L}(\tfrac{j}{k}, \gamma_{\boldsymbol{q}}(\tfrac{j}{k}), \dot\gamma_{\boldsymbol{q}}(\tfrac{j^-}{k})) - \partial_v \mathscr{L}(\tfrac{j}{k}, \gamma_{\boldsymbol{q}}(\tfrac{j}{k}), \dot\gamma_{\boldsymbol{q}}(\tfrac{j^+}{k})), \sigma_{\boldsymbol{q},\boldsymbol{v}}(\tfrac{j}{k})\rangle.
\end{aligned}
$$

Thus, by definition of the embedding λ_k, we have

$$
\sigma_{\boldsymbol{q},\boldsymbol{v}}(\tfrac{j}{k}) = \frac{\mathrm{d}}{\mathrm{d}s}\Big|_{s=0} \gamma_{\boldsymbol{q}+s\boldsymbol{v}}(\tfrac{j}{k}) = \frac{\mathrm{d}}{\mathrm{d}s}\Big|_{s=0} (q_j + s v_j) = v_j, \qquad \forall j \in \mathbb{Z}_k,
$$

and the claim follows. □

Corollary 4.4.2. *If $\boldsymbol{q} \in W_k$ is a critical point of $\mathscr{A}_k$, then the corresponding loop $\gamma_{\boldsymbol{q}} = \lambda_k(\boldsymbol{q})$ is a smooth solution of the Euler-Lagrange system of $\mathscr{L}$, and in particular it is a critical point of the action functional $\mathscr{A}$.*

Proof. By Proposition 4.4.1, $\boldsymbol{q}$ is a critical point of $\mathscr{A}_k$ if and only if

$$
\partial_v \mathscr{L}(\tfrac{j}{k}, \gamma_{\boldsymbol{q}}(\tfrac{j}{k}), \dot\gamma_{\boldsymbol{q}}(\tfrac{j^-}{k})) - \partial_v \mathscr{L}(\tfrac{j}{k}, \gamma_{\boldsymbol{q}}(\tfrac{j}{k}), \dot\gamma_{\boldsymbol{q}}(\tfrac{j^+}{k})) = 0, \qquad \forall j \in \mathbb{Z}_k. \tag{4.13}
$$

Assumption **(Q1)** (see Section 3.3) implies that the maps

$$
\partial_v \mathscr{L}(\tfrac{j}{k}, \gamma_{\boldsymbol{q}}(\tfrac{j}{k}), \cdot) : \mathbb{R}^m \to \mathbb{R}^m, \qquad \forall j \in \mathbb{Z}_k
$$

are diffeomorphisms, as we have proved in Section 1.2. Hence the equality in (4.13) holds if and only if $\dot\gamma_{\boldsymbol{q}}(\tfrac{j^-}{k}) = \dot\gamma_{\boldsymbol{q}}(\tfrac{j^+}{k})$ for each $j \in \mathbb{Z}_k$, that is if and only if $\gamma_{\boldsymbol{q}}$ is C^1. This implies that $\gamma_{\boldsymbol{q}} : \mathbb{T} \to \mathbb{R}^m$ satisfies the Euler-Lagrange system of $\mathscr{L}$ on the whole $\mathbb{T}$, and therefore it is smooth. □

Now, let us fix a critical point $\boldsymbol{q}' \in W_k$ of the discrete action functional $\mathscr{A}_k$. The tangent space of Λ_k at the smooth loop $\gamma_{\boldsymbol{q}'}$, that is the image of the differential

$$
\mathrm{d}\lambda_k(\boldsymbol{q}') : \mathrm{T}_{\boldsymbol{q}'}\Delta_k \xrightarrow{\simeq} \mathrm{T}_{\gamma_{\boldsymbol{q}'}}\Lambda_k,
$$

can be characterized as follows. We define a Lagrangian $L : \mathbb{T} \times \mathbb{R}^m \times \mathbb{R}^m \to \mathbb{R}$ by

$$L(t, q, v) = \frac{1}{2} \langle a(t)v, v \rangle + \langle b(t)q, v \rangle + \frac{1}{2} \langle c(t)q, q \rangle, \qquad \forall (t, q, v) \in \mathbb{T} \times \mathbb{R}^m \times \mathbb{R}^m, \tag{4.14}$$

where, for each $t \in \mathbb{T}$, $a(t)$, $b(t)$ and $c(t)$ are the $m \times m$ matrices given by

$$\begin{aligned} a_{ij}(t) &= \frac{\partial^2 \mathscr{L}}{\partial v^i \, \partial v^j}(t, \gamma_{\boldsymbol{q}'}(t), \dot{\gamma}_{\boldsymbol{q}'}(t)), \\ b_{ij}(t) &= \frac{\partial^2 \mathscr{L}}{\partial v^i \, \partial q^j}(t, \gamma_{\boldsymbol{q}'}(t), \dot{\gamma}_{\boldsymbol{q}'}(t)), \\ c_{ij}(t) &= \frac{\partial^2 \mathscr{L}}{\partial q^i \, \partial q^j}(t, \gamma_{\boldsymbol{q}'}(t), \dot{\gamma}_{\boldsymbol{q}'}(t)). \end{aligned}$$

A straightforward computation shows that the Euler-Lagrange system associated to L is given by the following linear system of ordinary differential equations for curves σ on $\mathbb{R}^m$,

$$a\, \ddot{\sigma} + (b + \dot{a} - b^T)\, \dot{\sigma} + (\dot{b} - c)\, \sigma = 0. \tag{4.15}$$

This is precisely the linearization of the Euler-Lagrange system of $\mathscr{L}$ along the periodic solution $\gamma_{\boldsymbol{q}'}$. The 1-periodic solutions $\sigma : \mathbb{T} \to \mathbb{R}^m$ of (4.15) are precisely the critical points of the action functional $A : W^{1,2}(\mathbb{T}; \mathbb{R}^m) \to \mathbb{R}$ associated to L, given as usual by

$$A(\xi) = \int_0^1 L(t, \xi(t), \dot{\xi}(t))\, dt, \qquad \forall \xi \in W^{1,2}(\mathbb{T}; \mathbb{R}^m).$$

Lemma 4.4.3. *The tangent space* $\mathrm{T}_{\gamma_{\boldsymbol{q}'}} \Lambda_k$ *is the space of continuous and piecewise smooth loops* $\sigma : \mathbb{T} \to \mathbb{R}^m$ *such that, for each* $j \in \mathbb{Z}_k$, *the restriction* $\sigma|_{[j/k,(j+1)/k]}$ *is a solution of the Euler-Lagrange system* (4.15).

Proof. By definition of tangent space, $\mathrm{T}_{\gamma_{\boldsymbol{q}'}} \Lambda_k$ consists of those continuous loops $\sigma : \mathbb{T} \to \mathbb{R}^m$ given by

$$\sigma(t) = \left.\frac{\partial}{\partial s}\right|_{s=0} \Sigma(s, t), \qquad \forall t \in \mathbb{T}, \tag{4.16}$$

for some continuous $\Sigma : (-\varepsilon, \varepsilon) \times \mathbb{T} \to \mathbb{R}^m$ such that:

- the restriction $\Sigma|_{(-\varepsilon,\varepsilon)\times[j/k,(j+1)/k]}$ is smooth for all $j \in \mathbb{Z}_k$,
- $\Sigma(s, \cdot) \in \Lambda_k$ for all $s \in (-\varepsilon, \varepsilon)$,
- $\Sigma(0, \cdot) = \gamma_{\boldsymbol{q}'}$.

Namely, Σ is a piecewise smooth variation of $\gamma_{\boldsymbol{q}'}$ such that the loops $\Sigma_s = \Sigma(s,\cdot)$ satisfy the Euler-Lagrange equations associated to $\mathscr{L}$ on the intervals $[\frac{j}{k},\frac{j+1}{k}]$ for all $j \in \mathbb{Z}_k$, i.e.,

$$\partial^2_{vv}\mathscr{L}(t,\Sigma_s,\dot\Sigma_s)\,\ddot\Sigma_s + \partial^2_{vq}\mathscr{L}(t,\Sigma_s,\dot\Sigma_s)\,\dot\Sigma_s + \partial^2_{vt}\mathscr{L}(t,\Sigma_s,\dot\Sigma_s) - \partial_q\mathscr{L}(t,\Sigma_s,\dot\Sigma_s) = 0, \qquad \forall t \in [j/k,(j+1)/k].$$

By differentiating the above equation with respect to s at $s = 0$, we obtain the Euler-Lagrange system (4.15) for the loop σ (as before, satisfied on the intervals $[\frac{j}{k},\frac{j+1}{k}]$ for all $j \in \mathbb{Z}_k$). Conversely, a continuous loop $\sigma : \mathbb{T} \to \mathbb{R}^m$ whose restrictions $\sigma|_{[j/k,(j+1)/k]}$ satisfy (4.15) is of the form (4.16) for some Σ as above, and therefore it is an element of $\mathrm{T}_{\gamma_{\boldsymbol{q}'}}\Lambda_k$. □

Now, we want to investigate the relationship between the Morse indices of the functionals $\mathscr{A}_k$ and $\mathscr{A}$ at the corresponding critical points $\boldsymbol{q}'$ and $\gamma_{\boldsymbol{q}'}$. We begin by characterizing the null-space of the Hessian of $\mathscr{A}$ at $\gamma_{\boldsymbol{q}'}$.

Lemma 4.4.4. *The null-space of* $\mathrm{Hess}\mathscr{A}(\gamma_{\boldsymbol{q}'})$ *consists of smooth loops* $\sigma : \mathbb{T} \to \mathbb{R}^m$ *that are solutions of the Euler-Lagrange system* (4.15) *on the whole* $\mathbb{T}$.

Proof. For every $\sigma, \xi \in W^{1,2}(\mathbb{T};\mathbb{R}^m)$ we have

$$\mathrm{Hess}\mathscr{A}(\gamma_{\boldsymbol{q}'})[\sigma,\xi] = \int_0^1 \Big(\langle a\,\dot\sigma,\dot\xi\rangle + \langle b\,\sigma,\dot\xi\rangle + \langle b^T\,\dot\sigma,\xi\rangle + \langle c\,\sigma,\xi\rangle\Big)\,\mathrm{d}t = \mathrm{d}A(\sigma)\xi.$$

Therefore σ is in the null-space of $\mathrm{Hess}\mathscr{A}(\gamma_{\boldsymbol{q}'})$ if and only if it is a critical point of A, that is if and only if it is a (smooth) solution of the Euler-Lagrange system (4.15). □

Remark 4.4.1. In the case where $\mathscr{L}$ is the autonomous Lagrangian given by

$$\mathscr{L}(q,v) = |v|^2_q, \qquad \forall (q,v) \in \mathrm{T}M,$$

the null-space of $\mathrm{Hess}\mathscr{A}(\gamma_{\boldsymbol{q}'})$ is given by the 1-periodic **Jacobi vector fields** along the closed geodesics $\gamma_{\boldsymbol{q}'}$, and the Euler-Lagrange system (4.15) is called the **Jacobi system**. This latter can also be intrinsically expressed as

$$\nabla_t^2\sigma + R(\sigma,\dot\gamma_{\boldsymbol{q}'})\dot\gamma_{\boldsymbol{q}'} = 0,$$

where ∇_t and R are respectively the covariant derivative and the Riemann tensor of the Riemannian manifold $(M,\langle\cdot,\cdot\rangle_\cdot)$. □

As a consequence of Lemmas 4.4.3 and 4.4.4, the null-space of $\mathrm{Hess}\mathscr{A}(\gamma_{\boldsymbol{q}'})$ is contained in $\mathrm{T}_{\gamma_{\boldsymbol{q}'}}\Lambda_k$, and therefore it is contained in the null-space of the Hessian of the restricted action $\mathrm{Hess}\mathscr{A}|_{\Lambda_k}(\gamma_{\boldsymbol{q}'})$. This inclusion is actually an equality, as shown by the following.

Lemma 4.4.5. $\mathrm{Hess}\mathscr{A}(\gamma_{\boldsymbol{q}'})$ *and* $\mathrm{Hess}\mathscr{A}|_{\Lambda_k}(\gamma_{\boldsymbol{q}'})$ *have the same null-space, and in particular* $\mathrm{nul}(\mathscr{A},\gamma_{\boldsymbol{q}'}) = \mathrm{nul}(\mathscr{A}|_{\Lambda_k},\gamma_{\boldsymbol{q}'})$.

Proof. We only need to show that any loop $\sigma \in \mathrm{T}_{\gamma_{q'}}\Lambda_k$ that is not everywhere smooth cannot be in the null-space of $\mathrm{Hess}\mathscr{A}|_{\Lambda_k}(\gamma_{q'})$. In fact, since σ is always smooth outside the points j/k (for $j \in \mathbb{Z}_k$), for each $\xi \in \mathrm{T}_{\gamma_{q'}}\Lambda_k$ we have

$$\begin{aligned}
\mathrm{Hess}\mathscr{A}|_{\Lambda_k}(\gamma_{q'})[\sigma,\xi] &= \sum_{j=0}^{k-1}\int_{j/k}^{(j+1)/k} \Big(\langle a\,\dot\sigma,\dot\xi\rangle + \langle b\,\sigma,\dot\xi\rangle + \langle b^T\dot\sigma,\xi\rangle + \langle c\,\sigma,\xi\rangle\Big)\,\mathrm{d}t\\
&= \sum_{j=0}^{k-1}\int_{j/k}^{(j+1)/k} \langle \underbrace{-a\,\ddot\sigma - b\,\dot\sigma - \dot a\,\dot\sigma - \dot b\,\sigma + b^T\dot\sigma + c\,\sigma}_{=0},\xi\rangle\,\mathrm{d}t\\
&\quad + \sum_{j=0}^{k-1}\langle a\,\dot\sigma + b\,\sigma,\xi\rangle\Big|_{(j/k)^+}^{((j+1)/k)^-}\\
&= \sum_{j=0}^{k-1}\langle a(\tfrac{j}{k})[\dot\sigma(\tfrac{j}{k}^-) - \dot\sigma(\tfrac{j}{k}^+)],\xi(\tfrac{j}{k})\rangle.
\end{aligned} \tag{4.17}$$

By assumption we have $\dot\sigma(\tfrac{j}{k}^+) \neq \dot\sigma(\tfrac{j}{k}^-)$ for some $j \in \mathbb{Z}_k$ and therefore

$$a(\tfrac{j}{k})[\dot\sigma(\tfrac{j}{k}^+) - \dot\sigma(\tfrac{j}{k}^-)] \neq \mathbf{0}.$$

Here we are using the fact that, by assumption **(Q1)**, the matrix $a(\tfrac{j}{k})$ is invertible. Now, consider $\xi \in \mathrm{T}_{\gamma_{q'}}\Lambda_k$ such that

$$\xi(\tfrac{h}{k}) = \begin{cases} a(\tfrac{j}{k})[\dot\sigma(\tfrac{j}{k}^+) - \dot\sigma(\tfrac{j}{k}^-)], & h = j,\\ \mathbf{0}, & h \neq j.\end{cases}$$

By (4.17) we have

$$\mathrm{Hess}\mathscr{A}|_{\Lambda_k}(\gamma_{q'})[\sigma,\xi] = \Big|a(\tfrac{j}{k})\big[\dot\sigma(\tfrac{j}{k}^+) - \dot\sigma(\tfrac{j}{k}^-)\big]\Big|^2 \neq 0,$$

and we conclude that σ is not in the null-space of $\mathrm{Hess}\mathscr{A}|_{\Lambda_k}(\gamma_{q'})$. □

Corollary 4.4.6. *The discrete action functional $\mathscr{A}_k$ and the full action functional $\mathscr{A}$ have the same nullity at the critical points $\boldsymbol{q}'$ and $\gamma_{\boldsymbol{q}'}$ respectively, i.e.,*

$$\mathrm{nul}(\mathscr{A}_k,\boldsymbol{q}') = \mathrm{nul}(\mathscr{A},\gamma_{\boldsymbol{q}'}).$$

Proof. First of all, notice that $\mathrm{d}\lambda_k(\boldsymbol{q}') : \mathbb{R}^m \to \mathrm{T}_{\gamma_{q'}}\Lambda_k$ is an isomorphism, and we have

$$\mathrm{Hess}\mathscr{A}_k(\boldsymbol{q}')[\boldsymbol{v},\boldsymbol{w}] = \mathrm{Hess}\mathscr{A}|_{\Lambda_k}(\gamma_{\boldsymbol{q}'})[\mathrm{d}\lambda_k(\boldsymbol{q}')\boldsymbol{v},\mathrm{d}\lambda_k(\boldsymbol{q}')\boldsymbol{w}], \qquad \forall \boldsymbol{v},\boldsymbol{w} \in \mathbb{R}^m.$$

This implies that $\mathscr{A}_k$ and $\mathscr{A}|_{\Lambda_k}$ have the same nullity at $\boldsymbol{q}'$ and $\gamma_{\boldsymbol{q}'}$ respectively, i.e., $\mathrm{nul}(\mathscr{A}_k,\boldsymbol{q}') = \mathrm{nul}(\mathscr{A}|_{\Lambda_k},\gamma_{\boldsymbol{q}'})$, and by Lemma 4.4.5 we obtain the assertion. □

So far we have proved that $\mathscr{A}_k$ and $\mathscr{A}$ have the same nullity at the corresponding critical points $\boldsymbol{q}'$ and $\gamma_{\boldsymbol{q}'}$. Now we want to prove that they also have the same Morse index provided that $k \in \mathbb{N}$ is sufficiently large. First of all, we need some preliminaries.

Let $\boldsymbol{E}$ be a real Hilbert space and $\mathscr{B} : \boldsymbol{E} \times \boldsymbol{E} \to \mathbb{R}$ a bounded symmetric bilinear form. We recall that the **Morse index** $\mathrm{ind}(\mathscr{B})$ of this form is the supremum of the dimension of the vector subspaces of $\boldsymbol{E}$ on which $\mathscr{B}$ is negative-definite. Now, let us fix an infinite sequence $\{\boldsymbol{E}_n \,|\, n \in \mathbb{N}\}$ of Hilbert subspaces of $\boldsymbol{E}$ such that

$$\boldsymbol{E}_1 \subset \boldsymbol{E}_2 \subset \boldsymbol{E}_3 \subset \cdots \subset \boldsymbol{E},$$

and that their union is dense in $\boldsymbol{E}$, i.e.,

$$\overline{\bigcup_{n\in\mathbb{N}} \boldsymbol{E}_n} = \boldsymbol{E}. \tag{4.18}$$

The following holds.

Lemma 4.4.7. *If the Morse index of $\mathscr{B}$ is finite, then it coincides with the Morse index of $\mathscr{B}$ restricted to $\boldsymbol{E}_n$ for all the sufficiently large $n \in \mathbb{N}$, i.e.,*

$$\mathrm{ind}(\mathscr{B}) = \mathrm{ind}(\mathscr{B}|_{\boldsymbol{E}_n \times \boldsymbol{E}_n}).$$

Proof. The inequality $\mathrm{ind}(\mathscr{B}) \geq \mathrm{ind}(\mathscr{B}|_{\boldsymbol{E}_n \times \boldsymbol{E}_n})$ is trivial, hence we only have to prove that $\mathrm{ind}(\mathscr{B}) \leq \mathrm{ind}(\mathscr{B}|_{\boldsymbol{E}_n \times \boldsymbol{E}_n})$. For each $n \in \mathbb{N}$, we denote by $P_n : \boldsymbol{E} \to \boldsymbol{E}_n$ the orthogonal projector onto $\boldsymbol{E}_n$. Let $\boldsymbol{V}$ be a vector subspace of $\boldsymbol{E}$ of dimension $\iota = \mathrm{ind}(\mathscr{B}) \in \mathbb{N}$, such that $\mathscr{B}$ is negative-definite on $\boldsymbol{V}$. We denote by $S(\boldsymbol{V})$ the $(\iota - 1)$-dimensional sphere in $\boldsymbol{V}$, i.e.,

$$S(\boldsymbol{V}) = \{\boldsymbol{e} \in \boldsymbol{V} \,|\, \|\boldsymbol{e}\|_{\boldsymbol{E}} = 1\}.$$

By (4.18) and since $\mathscr{B}$ is continuous, for each $\boldsymbol{e} \in S(\boldsymbol{V})$ there exists a positive integer $n_{\boldsymbol{e}} \in \mathbb{N}$ and a neighborhood $U_{\boldsymbol{e}} \subseteq S(\boldsymbol{V})$ of $\boldsymbol{e}$ such that

$$\mathscr{B}(P_n \boldsymbol{f}, P_n \boldsymbol{f}) < 0, \qquad \forall n \geq n_{\boldsymbol{e}},\ \boldsymbol{f} \in U_{\boldsymbol{e}}.$$

By compactness, $S(\boldsymbol{V})$ admits a finite cover $U_{\boldsymbol{e}_1}, \ldots, U_{\boldsymbol{e}_s}$. For every integer $n \geq \max\{n_{\boldsymbol{e}_1}, \ldots, n_{\boldsymbol{e}_s}\}$ and for every nonzero $\boldsymbol{v} = P_n \boldsymbol{w} \in P_n \boldsymbol{V}$ we have

$$\mathscr{B}(\boldsymbol{v}, \boldsymbol{v}) = \|\boldsymbol{w}\|^2_{\boldsymbol{E}} \underbrace{\mathscr{B}\left(P_n \tfrac{\boldsymbol{w}}{\|\boldsymbol{w}\|_{\boldsymbol{E}}}, P_n \tfrac{\boldsymbol{w}}{\|\boldsymbol{w}\|_{\boldsymbol{E}}}\right)}_{<0} < 0,$$

and therefore $\mathscr{B}$ is negative-definite on $P_n \boldsymbol{V}$. In order to conclude the proof we just need to show that $P_n \boldsymbol{V}$ still has dimension $\iota = \mathrm{ind}(\mathscr{B})$ provided n is sufficiently large. This is easily seen as follows. Let $\boldsymbol{v}_1, \ldots, \boldsymbol{v}_\iota$ be a basis for $\boldsymbol{V}$. By (4.18), for each $j = 1, \ldots, \iota$, we have that $P_n \boldsymbol{v}_i \to \boldsymbol{v}_i$ as $n \to \infty$. This implies that, for n sufficiently large, the vectors $P_n \boldsymbol{v}_1, \ldots, P_n \boldsymbol{v}_\iota$ are still linearly independent, and therefore $P_n \boldsymbol{V}$ has dimension ι. □

In order to apply this abstract lemma to our situation, we first need the following remark about the density of the spaces of broken affine loops in the Sobolev space $W^{1,2}(\mathbb{T};\mathbb{R}^m)$. For each $k \in \mathbb{N}$, we define the k-**broken affine loop space** of $\mathbb{R}^m$ as

$$\mathrm{Aff}_k(\mathbb{T};\mathbb{R}^m) := \Big\{ \sigma : \mathbb{T} \to \mathbb{R}^m \ \Big| \ \sigma(\tfrac{j+s}{k}) = (1-s)\,\sigma(\tfrac{j}{k}) + s\,\sigma(\tfrac{j+1}{k}) \quad \forall s \in [0,1],\ j \in \mathbb{Z}_k \Big\}.$$

This space is isomorphic to the k-fold product $\mathbb{R}^m \times \cdots \times \mathbb{R}^m$ by the linear map $\alpha_k : \mathrm{Aff}_k(\mathbb{T};\mathbb{R}^m) \to \mathbb{R}^m \times \cdots \times \mathbb{R}^m$ given by

$$\alpha_k(\sigma) = \big(\sigma(0), \sigma(\tfrac{1}{k}), \dots, \sigma(\tfrac{k-1}{k})\big), \qquad \forall \sigma \in \mathrm{Aff}_k(\mathbb{T};\mathbb{R}^m).$$

In particular, since $\mathrm{Aff}_k(\mathbb{T};\mathbb{R}^m)$ is a finite-dimensional vector space, it is a Hilbert subspace of $W^{1,2}(\mathbb{T};\mathbb{R}^m)$.

Lemma 4.4.8. *The union of the spaces* $\mathrm{Aff}_k(\mathbb{T};\mathbb{R}^m)$*, for all* $k \in \mathbb{N}$*, is dense in* $W^{1,2}(\mathbb{T};\mathbb{R}^m)$*, i.e.,*

$$\overline{\bigcup_{k\in\mathbb{N}} \mathrm{Aff}_k(\mathbb{T};\mathbb{R}^m)} = W^{1,2}(\mathbb{T};\mathbb{R}^m).$$

Proof. It is well known that $C^\infty(\mathbb{T};\mathbb{R}^m)$ is dense in $W^{1,2}(\mathbb{T};\mathbb{R}^m)$, see for instance [AF03, page 68]. Hence, all we have to do in order to prove the lemma is to show that, for an arbitrary $\gamma \in C^\infty(\mathbb{T};\mathbb{R}^m)$, there exists a sequence $\{\gamma_k \mid k \in \mathbb{N}\}$ such that $\gamma_k \in \mathrm{Aff}_k(\mathbb{T};\mathbb{R}^m)$ and $\gamma_k \to \gamma$ in $W^{1,2}(\mathbb{T};\mathbb{R}^m)$. A candidate for this sequence is built by defining γ_k to be the map in $\mathrm{Aff}_k(\mathbb{T};\mathbb{R}^m)$ such that $\gamma_k(\frac{j}{k}) = \gamma(\frac{j}{k})$ for each $j \in \mathbb{Z}_k$, see Figure 4.1.

Since γ is smooth and 1-periodic, if we fix an arbitrary $\varepsilon > 0$ there exists a positive $\delta > 0$ such that, for each $t \in \mathbb{T}$ and for each $\delta_0, \delta_1 > 0$ with $0 < \delta_0 + \delta_1 \le \delta$, we have

$$\left| \dot\gamma(t) - \frac{\gamma(t+\delta_1) - \gamma(t-\delta_0)}{\delta_1 + \delta_0} \right| \le \varepsilon. \tag{4.19}$$

Now, notice that, for each $k \in \mathbb{N}$, the periodic curve $\gamma_k : \mathbb{T} \to \mathbb{R}$ is differentiable outside the points j/k (for each $j \in \mathbb{Z}_k$) and we have

$$\dot\gamma_k(t) = k \left[\gamma\left(\tfrac{\lfloor kt \rfloor + 1}{k}\right) - \gamma\left(\tfrac{\lfloor kt \rfloor}{k}\right) \right], \qquad t \in \mathbb{T} \setminus \{\tfrac{j}{k} \mid j \in \mathbb{Z}_k\},$$

where $\lfloor \cdot \rfloor$ gives the integer part of its argument. In particular, for each irrational number $t \in \mathbb{R} \setminus \mathbb{Q}$, or more precisely for each $t \in (\mathbb{R} \setminus \mathbb{Q})/\mathbb{Z}$, we have

$$\lim_{k\to\infty} \dot\gamma_k(t) = \dot\gamma(t).$$

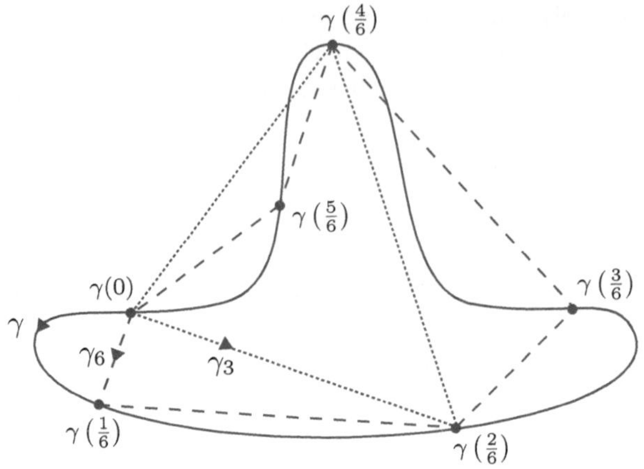

Figure 4.1. Example of a smooth loop γ (solid line) and the corresponding piecewise affine loops γ_3 (dotted line) and γ_6 (dashed line) of the sequence $\{\gamma_k \mid k \in \mathbb{N}\}$.

Since the rational numbers have Lebesgue measure 0, the above equation implies that $\dot{\gamma}_k$ converges to $\dot{\gamma}$ almost everywhere as $k \to \infty$. By (4.19), for each $\varepsilon > 0$ there exists $\delta > 0$ such that, for each $k \geq \delta^{-1}$ and for almost every $t \in \mathbb{T}$, we have

$$|\dot{\gamma}(t) - \dot{\gamma}_k(t)| = \left| \dot{\gamma}(t) - \frac{\gamma\left(\frac{\lfloor kt \rfloor + 1}{k}\right) - \gamma\left(\frac{\lfloor kt \rfloor}{k}\right)}{\frac{1}{k}} \right| \leq \varepsilon.$$

We can then apply the dominated convergence theorem to conclude that $\dot{\gamma}_k \to \dot{\gamma}$ in $L^2(\mathbb{T};\mathbb{R}^m)$ as $k \to \infty$, and therefore $\gamma_k \to \gamma$ in $W^{1,2}(\mathbb{T};\mathbb{R}^m)$ as $k \to \infty$. □

Once these preliminaries are established, we can prove the previously announced result about the Morse index of the discrete action functionals.

Lemma 4.4.9. *For all $k \in \mathbb{N}$ sufficiently large, the Morse index of $\mathscr{A}_k$ at $\mathbf{q}'$ coincides with the Morse index of $\mathscr{A}$ at $\gamma_{\mathbf{q}'}$, i.e.,*

$$\operatorname{ind}(\mathscr{A}_k, \mathbf{q}') = \operatorname{ind}(\mathscr{A}, \gamma_{\mathbf{q}'}).$$

Proof. Since the functionals $\mathscr{A}_k$ and $\mathscr{A}|_{\Lambda_k}$ have the same Morse index and nullity (see the proof of Corollary 4.4.6), we just need to prove that

$$\operatorname{ind}(\mathscr{A}|_{\Lambda_k}, \gamma_{\mathbf{q}'}) = \operatorname{ind}(\mathscr{A}, \gamma_{\mathbf{q}'})$$

for all $k \in \mathbb{N}$ sufficiently large.

By definition of Morse index, there exists an $\operatorname{ind}(\mathscr{A}, \gamma_{\mathbf{q}'})$-dimensional vector subspace $\boldsymbol{V} \subseteq W^{1,2}(\mathbb{T};\mathbb{R}^m)$ on which $\operatorname{Hess}\mathscr{A}(\gamma_{\mathbf{q}'})$ is negative-definite, i.e.,

$$\operatorname{Hess}\mathscr{A}(\gamma_{\mathbf{q}'})[\sigma,\sigma] < 0, \quad \forall \sigma \in \boldsymbol{V} \setminus \{\mathbf{0}\}. \tag{4.20}$$

By Lemmas 4.4.8 and 4.4.7, we can choose $\boldsymbol{V}$ to be a subspace of $\mathrm{Aff}_k(\mathbb{T};\mathbb{R}^m)$. Let us define a linear map $K:\boldsymbol{V}\to \mathrm{T}_{\gamma_{\boldsymbol{q}'}}\Lambda_k$ by $K(\sigma)=\tilde\sigma$, where $\tilde\sigma$ is the unique element in $\mathrm{T}_{\gamma_{\boldsymbol{q}'}}\Lambda_k$ such that

$$\sigma(\tfrac{j}{k})=\tilde\sigma(\tfrac{j}{k}),\qquad \forall j\in\mathbb{Z}_k.$$

Notice that K is injective. In fact, if $K(\sigma)=\boldsymbol{0}$, we have $\sigma(\tfrac{j}{k})=\boldsymbol{0}$ for each $j\in\mathbb{Z}_k$, and since $\sigma\in\mathrm{Aff}_k(\mathbb{T};\mathbb{R}^m)$ we conclude that $\sigma=\boldsymbol{0}$. Hence, $\tilde{\boldsymbol{V}}=K\boldsymbol{V}$ is an $\mathrm{ind}(\mathscr{A},\gamma_{\boldsymbol{q}'})$-dimensional vector subspace of $\mathrm{T}_{\gamma_{\boldsymbol{q}'}}\Lambda_k$.

In order to conclude we just have to show that $\mathrm{Hess}\mathscr{A}(\gamma_{\boldsymbol{q}'})$ is negative-definite on the vector space $\tilde{\boldsymbol{V}}$. First of all, notice that, for each $\sigma\in W^{1,2}(\mathbb{T};\mathbb{R}^m)$, we have

$$\begin{aligned}\mathrm{Hess}\mathscr{A}(\gamma_{\boldsymbol{q}'})[\sigma,\sigma]&=\int_0^1\big(\langle a\,\dot\sigma,\dot\sigma\rangle+\langle b\,\sigma,\dot\sigma\rangle+\langle b^T\,\dot\sigma,\sigma\rangle+\langle c\,\sigma,\sigma\rangle\big)\,\mathrm{d}t\\&=2\int_0^1 L(t,\sigma(t),\dot\sigma(t))\,\mathrm{d}t=2\,A(\sigma).\end{aligned}\tag{4.21}$$

Now, consider an arbitrary $\tilde\sigma\in\tilde{\boldsymbol{V}}\setminus\{\boldsymbol{0}\}$ and set $\sigma=K^{-1}(\tilde\sigma)\in\boldsymbol{V}\setminus\{\boldsymbol{0}\}$. For each $j\in\mathbb{Z}_k$ the curve $\tilde\sigma|_{[j/k,(j+1)/k]}$ is an action minimizer with respect to L, and therefore $A(\tilde\sigma)\leq A(\sigma)$. By (4.20) and (4.21) we conclude

$$\mathrm{Hess}\mathscr{A}|_{\Lambda_k}(\gamma_{\boldsymbol{q}'})[\tilde\sigma,\tilde\sigma]=2\,A(\tilde\sigma)\leq 2\,A(\sigma)=\mathrm{Hess}\mathscr{A}(\gamma_{\boldsymbol{q}'})[\sigma,\sigma]<0.\qquad\square$$

4.5 Homotopic approximation of the action sublevels

The discretization technique introduced in this chapter can also be used to build finite-dimensional homotopic approximations of the sublevels of the action functional. Let $\mathscr{L}:\mathbb{T}\times\mathrm{T}M\to\mathbb{R}$ be a 1-periodic convex quadratic-growth Lagrangian with associated action $\mathscr{A}:W^{1,2}(\mathbb{T};M)\to\mathbb{R}$, as in the previous sections, and consider the constants $\varepsilon_0=\varepsilon_0(\mathscr{L})>0$ and $\rho_0=\rho_0(\mathscr{L})>0$ given by Theorem 4.1.1. We will need the following statement.

Proposition 4.5.1. *For each $c\in\mathbb{R}$ there exists $\bar\varepsilon=\bar\varepsilon(\mathscr{L},c)>0$ such that, for each $\zeta\in W^{1,2}(\mathbb{T};M)$ with $\mathscr{A}(\zeta)<c$ and for each interval $[t_0,t_1]\subset\mathbb{R}$ with $0<t_1-t_0\leq\bar\varepsilon$ we have* $\mathrm{dist}(\zeta(t_0),\zeta(t_1))<\rho_0$.

Proof. Up to adding a positive constant to the convex quadratic-growth Lagrangian $\mathscr{L}$, we can always assume that there exists a constant $\underline{\ell}>0$ such that

$$\mathscr{L}(t,q,v)\geq\underline{\ell}\,|v|_q^2,\qquad\forall(t,q,v)\in\mathbb{T}\times\mathrm{T}M.$$

Let us consider an arbitrary loop $\zeta\in W^{1,2}(\mathbb{T};M)$ such that $\mathscr{A}(\zeta)<c$. For each

interval $[t_0, t_1] \subset \mathbb{R}$ with $0 < t_1 - t_0 \leq 1$ we have

$$\begin{aligned}\operatorname{dist}(\zeta(t_0), \zeta(t_1))^2 &\leq \left(\int_{t_0}^{t_1} |\dot\zeta(t)|_{\zeta(t)} \mathrm{d}t \right)^2 \leq (t_1 - t_0) \int_{t_0}^{t_1} |\dot\zeta(t)|^2_{\zeta(t)} \, \mathrm{d}t \\ &\leq (t_1 - t_0) \int_{t_0}^{t_1} \underline{\ell}^{-1} \mathscr{L}(t, \zeta(t), \dot\zeta(t)) \, \mathrm{d}t \leq (t_1 - t_0) \underline{\ell}^{-1} \mathscr{A}(\zeta) \\ &< (t_1 - t_0) \underline{\ell}^{-1} c.\end{aligned}$$

Hence, for $\bar\varepsilon = \bar\varepsilon(\mathscr{L}, c) := \rho_0^2 \, \underline{\ell} \, c^{-1}$, we obtain the claim. □

From now on, we will denote the open sublevels of the action and of the discrete action by

$$(\mathscr{A})_c := \mathscr{A}^{-1}(-\infty, c), \qquad (\mathscr{A}_k)_c := \mathscr{A}_k^{-1}(-\infty, c), \qquad \forall c \in \mathbb{R}.$$

Let us consider the integer

$$\bar k(\mathscr{L}, c) := \left\lceil \max \left\{ \frac{1}{\varepsilon_0(\mathscr{L})}, \frac{1}{\bar\varepsilon(\mathscr{L}, c)} \right\} \right\rceil \in \mathbb{N}.$$

We want to show that, for each $c \in \mathbb{R}$ and for each integer $k \geq \bar k(\mathscr{L}, c)$, the map $\lambda_k : \Delta_k \hookrightarrow W^{1,2}(\mathbb{T}; M)$ (see the definition after (4.11)) restricts to a homotopy equivalence

$$\lambda_k : (\mathscr{A}_k)_c \xhookrightarrow{\sim} (\mathscr{A})_c.$$

Since $\mathscr{A}_k = \mathscr{A} \circ \lambda_k$ and since λ_k maps Δ_k diffeomorphically onto Λ_k, this property can be equivalently expressed by saying that the inclusion of the open sublevel $(\mathscr{A}|_{\Lambda_k})_c := (\mathscr{A})_c \cap \Lambda_k$ into $(\mathscr{A})_c$ is a homotopy equivalence.

A candidate homotopy inverse of λ_k is given by the map $r_k : (\mathscr{A})_c \to (\mathscr{A}_k)_c$ defined by

$$r_k(\zeta) = \left(\zeta(0), \zeta(\tfrac{1}{k}), \ldots, \zeta(\tfrac{k-1}{k})\right), \qquad \forall \zeta \in (\mathscr{A})_c. \tag{4.22}$$

Notice that, by Proposition 4.5.1, each $\zeta \in (\mathscr{A})_c$ satisfies $\operatorname{dist}(\zeta(\frac{j}{k}), \zeta(\frac{j+1}{k})) < \rho_0$ for each $j \in \mathbb{Z}_k$. Hence the map r_k is well defined. The composition $r_k \circ \lambda_k$ is the identity on $(\mathscr{A}_k)_c$. As for the inverse composition, $\lambda_k \circ r_k$, we build a homotopy $R_k : [0,1] \times (\mathscr{A})_c \to (\mathscr{A})_c$ as follows: for each $j \in \mathbb{Z}_k$, $s \in [\frac{j}{k}, \frac{j+1}{k}]$ and $\zeta \in (\mathscr{A})_c$, the loop $R_k(s, \zeta)$ is defined by setting

$$\begin{aligned} R_k(s, \zeta)|_{[0, j/k]} &:= \lambda_k \circ r_k(\zeta)|_{[0, j/k]}, \\ R_k(s, \zeta)|_{[s,1]} &:= \zeta|_{[s,1]}, \end{aligned}$$

and by setting $R_k(s, \zeta)|_{[j/k, s]}$ to be the unique action minimizer with endpoints $\zeta(\frac{j}{k})$ and $\zeta(s)$, see Figure 4.2. Then $R_k(0, \cdot)$ is the identity on $(\mathscr{A})_c$, $R_k(1, \cdot)$ is the composition $\lambda_k \circ r_k$, and

$$\mathscr{A}(R_k(s, \zeta)) \leq \mathscr{A}(\zeta) \qquad \forall \zeta \in (\mathscr{A})_c, \ s \in [0,1].$$

Therefore R_k is a well-defined homotopy from the identity on $(\mathscr{A})_c$ to $\lambda_k \circ r_k$.

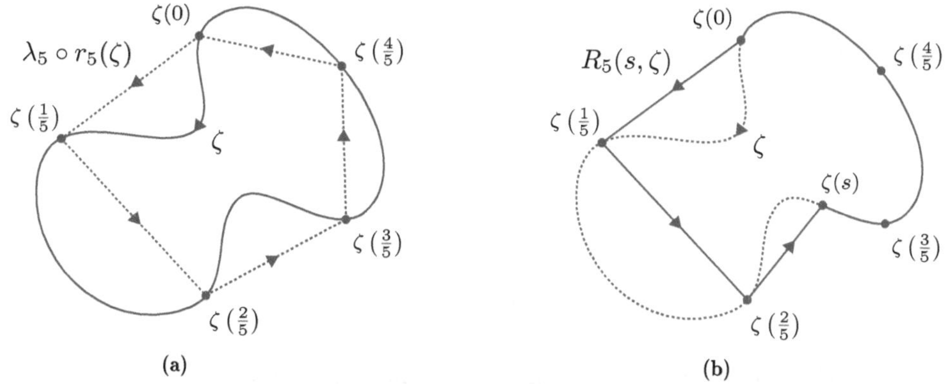

Figure 4.2. (a) Example of a loop ζ (solid line) and the corresponding $\lambda_5 \circ r_5(\zeta)$ (dotted line), for the case of the autonomous Lagrangian generating the geodesic flow on the flat $\mathbb{R}^2$, i.e., $\mathscr{L}(t,q,v) = \mathscr{L}(v) = v_1^2 + v_2^2$. (b) Homotoped curve $R_5(s,\zeta)$ (solid line).

Remark 4.5.1. Actually, R_k is even a strong deformation retraction. In fact, for each $\zeta \in (\mathscr{A})_c$, we have that $R_k(s,\zeta) = \zeta$ for all $s \in [0,1]$ if and only if $\zeta \in \Lambda_k = \lambda_k(\Delta_k)$. □

If $c_1 < c_2 \le c$, the same homotopy R_k can be used to show that the pair $((\mathscr{A})_{c_2}, (\mathscr{A})_{c_1})$ deformation retracts strongly onto $((\mathscr{A}|_{\Lambda_k})_{c_2}, (\mathscr{A}|_{\Lambda_k})_{c_1})$. Furthermore, if $\gamma \in W^{1,2}(\mathbb{T}; M)$ is a critical point of $\mathscr{A}$ with $\mathscr{A}(\gamma) = c$, up to increasing k we have that $\gamma \in \Lambda_k$ and R_k gives a deformation retraction of the pair $((\mathscr{A})_c \cup \{\gamma\}, (\mathscr{A})_c)$ onto $((\mathscr{A}|_{\Lambda_k})_c \cup \{\gamma\}, (\mathscr{A}|_{\Lambda_k})_c)$.

Summing up, we have obtained the following.

Lemma 4.5.2.

(i) *Let $c_1 < c_2 < \infty$. Then there exists a positive integer $\bar{k} = \bar{k}(\mathscr{L}, c_2)$ such that, for every integer $k \ge \bar{k}$, the embedding λ_k restricts to a homotopy equivalence of topological pairs*

$$\lambda_k : ((\mathscr{A}_k)_{c_2}, (\mathscr{A}_k)_{c_1}) \xhookrightarrow{\simeq} ((\mathscr{A})_{c_2}, (\mathscr{A})_{c_1}).$$

(ii) *Let $\boldsymbol{q} \in \Delta_k$ be a critical point of $\mathscr{A}_k$ such that $\mathscr{A}_k(\boldsymbol{q}) = c$. Then there exists a positive integer $\bar{k} = \bar{k}(\mathscr{L}, c)$ such that, for every integer $k \ge \bar{k}$, the embedding λ_k restricts to a homotopy equivalence of topological pairs*

$$\lambda_k : ((\mathscr{A}_k)_c \cup \{\boldsymbol{q}\}, (\mathscr{A}_k)_c) \xhookrightarrow{\simeq} ((\mathscr{A})_c \cup \{\gamma_{\boldsymbol{q}}\}, (\mathscr{A})_c),$$

where $\gamma_{\boldsymbol{q}} = \lambda_k(\boldsymbol{q})$. □

In the forthcoming chapters, we will mainly apply the above lemma to show that λ_k induces isomorphisms between invariant groups that are fundamental in

Morse theory: the **local homology groups** (see Section A.4). We recall that the local homology of $\mathscr{A}_k$ at a critical point $\boldsymbol{q}$ is defined as

$$\mathrm{C}_*(\mathscr{A}_k, \boldsymbol{q}) := \mathrm{H}_*\left((\mathscr{A}_k)_c \cup \{\boldsymbol{q}\}, (\mathscr{A}_k)_c\right),$$

where H_* in the right-hand side denotes the singular homology functor (with any coefficient group). The local homology of the full action functional $\mathscr{A}$ at the corresponding critical point $\gamma_{\boldsymbol{q}} = \lambda_k(\boldsymbol{q})$ is defined analogously as

$$\mathrm{C}_*(\mathscr{A}, \gamma_{\boldsymbol{q}}) := \mathrm{H}_*\left((\mathscr{A})_c \cup \{\gamma_{\boldsymbol{q}}\}, (\mathscr{A})_c\right).$$

Hence, point (ii) of the above Lemma 4.5.2 has the following immediate consequence.

Corollary 4.5.3. *For each integer $k >\geq k(\mathscr{L}, c)$ the embedding λ_k induces the homology isomorphism*

$$\mathrm{H}_*(\lambda_k) : \mathrm{C}_*(\mathscr{A}_k, \boldsymbol{q}) \xrightarrow{\simeq} \mathrm{C}_*(\mathscr{A}, \gamma_{\boldsymbol{q}}).$$ □

It is well known that the local homology groups of a C^2 functional at a critical point are trivial in dimension that is smaller than the Morse index or larger than the sum of the Morse index and the nullity (Corollary A.5.4). This result can be recovered for the C^1 action functional $\mathscr{A} : W^{1,2}(\mathbb{T}; M) \to \mathbb{R}$.

Corollary 4.5.4. *Let γ be an isolated critical point of $\mathscr{A}$. Then the local homology group $\mathrm{C}_j(\mathscr{A}, \gamma)$ is trivial if $j < \mathrm{ind}(\mathscr{A}, \gamma)$ or $j > \mathrm{ind}(\mathscr{A}, \gamma) + \mathrm{nul}(\mathscr{A}, \gamma)$.*

Proof. For each sufficiently large $k \in \mathbb{N}$, there exists $\boldsymbol{q} \in \Delta_k$ such that

$$\gamma = \gamma_{\boldsymbol{q}} = \lambda_k(\boldsymbol{q})$$

and $\boldsymbol{q}$ is an isolated critical point of the discrete action functional $\mathscr{A}_k$ (see Section 4.4). By Corollary 4.4.6 and Lemma 4.4.9, up to increasing k we have that

$$\mathrm{ind}(\mathscr{A}_k, \boldsymbol{q}) = \mathrm{ind}(\mathscr{A}, \gamma_{\boldsymbol{q}}),$$
$$\mathrm{nul}(\mathscr{A}_k, \boldsymbol{q}) = \mathrm{nul}(\mathscr{A}, \gamma_{\boldsymbol{q}}).$$

By the above Corollary 4.5.3, up to further increasing k we have

$$\mathrm{C}_*(\mathscr{A}_k, \boldsymbol{q}) \simeq \mathrm{C}_*(\mathscr{A}, \gamma_{\boldsymbol{q}}).$$

Therefore our claim follows from Corollary A.5.4 applied to the local homology group $\mathrm{C}_*(\mathscr{A}_k, \boldsymbol{q})$. □

4.6 Multiplicity of periodic orbits with prescribed period

In this section we apply the discretization technique that we have discussed so far in order to prove a multiplicity result, originally due to Benci [Ben86], for 1-periodic orbits of the Euler-Lagrange system associated to a convex quadratic-growth Lagrangian $\mathscr{L} : \mathbb{T} \times \mathrm{T}M \to \mathbb{R}$. Following the notation of Section A.6, for each homology class a we denote by $\mathfrak{S}_a$ the family of the supports of those cycles representing a. If $a \in \mathrm{H}_*(W^{1,2}(\mathbb{T}; M))$, we recall that the **minimax** of $\mathscr{A}$ over $\mathfrak{S}_a$ is the quantity

$$\underset{\mathfrak{S}_a}{\operatorname{minimax}} \mathscr{A} := \inf_{\mathscr{K} \in \mathfrak{S}_a} \sup_{\gamma \in \mathscr{K}} \{\mathscr{A}(\gamma)\} \in \mathbb{R}.$$

Theorem 4.6.1. *Let M be a smooth closed manifold of positive dimension, and consider a 1-periodic convex quadratic-growth Lagrangian $\mathscr{L} : \mathbb{T} \times \mathrm{T}M \to \mathbb{R}$ with associated action functional $\mathscr{A} : W^{1,2}(\mathbb{T}; M) \to \mathbb{R}$.*

(i) *If $a \in \mathrm{H}_d(W^{1,2}(\mathbb{T}; M))$ is a nonzero homology class, then $\operatorname{minimax}_{\mathfrak{S}_a} \mathscr{A}$ is a critical value of $\mathscr{A}$. Moreover, if the critical points of $\mathscr{A}$ corresponding to this minimax critical value are isolated, there exists a critical point γ of $\mathscr{A}$ such that*

$$\begin{aligned} \mathscr{A}(\gamma) &= \underset{\mathfrak{S}_a}{\operatorname{minimax}} \mathscr{A}, \\ \operatorname{ind}(\mathscr{A}, \gamma) \leq d &\leq \operatorname{ind}(\mathscr{A}, \gamma) + \operatorname{nul}(\mathscr{A}, \gamma). \end{aligned} \tag{4.23}$$

(ii) *If[2] $a \prec b$ in $\mathrm{H}_*(W^{1,2}(\mathbb{T}; M))$, then*

$$\underset{\mathfrak{S}_a}{\operatorname{minimax}} \mathscr{A} \leq \underset{\mathfrak{S}_b}{\operatorname{minimax}} \mathscr{A}$$

and either the inequality above is strict or $\mathscr{A}$ has infinitely many critical points corresponding to the critical value $\operatorname{minimax}_{\mathfrak{S}_a} \mathscr{A} = \operatorname{minimax}_{\mathfrak{S}_b} \mathscr{A}$.

(iii) *The number of critical points of $\mathscr{A}$ is at least $\mathrm{CL}(W^{1,2}(\mathbb{T}; M)) + 1$, where $\mathrm{CL}(W^{1,2}(\mathbb{T}; M))$ is the cup-length of $W^{1,2}(\mathbb{T}; M)$.*

Thus, the multiplicity of the 1-periodic orbits is guaranteed by the richness of the homology of the free loop space of M. From Theorem 3.2.5, we know that the homology of the free loop space is rich whenever the configuration space is simply connected. More generally, we have the following.

Corollary 4.6.2. *If $\pi_1(M)$ is finite, then the Euler-Lagrange system of $\mathscr{L}$ admits infinitely many contractible 1-periodic orbits. If they are all isolated, then there exists an infinite sequence of contractible 1-periodic orbits $\{\gamma_n \mid n \in \mathbb{N}\}$ such that*

$$\lim_{n \to \infty} \mathscr{A}(\gamma_n) = \infty,$$
$$\lim_{n \to \infty} \operatorname{ind}(\mathscr{A}, \gamma_n) = \infty.$$

[2]See Section A.6 for the definition of the homological relation "$\prec$".

Proof. Let us assume that the contractible 1-periodic orbits of the Euler-Lagrange system of $\mathscr{L}$ are all isolated (otherwise there is nothing to prove). Let $\pi : \widetilde{M} \to M$ be the universal covering of M. Since $\pi_1(M)$ is finite, $\widetilde{M}$ has finite degree and in particular it is compact. We can lift $\mathscr{L}$ to a convex quadratic-growth Lagrangian $\widetilde{\mathscr{L}} : \mathbb{T} \times \mathrm{T}\widetilde{M} \to \mathbb{R}$ by setting

$$\widetilde{\mathscr{L}}(t, \widetilde{q}, \widetilde{v}) := \mathscr{L}(t, \pi(\widetilde{q}), \mathrm{d}\pi(\widetilde{q})\widetilde{v}), \qquad \forall (t, \widetilde{q}, \widetilde{v}) \in \mathbb{T} \times \mathrm{T}\widetilde{M}.$$

Notice that, if $\widetilde{\gamma} : \mathbb{T} \to \widetilde{M}$ is a (contractible) 1-periodic solution of the Euler-Lagrange system of $\widetilde{\mathscr{L}}$, then its projection $\gamma := \pi \circ \widetilde{\gamma}$ is a contractible 1-periodic solution of the Euler-Lagrange system of $\mathscr{L}$ and, if we denote the action of $\widetilde{\mathscr{L}}$ by $\widetilde{\mathscr{A}}$, we have

$$\begin{aligned} \widetilde{\mathscr{A}}(\widetilde{\gamma}) &= \mathscr{A}(\gamma), \\ \mathrm{ind}(\widetilde{\mathscr{A}}, \widetilde{\gamma}) &= \mathrm{ind}(\mathscr{A}, \gamma), \\ \mathrm{nul}(\widetilde{\mathscr{A}}, \widetilde{\gamma}) &= \mathrm{nul}(\mathscr{A}, \gamma). \end{aligned}$$

Hence, we only need to prove the statement for the Euler-Lagrange system associated to $\widetilde{\mathscr{L}}$.

By Theorem 3.2.5, there exists an infinite sequence of homology classes

$$\left\{ a_n \in \mathrm{H}_{d_n}(W^{1,2}(\mathbb{T}; \widetilde{M})) \,|\, n \in \mathbb{N} \right\}$$

such that $d_n \to \infty$ as $n \to \infty$. By Theorem 4.6.1, there exists an associated sequence $\{\widetilde{\gamma}_n \,|\, n \in \mathbb{N}\}$ of 1-periodic solutions of the Euler-Lagrange system of $\widetilde{\mathscr{L}}$ such that

$$\lim_{n\to\infty} \left[\mathrm{ind}(\widetilde{\mathscr{A}}, \widetilde{\gamma}_n) + \mathrm{nul}(\widetilde{\mathscr{A}}, \widetilde{\gamma}_n) \right] = \infty.$$

This, together with Proposition 2.1.4, implies that

$$\lim_{n\to\infty} \widetilde{\mathscr{A}}(\widetilde{\gamma}_n) = \infty.$$

Finally, since the nullity of the action functional at a critical point is always less than or equal to $2\dim(M) = 2\dim(\widetilde{M})$ (see Lemma 4.4.4), we obtain

$$\lim_{n\to\infty} \mathrm{ind}(\widetilde{\mathscr{A}}, \widetilde{\gamma}_n) = \infty. \qquad \square$$

Corollary 4.6.3. *If $\pi_1(M)$ is abelian (e.g., if M is a Lie group or, more generally, an H-space*[3]*), then the Euler-Lagrange system of $\mathscr{L}$ admits infinitely many 1-periodic orbits.*

[3]We recall that a topological space X is an **H-space**, "H" standing for "Hopf", when there is a multiplication map $* : X \times X \to X$ and an identity element $e \in X$ such that the two maps $X \to X$ given by $x \mapsto x * e$ and $x \mapsto e * x$ are homotopic to the identity on X with homotopies relative to $\{e\}$.

Proof. If $\pi_1(M)$ is finite, then the statement follows from Corollary 4.6.2. If $\pi_1(M)$ is infinite and abelian, then it has infinitely many conjugacy classes (each element of $\pi_1(M)$ is a conjugacy class) and the statement follows from Theorem 3.5.5. □

In the remainder of this section we will prove Theorem 4.6.1. As the reader might guess, we want to apply the abstract minimax statements from section A.6. However, in order to get the estimate (4.23) on the Morse index and nullity pair of the periodic orbits, we cannot immediately apply Theorem A.6.3 since the action functional $\mathscr{A}$ is not C^2. The idea here is to apply the homological minimax theorem to the (smooth) discrete action functional. First of all, we show that the minimax values of the action functional and of the discrete action functional are the same. Consider a nonzero homology class $a \in \mathrm{H}_*(W^{1,2}(\mathbb{T};M))$ represented by the cycle α. We denote by $\mathscr{K}_\alpha \subset W^{1,2}(\mathbb{T};M)$ the (compact) support of α, and we choose a constant $c \in \mathbb{R}$ such that

$$c > \max_{\gamma\in\mathscr{K}_\alpha} \{\mathscr{A}(\gamma)\} \in \mathbb{R}. \tag{4.24}$$

Notice that α also represents a nonzero homology class $a' \in \mathrm{H}_*((\mathscr{A})_c)$, and

$$\operatorname*{minimax}_{\mathfrak{S}_a} \mathscr{A} = \operatorname*{minimax}_{\mathfrak{S}_{a'}} \mathscr{A}. \tag{4.25}$$

Now, consider the constant $\bar{k}(\mathscr{L},c) \in \mathbb{N}$ given by Lemma 4.5.2. For each integer $k \geq \bar{k}(\mathscr{L},c)$, the embedding $\lambda_k : (\mathscr{A}_k)_c \hookrightarrow (\mathscr{A})_c$ is a homotopy equivalence, with a homotopic inverse $r_k : (\mathscr{A})_c \to (\mathscr{A}_k)_c$ given by (4.22), and therefore we can define a nonzero homology class

$$a_k := \mathrm{H}_*(r_k)\, a' \in \mathrm{H}_*((\mathscr{A}_k)_c).$$

Lemma 4.6.4. $\operatorname*{minimax}_{\mathfrak{S}_a} \mathscr{A} = \operatorname*{minimax}_{\mathfrak{S}_{a_k}} \mathscr{A}_k.$

Proof. By (4.25), we only need to prove that

$$\operatorname*{minimax}_{\mathfrak{S}_{a'}} \mathscr{A} = \operatorname*{minimax}_{\mathfrak{S}_{a_k}} \mathscr{A}_k.$$

For each $\mathscr{K} \in \mathfrak{S}_{a_k}$, we have $\lambda_k(\mathscr{K}) \in \mathfrak{S}_{a'}$, therefore

$$\operatorname*{minimax}_{\mathfrak{S}_{a'}} \mathscr{A} \leq \operatorname*{minimax}_{\mathfrak{S}_{a_k}} \mathscr{A}_k.$$

On the other hand, for each $\gamma \in (\mathscr{A})_c$ we have that $\mathscr{A}(r_k(\gamma)) \leq \mathscr{A}(\gamma)$ and, for each $\mathscr{K}' \in \mathfrak{S}_{a'}$, we have that $r_k(\mathscr{K}') \in \mathfrak{S}_{a_k}$. Therefore

$$\operatorname*{minimax}_{\mathfrak{S}_{a'}} \mathscr{A} = \inf_{\mathscr{K}'\in\mathfrak{S}_{a'}} \sup_{\gamma\in\mathscr{K}'} \{\mathscr{A}(\gamma)\} \geq \inf_{\mathscr{K}'\in\mathfrak{S}_{a'}} \sup_{\gamma\in\mathscr{K}'} \{\mathscr{A}_k(r_k(\gamma))\} = \operatorname*{minimax}_{\mathfrak{S}_{a_k}} \mathscr{A}_k.$$

□

Proof of Theorem 4.6.1. Everything beside the estimate (4.23) follows from the abstract Theorems A.6.2, A.6.5 and from Corollary A.6.6. As for (4.23), consider a cycle α representing the homology class $a \in \mathrm{H}_d(W^{1,2}(\mathbb{T}; M))$, and a real constant c satisfying (4.24). The cycle α also represents a nonzero $a' \in \mathrm{H}_d((\mathscr{A})_c)$ and, for each integer $k \geq \bar{k}(\mathscr{L}, c)$, we obtain a nonzero $a_k := \mathrm{H}_d(r_k)\, a' \in \mathrm{H}_d((\mathscr{A}_k)_c)$ such that, by Lemma 4.6.4,

$$\underset{\mathfrak{S}_a}{\operatorname{minimax}} \mathscr{A} = \underset{\mathfrak{S}_{a_k}}{\operatorname{minimax}} \mathscr{A}_k.$$

Now, assume that the critical level $\operatorname{minimax}_{\mathfrak{S}_a} \mathscr{A}$ contains only isolated critical points of $\mathscr{A}$. Since $\mathscr{A}$ satisfies the Palais-Smale condition (Proposition 3.5.2), there exist only finitely many critical points $\gamma_1, \ldots, \gamma_s$ of $\mathscr{A}$ with critical value $\operatorname{minimax}_{\mathfrak{S}_a} \mathscr{A}$. This implies that there exist only finitely many critical points $\boldsymbol{q}_1, \ldots, \boldsymbol{q}_s$ of $\mathscr{A}_k$ with critical value $\operatorname{minimax}_{\mathfrak{S}_{a_k}} \mathscr{A}_k = \operatorname{minimax}_{\mathfrak{S}_a} \mathscr{A}$. Notice that, up to increasing k, by Corollary 4.4.6 and Lemma 4.4.9 we have

$$\operatorname{ind}(\mathscr{A}_k, \boldsymbol{q}_j) = \operatorname{ind}(\mathscr{A}, \gamma_j),$$
$$\operatorname{nul}(\mathscr{A}_k, \boldsymbol{q}_j)) = \operatorname{nul}(\mathscr{A}, \gamma_j),$$

for all $j = 1, \ldots, s$. Now, since the discrete action $\mathscr{A}_k$ is C^∞, by Theorem A.6.3 there exists $\boldsymbol{q} \in \{\boldsymbol{q}_1, \ldots, \boldsymbol{q}_s\}$ such that

$$\mathscr{A}_k(\boldsymbol{q}) = \underset{\mathfrak{S}_{a_k}}{\operatorname{minimax}} \mathscr{A}_k,$$
$$\operatorname{ind}(\mathscr{A}_k, \boldsymbol{q}) \leq d \leq \operatorname{ind}(\mathscr{A}_k, \boldsymbol{q}) + \operatorname{nul}(\mathscr{A}_k, \boldsymbol{q}).$$

□

4.7 Discretizations in higher period

So far we have only dealt with 1-periodic loop spaces, but our arguments extend to every period $n \in \mathbb{N}$ as follows. Consider the smooth 1-periodic convex quadratic-growth Lagrangian $\mathscr{L} : \mathbb{T} \times \mathrm{T}M \to \mathbb{R}$ of the previous sections. For each $k \in \mathbb{N}$ we define $C^\infty_k(\mathbb{T}^{[n]}; M)$ as the space of continuous n-periodic curves $\zeta : \mathbb{T}^{[n]} \to M$ such that, for each $j \in \mathbb{Z}_{nk}$, the restriction $\zeta|_{[j/k,(j+1)/k]}$ is smooth. We endow this space with the broken smooth topology that turns it into a Fréchet manifold, as we did for $C^\infty_k(\mathbb{T}; M)$. If $k \geq 1/\varepsilon_0(\mathscr{L})$, we define a smooth embedding

$$\lambda^{[n]}_k = \lambda^{[n]}_{k,\mathscr{L}} : \Delta_{nk} \hookrightarrow C^\infty_k(\mathbb{T}^{[n]}; M)$$

in the following way: for each $\boldsymbol{q} = (q_0, \ldots, q_{nk-1}) \in \Delta_{nk}$ we set $\lambda^{[n]}_k(\boldsymbol{q}) := \gamma_{\boldsymbol{q}}$, where $\gamma_{\boldsymbol{q}}$ is the unique loop whose restrictions $\gamma_{\boldsymbol{q}}|_{[j/k,(j+1)/k]}$ are the unique action minimizers (with respect to $\mathscr{L}$) with endpoints q_j and q_{j+1}, for each $j \in \mathbb{Z}_{nk}$. We define the **k-broken n-periodic Euler-Lagrange loop space** (with respect to $\mathscr{L}$) as the image

$$\Lambda^{[n]}_k = \Lambda^{[n]}_{k,\mathscr{L}} := \lambda^{[n]}_k(\Delta_{nk})$$

of the embedding $\lambda_k^{[n]}$. As for the 1-periodic case, $\Lambda_k^{[n]}$ is a smooth submanifold of $C_k^\infty(\mathbb{T}^{[n]};M)$ (and of $W^{1,2}(\mathbb{T}^{[n]};M)$) with finite dimension nkm, where m is the dimension of M. Finally, we define the **discrete mean action functional**

$$\mathscr{A}_k^{[n]} : \Delta_{nk} \to \mathbb{R}$$

as the composition of the mean action functional $\mathscr{A}^{[n]}$ of $\mathscr{L}$ (see Section 3.6) with $\lambda_k^{[n]}$, namely

$$\mathscr{A}_k^{[n]}(\boldsymbol{q}) = \frac{1}{n}\int_0^n \mathscr{L}(t,\gamma_{\boldsymbol{q}}(t),\dot\gamma_{\boldsymbol{q}}(t))\,\mathrm{d}t, \qquad \forall \boldsymbol{q} \in \Delta_{nk}.$$

All the results of Sections 4.3, 4.4, 4.5 and 4.6 still hold if we replace $\mathscr{A}$ and $\mathscr{A}_k$ with the mean versions $\mathscr{A}^{[n]}$ and $\mathscr{A}_k^{[n]}$, for each $n \in \mathbb{N}$.

The nth-iteration map $\psi^{[n]} : W^{1,2}(\mathbb{T};M) \hookrightarrow W^{1,2}(\mathbb{T}^{[n]};M)$ restricts to a continuous embedding of k-broken Euler-Lagrange loop spaces

$$\psi^{[n]} : \Lambda_k \hookrightarrow \Lambda_k^{[n]}.$$

Moreover, by means of the diffeomorphisms λ_k and $\lambda_k^{[n]}$ we can define the **discrete nth-iteration map**

$$\psi_k^{[n]} : \Delta_k \hookrightarrow \Delta_{nk},$$

in such a way that the following diagram commutes.

$$\begin{array}{ccc}
\Delta_k & \xrightarrow{\ \psi_k^{[n]}\ } & \Delta_{nk} \\
{\scriptstyle \lambda_k}\downarrow{\scriptstyle \simeq} & & {\scriptstyle \simeq}\downarrow{\scriptstyle \lambda_k^{[n]}} \\
\Lambda_k & \xrightarrow{\ \psi^{[n]}\ } & \Lambda_k^{[n]}
\end{array}$$

Namely, for each $\boldsymbol{q} \in \Delta_k$, we have $\psi_k^{[n]}(\boldsymbol{q}) = \boldsymbol{q}^{[n]}$, where

$$\boldsymbol{q}^{[n]} := (\underbrace{\boldsymbol{q},\dots,\boldsymbol{q}}_{n \text{ times}}).$$

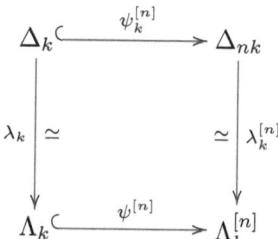

[illegible] As for the [illegible] [illegible] Finally we define the discrete loop action functional

[illegible]

[illegible]

[illegible]

Chapter 5

Local Homology and Hilbert Subspaces

In a Lagrangian system associated to a Lagrangian function $\mathscr{L} : \mathbb{R}/\mathbb{Z} \times \mathrm{T}M \to \mathbb{R}$, a periodic orbit $\gamma : \mathbb{R}/\mathbb{Z} \to M$ produces a critical point of the action functional in any period $n \in \mathbb{N}$. Namely, the iterated curve $\gamma^{[n]}$ is a critical point of $\mathscr{A}^{[n]}$. As we already saw in Section 4.6, critical point theory allows us to investigate the multiplicity of periodic orbits with prescribed period n. On the other hand, if one is interested in the multiplicity of periodic orbits with any period, some additional arguments are needed in order to recognize when an n-periodic orbit is the iteration of a periodic orbit of smaller period. Bott's theory, which we discussed in Section 2.2, served this purpose by studying the behavior of the Morse index and nullity of iterated periodic orbits. Now, we wish to go one step further, and investigate the behavior of the local homology of periodic orbits under iteration.

The first result in this direction is due to Gromoll and Meyer [GM69b, Theorem 3]. In the setting of closed geodesics (i.e., when the Lagrangian is given by a squared Riemannian norm), they proved that the local homology of an isolated closed geodesic is unchanged under an iteration that does not change the nullity, up to a shift in the degree that is given by the shift of the Morse index of the iterated geodesic. For mechanical Lagrangians (i.e., fiberwise quadratic Lagrangians of the form kinetic plus potential energy), Long [Lon00, Theorem 3.7] proved that, if an isolated periodic orbit γ and its iteration $\gamma^{[n]}$ have the same Morse index and nullity, the isomorphism between the local homology of γ and the local homology of $\gamma^{[n]}$ is induced by the iteration map. This fact turns out to be crucial in the study of the multiplicity of periodic orbits with unprescribed period, as we will see in Section 6.5 of the next chapter.

In this chapter, following [Maz11, Section 4], we carry out Long's argument in an abstract setting: given a C^2 function defined on an open set of a separable Hilbert space, the local homology of the function at an isolated critical point does not change if we restrict the function to a Hilbert subspace containing the critical point, provided the Morse index and the nullity of the critical point do not change under the restriction. In Section 5.1 we state this result. The proof will be carried out in Section 5.4, after establishing several preliminaries in Sections 5.2 and 5.3. Finally, in Section 5.5 we discuss applications to the Lagrangian action functional, proving Long's result for the more general case of Lagrangian systems associated to convex quadratic-growth Lagrangians. We refer the reader to the Appendix for the background needed from Morse theory.

5.1 The abstract result

Let $\mathscr{U}$ be an open set of a real separable Hilbert space $\boldsymbol{E}$, and $\mathscr{F} : \mathscr{U} \to \mathbb{R}$ a C^2 function that satisfies the Palais-Smale condition. Assume that $\boldsymbol{x} \in \mathscr{U}$ is an isolated critical point of $\mathscr{F}$ with finite Morse index and assume that the Hessian of $\mathscr{F}$ at $\boldsymbol{x}$ is Fredholm. We recall that this latter condition means that the associated bounded self-adjoint operator H on $\boldsymbol{E}$ is Fredholm, where

$$\mathrm{Hess}\mathscr{F}(\boldsymbol{x})[\boldsymbol{v},\boldsymbol{w}] = \langle\langle H\boldsymbol{v},\boldsymbol{w}\rangle\rangle_{\boldsymbol{E}}, \qquad \forall \boldsymbol{v},\boldsymbol{w} \in \boldsymbol{E}. \tag{5.1}$$

In particular the nullity of $\mathscr{F}$ at $\boldsymbol{x}$ is finite, being the dimension of the kernel of H.

Now, let $\boldsymbol{E}_\bullet$ be a Hilbert subspace of $\boldsymbol{E}$ containing the critical point $\boldsymbol{x}$. We assume that $\mathscr{U}_\bullet := \mathscr{U} \cap \boldsymbol{E}_\bullet$ is invariant under the gradient flow of $\mathscr{F}$. This requirement can be expressed in terms of the isometric inclusion $J : \mathscr{U}_\bullet \hookrightarrow \mathscr{U}$ as

$$(\mathrm{Grad}\mathscr{F}) \circ J = J \circ \mathrm{Grad}(\mathscr{F} \circ J). \tag{5.2}$$

Throughout this chapter, for simplicity, all the homology groups are assumed to have coefficients in a field $\mathbb{F}$ (in this way we avoid the torsion terms that appear in the Künneth formula). We recall that the **local homology** of the function $\mathscr{F}$ at $\boldsymbol{x}$ is defined as

$$\mathrm{C}_*(\mathscr{F},\boldsymbol{x}) := \mathrm{H}_*\left((\mathscr{F})_c \cup \{\boldsymbol{x}\}, (\mathscr{F})_c\right),$$

where $c = \mathscr{F}(\boldsymbol{x})$ and $(\mathscr{F})_c := \mathscr{F}^{-1}(-\infty, c)$. If we denote by $\mathscr{F}_\bullet : \mathscr{U}_\bullet \to \mathbb{R}$ the restricted function $\mathscr{F}|_{\mathscr{U}_\bullet}$, then $\boldsymbol{x}$ is a critical point of $\mathscr{F}_\bullet$ as well and the local homology $\mathrm{C}_*(\mathscr{F}_\bullet,\boldsymbol{x})$ is defined analogously as $\mathrm{H}_*((\mathscr{F}_\bullet)_c \cup \{\boldsymbol{x}\}, (\mathscr{F}_\bullet)_c)$. The inclusion J restricts to a continuous map of pairs

$$J : ((\mathscr{F}_\bullet)_c \cup \{\boldsymbol{x}\}, (\mathscr{F}_\bullet)_c) \hookrightarrow ((\mathscr{F})_c \cup \{\boldsymbol{x}\}, (\mathscr{F})_c),$$

and therefore induces the homology homomorphism

$$\mathrm{H}_*(J) : \mathrm{C}_*(\mathscr{F}_\bullet,\boldsymbol{x}) \to \mathrm{C}_*(\mathscr{F},\boldsymbol{x}).$$

The main result of this chapter is the following.

Theorem 5.1.1. *If the Morse indices and nullity of $\mathscr{F}$ and $\mathscr{F}_\bullet$ at $\boldsymbol{x}$ coincide, i.e.,*

$$\begin{aligned}\mathrm{ind}(\mathscr{F},\boldsymbol{x}) &= \mathrm{ind}(\mathscr{F}_\bullet,\boldsymbol{x}),\\ \mathrm{nul}(\mathscr{F},\boldsymbol{x}) &= \mathrm{nul}(\mathscr{F}_\bullet,\boldsymbol{x}),\end{aligned}$$

then $\mathrm{H}_*(J)$ is an isomorphism of local homology groups.

The proof of this theorem will be carried out in Section 5.4, after several preliminaries.

Remark 5.1.1. One might ask if Theorem 5.1.1 still holds without the assumption (5.2). This is true in the case where $\boldsymbol{x}$ is a non-degenerate critical point: a relative cycle that represents a generator of $\mathrm{C}_{\mathrm{ind}(\mathscr{F},\boldsymbol{x})}(\mathscr{F}_\bullet,\boldsymbol{x})$ also represents a generator of $\mathrm{C}_{\mathrm{ind}(\mathscr{F},\boldsymbol{x})}(\mathscr{F},\boldsymbol{x})$, and in every other degree $j \neq \mathrm{ind}(\mathscr{F},\boldsymbol{x}) = \mathrm{ind}(\mathscr{F}_\bullet,\boldsymbol{x})$ the local homology groups $\mathrm{C}_j(\mathscr{F}_\bullet,\boldsymbol{x})$ and $\mathrm{C}_j(\mathscr{F},\boldsymbol{x})$ are trivial (see Theorem A.4.1). However, in the general case, assumption (5.2) cannot be dropped, as is shown by the following simple example. Consider the function $\mathscr{F}:\mathbb{R}^2\to\mathbb{R}$ from Example A.4 in the Appendix, i.e.,

$$\mathscr{F}(x,y) = (y-x^2)(y-2x^2), \qquad \forall (x,y)\in\mathbb{R}^2.$$

The origin $\mathbf{0}$ is clearly an isolated critical point of $\mathscr{F}$, and the corresponding Hessian is given in matrix form by

$$\mathrm{Hess}\mathscr{F}(0,0) = \begin{bmatrix} 0 & 0\\ 0 & 2\end{bmatrix}. \tag{5.3}$$

Now, let us consider the inclusion $J:\mathbb{R}\hookrightarrow\mathbb{R}^2$ given by $J(x)=(x,0)$, namely the inclusion of the x-axis in $\mathbb{R}^2$. Then

$$\begin{aligned}\mathrm{ind}(\mathscr{F},\mathbf{0}) &= \mathrm{ind}(\mathscr{F}\circ J,0) = 0,\\ \mathrm{nul}(\mathscr{F},\mathbf{0}) &= \mathrm{nul}(\mathscr{F}\circ J,0) = 1.\end{aligned} \tag{5.4}$$

However the gradient of $\mathscr{F}$ on the x-axis is given by

$$\mathrm{Grad}\mathscr{F}(x,0) = (8x^3,-3x^2), \qquad \forall x\in\mathbb{R}.$$

Hence condition (5.2) is not satisfied for the inclusion J. The local homology groups of $\mathscr{F}$ and $\mathscr{F}\circ J$ at the origin are not isomorphic (and consequently $\mathrm{H}_*(J)$ is not an isomorphism). In fact, by examining the sublevel $(\mathscr{F})_0$ (see Figure A.5 in the Appendix) it is clear that the origin is a saddle point of $\mathscr{F}$ and a minimum of $\mathscr{F}\circ J$. Therefore we have

$$\mathrm{C}_j(\mathscr{F},\mathbf{0}) = \begin{cases}\mathbb{F} & j=1,\\ 0 & j\neq 1,\end{cases} \qquad \mathrm{C}_j(\mathscr{F}\circ J,0) = \begin{cases}\mathbb{F} & j=0,\\ 0 & j\neq 0.\end{cases} \qquad \square$$

Remark 5.1.2. The C^2 regularity of the involved function is also an essential assumption in Theorem 5.1.1. In fact, let us modify the function $\mathscr{F}$ of the previous remark in the following way:

$$\mathscr{F}(x,y) = (y - x^2)(y - 2x^2) + 3x^6 \arctan\left(\frac{y}{x^4}\right), \qquad \forall (x,y) \in \mathbb{R}^2.$$

This function is C^1 and twice Gâteaux differentiable, but it is not C^2 at the origin, which is an isolated critical point. The Hessian of $\mathscr{F}$ at the origin is still as in equation (5.3). The gradient of $\mathscr{F}$ on the x-axis is given by

$$\mathrm{Grad}\mathscr{F}(x,0) = (8x^3, 0), \qquad \forall x \in \mathbb{R}.$$

Hence, condition (5.2) is satisfied, i.e.,

$$(\mathrm{Grad}\mathscr{F}) \circ J = \mathrm{Grad}(\mathscr{F} \circ J),$$

where $J : \mathbb{R} \hookrightarrow \mathbb{R}^2$ is given by $J(x) = (x,0)$. The Morse index and nullity of $\mathscr{F}$ and $\mathscr{F}$ at the origin are still as in (5.4). Moreover, 0 is a local minimum of $\mathscr{F} \circ J$, which implies

$$\mathrm{C}_j(\mathscr{F} \circ J, 0) = \begin{cases} \mathbb{F} & j = 0, \\ 0 & j \neq 0. \end{cases}$$

However, the origin $\mathbf{0} \in \mathbb{R}^2$ is not a local minimum of the function $\mathscr{F}$. In fact, a straightforward computation shows that $\mathbf{0}$ is a local maximum of the function $\mathscr{F}$ restricted to the parabola $y = \frac{3}{2}x^2$, namely $0 \in \mathbb{R}$ is a local maximum of the function

$$x \mapsto \mathscr{F}\left(x, \tfrac{3}{2}x^2\right) = -\frac{1}{4}x^4 + 3x^6 \arctan\left(\frac{3}{2x^2}\right).$$

This readily implies that $\mathrm{C}_0(\mathscr{F}, \mathbf{0})$ is trivial. □

5.2 The generalized Morse lemma revisited

In order to prove Theorem 5.1.1 we need to reconsider the generalized Morse lemma (Lemma A.2.1): more specifically, for reasons that will become clear in the next two sections, we need to specify how the map ϕ and the function $\mathscr{F}^0$ appearing in the statement are defined. This can be done by inspecting the original proof in [GM69a] or, alternatively, in [MW89, page 189] and [Cha93, page 44]. For the reader's convenience, we will include the proof here with all the details.

In order to simplify the notation, from now on we will assume, without loss of generality, that $\boldsymbol{x} = \mathbf{0} \in \boldsymbol{E}$. Consider the operator H associated to the Hessian of $\mathscr{F}$ at the critical point $\mathbf{0}$ defined in (5.1). According to the spectral theorem, we have an orthogonal splitting $\boldsymbol{E} = \boldsymbol{E}^+ \oplus \boldsymbol{E}^- \oplus \boldsymbol{E}^0$, where $\boldsymbol{E}^+$ [resp. $\boldsymbol{E}^-$] is a vector subspace of $\boldsymbol{E}$ on which H is positive-definite [resp. negative-definite], while $\boldsymbol{E}^0$ is the finite-dimensional kernel of H. We will write every vector $\boldsymbol{v} \in \boldsymbol{E}$ as

$$\boldsymbol{v} = \boldsymbol{v}^0 + \boldsymbol{v}^{\pm},$$

where $\boldsymbol{v}^{\pm} \in \boldsymbol{E}^{\pm} := \boldsymbol{E}^{+} \oplus \boldsymbol{E}^{-}$ and $\boldsymbol{v}^{0} \in \boldsymbol{E}^{0}$. On $\boldsymbol{E}^{\pm} \setminus \{\mathbf{0}\}$ we introduce the local flow Θ_H defined by $\Theta_H(s, \sigma(0)) = \sigma(s)$, where $\sigma : (s_0, s_1) \to \boldsymbol{E}^{\pm} \setminus \{\mathbf{0}\}$ (with $s_0 < 0 < s_1$) is a curve that satisfies

$$\dot{\sigma}(s) = -\frac{H\sigma(s)}{\|H\sigma(s)\|_{\boldsymbol{E}}}, \qquad \forall s \in (s_0, s_1). \tag{5.5}$$

We also set $\Theta_H(0, \mathbf{0}) := \mathbf{0}$. Notice that Θ_H is a reparametrization of the anti-gradient flow of the function $\mathscr{F}^{\pm} : \boldsymbol{E}^{\pm} \to \mathbb{R}$ defined by

$$\mathscr{F}^{\pm}(\boldsymbol{v}^{\pm}) := \tfrac{1}{2}\mathrm{Hess}\mathscr{F}(\mathbf{0})[\boldsymbol{v}^{\pm}, \boldsymbol{v}^{\pm}] = \tfrac{1}{2}\langle\!\langle H\boldsymbol{v}^{\pm}, \boldsymbol{v}^{\pm}\rangle\!\rangle_{\boldsymbol{E}}, \qquad \forall \boldsymbol{v}^{\pm} \in \boldsymbol{E}^{\pm}. \tag{5.6}$$

The generalized Morse lemma can be restated as follows.

Lemma 5.2.1 (Generalized Morse Lemma revisited). *If $\mathscr{V} \subseteq \boldsymbol{E}$ is a sufficiently small open neighborhood of $\mathbf{0}$, there exist*

(i) *a C^1 map $\psi : (\mathscr{V} \cap \boldsymbol{E}^0, \mathbf{0}) \to (\boldsymbol{E}^{\pm}, \mathbf{0})$ implicitly defined by*

$$\mathrm{d}\mathscr{F}(\boldsymbol{v}^0 + \psi(\boldsymbol{v}^0))\big|_{\boldsymbol{E}^{\pm}} = \mathbf{0}, \qquad \forall \boldsymbol{v}^0 \in \mathscr{V} \cap \boldsymbol{E}^0, \tag{5.7}$$

(ii) *a map $\phi : (\mathscr{V}, \mathbf{0}) \to (\mathscr{U}, \mathbf{0})$ that is a homeomorphism onto its image, defined by*

$$\begin{aligned}\phi(\boldsymbol{v}) = \boldsymbol{v}^0 + \Theta_H(\tau(\boldsymbol{v} - \psi(\boldsymbol{v}^0)), \boldsymbol{v}^{\pm} - \psi(\boldsymbol{v}^0)),&\\ \forall \boldsymbol{v} = \boldsymbol{v}^0 + \boldsymbol{v}^{\pm} \in \mathscr{V},&\end{aligned} \tag{5.8}$$

where τ is a continuous function implicitly defined by

$$\begin{aligned}&|\tau(\boldsymbol{v})| \leq \tfrac{1}{2}\|\boldsymbol{v}^{\pm}\|_{\boldsymbol{E}},\\ &\mathscr{F}^{\pm}(\Theta_H(\tau(\boldsymbol{v}), \boldsymbol{v}^{\pm})) = \mathscr{F}\left(\boldsymbol{v} + \psi(\boldsymbol{v}^0)\right) - \mathscr{F}\left(\boldsymbol{v}^0 + \psi(\boldsymbol{v}^0)\right).\end{aligned}$$

Then we have

$$\mathscr{F} \circ \phi^{-1}(\boldsymbol{v}) = \mathscr{F}^0(\boldsymbol{v}^0) + \mathscr{F}^{\pm}(\boldsymbol{v}^{\pm}), \qquad \forall \boldsymbol{v} = \boldsymbol{v}^0 + \boldsymbol{v}^{\pm} \in \phi(\mathscr{V}), \tag{5.9}$$

where $\mathscr{F}^0 : \mathscr{V} \cap \boldsymbol{E}^0 \to \mathbb{R}$ is the C^2 function

$$\mathscr{F}^0(\boldsymbol{v}^0) = \mathscr{F}(\boldsymbol{v}^0 + \psi(\boldsymbol{v}^0)), \qquad \forall \boldsymbol{v}^0 \in \mathscr{V} \cap \boldsymbol{E}^0. \tag{5.10}$$

Moreover, the origin is a totally degenerate isolated critical point of $\mathscr{F}^0$ and a non-degenerate critical point of $\mathscr{F}^{\pm}$.

Proof. Since the Hessian of $\mathscr{F}$ at $\mathbf{0}$ is non-degenerate when restricted to $\boldsymbol{E}^{\pm} \times \boldsymbol{E}^{\pm}$, by the implicit function theorem we can find an open neighborhood $\mathscr{V}^0 \subset \mathscr{U} \cap \boldsymbol{E}^0$ of $\mathbf{0}$ and a C^1 map $\psi : (\mathscr{V}^0, \mathbf{0}) \to (\boldsymbol{E}^{\pm}, \mathbf{0})$ that is implicitly defined as in (5.7).

Consider the function $\mathscr{F}^0 : \mathscr{V}^0 \to \mathbb{R}$ defined as in (5.10). The origin is an isolated critical point of $\mathscr{F}^0$, for $\psi(\mathbf{0}) = \mathbf{0}$ and

$$\begin{aligned} \mathrm{d}\mathscr{F}^0(\boldsymbol{v}^0) &= \mathrm{d}\mathscr{F}(\boldsymbol{v}^0 + \psi(\boldsymbol{v}^0)) + \mathrm{d}\mathscr{F}(\boldsymbol{v}^0 + \psi(\boldsymbol{v}^0))\,\mathrm{d}\psi(\boldsymbol{v}^0) \\ &= \mathrm{d}\mathscr{F}(\boldsymbol{v}^0 + \psi(\boldsymbol{v}^0)). \end{aligned}$$

Moreover,

$$\mathrm{Hess}\mathscr{F}^0(\boldsymbol{v}^0) = \mathrm{Hess}\mathscr{F}(\boldsymbol{v}^0 + \psi(\boldsymbol{v}^0))[(I + \mathrm{d}\psi(\boldsymbol{v}^0))\cdot, \cdot],$$

where I denotes the identity on $\boldsymbol{E}^0$. This shows that $\mathscr{F}^0$ is a C^2 function whose Hessian vanishes at the origin (i.e., the origin is a totally degenerate critical point of $\mathscr{F}^0$). As for the function $\mathscr{F}^\pm : \boldsymbol{E}^\pm \to \mathbb{R}$ defined in (5.6), the origin is clearly a non-degenerate critical point of it. In order to conclude the proof, we need to show that we can define a map ϕ as in point (ii) of the statement such that (5.9) holds.

Consider the flow Θ_H introduced above. We claim that Θ_H is well defined on

$$\left\{(s, \boldsymbol{v}^\pm) \in \mathbb{R} \times \boldsymbol{E}^\pm \,\middle|\, |s| < \|\boldsymbol{v}^\pm\|_{\boldsymbol{E}}\right\}.$$

In fact, for any nonzero $\boldsymbol{v}^\pm \in \boldsymbol{E}^\pm$, consider the maximal flow line

$$\sigma : (s_0, s_1) \to \boldsymbol{E}^\pm \setminus \{\mathbf{0}\}$$

of Θ_H with $s_0 < 0 < s_1$ and $\sigma(0) = \boldsymbol{v}^\pm$ (see equation (5.5)). Assume by contradiction that $s_1 < \|\boldsymbol{v}^\pm\|_{\boldsymbol{E}}$. Since $\|\dot\sigma(s)\|_{\boldsymbol{E}} = 1$ for each $s \in (s_0, s_1)$, we have

$$\|\sigma(s) - \boldsymbol{v}^\pm\|_{\boldsymbol{E}} = \left\|\int_0^s \dot\sigma(r)\,\mathrm{d}r\right\|_{\boldsymbol{E}} \le \int_0^s \|\dot\sigma(r)\|_{\boldsymbol{E}}\,\mathrm{d}r = |s|.$$

Therefore $0 < \|\sigma(s_1)\|_{\boldsymbol{E}} < 2\|\boldsymbol{v}^\pm\|_{\boldsymbol{E}}$ and $\dot\sigma(s_1) = -H\sigma(s_1)/\|H\sigma(s_1)\|_{\boldsymbol{E}}$ is well defined, contradicting the maximality of s_1. An analogous argument shows that $s_0 \le -\|\boldsymbol{v}^\pm\|_{\boldsymbol{E}}$.

Since H is a Fredholm operator, 0 is an isolated point in the spectrum of H (see the proof of Lemma 2.1.1). Fix $\varepsilon > 0$ such that the interval $[-\varepsilon, \varepsilon]$ intersects the spectrum of H only in 0. For each $\boldsymbol{v}^\pm \in \boldsymbol{E}^\pm \setminus \{\mathbf{0}\}$, the function $s \mapsto \mathscr{F}^\pm \circ \Theta_H(s, \boldsymbol{v}^\pm)$ is strictly decreasing, and

$$\begin{aligned} |\mathscr{F}^\pm(\Theta_H(s, \boldsymbol{v}^\pm)) - \mathscr{F}^\pm(\boldsymbol{v}^\pm)| &= \int_0^{|s|} \|H\,\Theta_H(r, \boldsymbol{v}^\pm)\|_{\boldsymbol{E}}\,\mathrm{d}r \\ &\ge \varepsilon \int_0^{|s|} \|\Theta_H(r, \boldsymbol{v}^\pm)\|_{\boldsymbol{E}}\,\mathrm{d}r \\ &\ge \varepsilon \int_0^{|s|} \left(\|\boldsymbol{v}^\pm\|_{\boldsymbol{E}} - r\right)\mathrm{d}r \\ &\ge \varepsilon \left(\|\boldsymbol{v}^\pm\|_{\boldsymbol{E}}\,|s| - \frac{s^2}{2}\right). \end{aligned}$$

In particular, for $s = \frac{1}{2}\|\boldsymbol{v}^{\pm}\|_{\boldsymbol{E}}$, we have

$$\begin{aligned} \mathscr{F}^{\pm}(\Theta_H(-s,\boldsymbol{v}^{\pm})) - \mathscr{F}^{\pm}(\boldsymbol{v}^{\pm}) &\geq \varepsilon'\|\boldsymbol{v}^{\pm}\|^2_{\boldsymbol{E}}, \\ \mathscr{F}^{\pm}(\boldsymbol{v}^{\pm}) - \mathscr{F}^{\pm}(\Theta_H(s,\boldsymbol{v}^{\pm})) &\geq \varepsilon'\|\boldsymbol{v}^{\pm}\|^2_{\boldsymbol{E}}, \end{aligned} \tag{5.11}$$

where $\varepsilon' := (\frac{1}{2} - \frac{1}{8})\varepsilon > 0$. Now, consider an open neighborhood of $\mathscr{V}^{\pm} \subset \boldsymbol{E}^{\pm}$ of $\boldsymbol{0}$ such that

$$\mathscr{V} := \mathscr{V}^0 \oplus \mathscr{V}^{\pm} \subset \mathscr{U}.$$

For each $\boldsymbol{v} = \boldsymbol{v}^{\pm} + \boldsymbol{v}^0 \in \mathscr{V}$ we have

$$\begin{aligned} &\mathscr{F}(\boldsymbol{v} + \psi(\boldsymbol{v}^0)) - \mathscr{F}(\boldsymbol{v}^0 + \psi(\boldsymbol{v}^0)) \\ &\quad = \int_0^1 \mathrm{d}\mathscr{F}(s\boldsymbol{v}^{\pm} + \boldsymbol{v}^0 + \psi(\boldsymbol{v}^0))\boldsymbol{v}^{\pm}\,\mathrm{d}s \\ &\quad = \int_0^1\!\!\int_0^1 s\,\mathrm{Hess}\mathscr{F}(rs\boldsymbol{v}^{\pm} + \boldsymbol{v}^0 + \psi(\boldsymbol{v}^0))[\boldsymbol{v}^{\pm},\boldsymbol{v}^{\pm}]\,\mathrm{d}r\,\mathrm{d}s. \end{aligned}$$

Since the function $\mathscr{F}$ is C^2, up to shrinking the neighborhood $\mathscr{V}$ of $\boldsymbol{0}$, for each $\boldsymbol{v} = \boldsymbol{v}^{\pm} + \boldsymbol{v}^0 \in \mathscr{V}$ we have

$$\begin{aligned} &|\mathscr{F}(\boldsymbol{v} + \psi(\boldsymbol{v}^0)) - \mathscr{F}(\boldsymbol{v}^0 + \psi(\boldsymbol{v}^0)) - \mathscr{F}^{\pm}(\boldsymbol{v}^{\pm})| \\ &\quad = \left| \int_0^1\!\!\int_0^1 s\left(\mathrm{Hess}\mathscr{F}(rs\boldsymbol{v}^{\pm} + \boldsymbol{v}^0)[\boldsymbol{v}^{\pm},\boldsymbol{v}^{\pm}] - \mathrm{Hess}\mathscr{F}(\boldsymbol{0})[\boldsymbol{v}^{\pm},\boldsymbol{v}^{\pm}]\right)\mathrm{d}r\,\mathrm{d}s \right| \\ &\quad \leq \tfrac{1}{2}\varepsilon'\|\boldsymbol{v}^{\pm}\|^2_{\boldsymbol{E}}. \end{aligned}$$

Set $s = \frac{1}{2}\|\boldsymbol{v}^{\pm}\|_{\boldsymbol{E}}$. This estimate, together with the ones in (5.11), implies that

$$\begin{aligned} \big|\mathscr{F}(\boldsymbol{v} + \psi(\boldsymbol{v}^0)) &- \mathscr{F}(\boldsymbol{v}^0 + \psi(\boldsymbol{v}^0)) - \mathscr{F}^{\pm}(\Theta_H(s,\boldsymbol{v}^{\pm}))\big| \\ &\geq \big|\mathscr{F}^{\pm}(\boldsymbol{v}^{\pm}) - \mathscr{F}^{\pm}(\Theta_H(s,\boldsymbol{v}^{\pm}))\big| \\ &\quad - \big|\mathscr{F}(\boldsymbol{v} + \psi(\boldsymbol{v}^0)) - \mathscr{F}(\boldsymbol{v}^0 + \psi(\boldsymbol{v}^0)) - \mathscr{F}^{\pm}(\boldsymbol{v}^{\pm})\big| \\ &\geq \tfrac{1}{2}\varepsilon'\|\boldsymbol{v}^{\pm}\|^2_{\boldsymbol{E}} \\ &\geq 0, \end{aligned}$$

and actually

$$\mathscr{F}^{\pm}(\Theta_H(s,\boldsymbol{v}^{\pm})) \leq \mathscr{F}(\boldsymbol{v} + \psi(\boldsymbol{v}^0)) - \mathscr{F}(\boldsymbol{v}^0 + \psi(\boldsymbol{v}^0)) \leq \mathscr{F}^{\pm}(\Theta_H(-s,\boldsymbol{v}^{\pm})). \tag{5.12}$$

Notice that these inequalities become equalities in case $\boldsymbol{v}^{\pm} = \boldsymbol{0}$. If $\boldsymbol{v}^{\pm} \neq \boldsymbol{0}$ and $|r| < \|\boldsymbol{v}^{\pm}\|_{\boldsymbol{E}}$, we have

$$\frac{\mathrm{d}}{\mathrm{d}r}\mathscr{F}^{\pm}(\Theta_H(r,\boldsymbol{v}^{\pm})) = -\|H\Theta_H(r,\boldsymbol{v}^{\pm})\|_{\boldsymbol{E}} < 0,$$

and by the implicit function theorem there exists a C^1 function

$$\tau : \mathscr{V} \setminus (\mathscr{V}^0 \oplus \{\boldsymbol{0}\}) \to \mathbb{R}$$

such that

$$\mathscr{F}^{\pm}(\Theta_H(\tau(\boldsymbol{v}),\boldsymbol{v}^{\pm})) = \mathscr{F}(\boldsymbol{v}+\psi(\boldsymbol{v}^0)) - \mathscr{F}(\boldsymbol{v}^0+\psi(\boldsymbol{v}^0)), \qquad \forall \boldsymbol{v}=\boldsymbol{v}^0+\boldsymbol{v}^{\pm}\in\mathscr{V}\setminus\{\boldsymbol{0}\}. \tag{5.13}$$

Notice that, by the estimates (5.12), we have $|\tau(\boldsymbol{v})| \leq \frac{1}{2}\|\boldsymbol{v}^{\pm}\|_{\boldsymbol{E}}$. Therefore we can extend τ to a continuous function defined on all of $\mathscr{V}$.

Once more, since the function $\mathscr{F}$ is C^2, up to further shrinking the neighborhood $\mathscr{V}$ of $\boldsymbol{0}$, for each $\boldsymbol{v}=\boldsymbol{v}^{\pm}+\boldsymbol{v}^0\in\mathscr{V}$ with $\boldsymbol{v}^{\pm}\neq 0$ and $|r|<\|\boldsymbol{v}^{\pm}\|_{\boldsymbol{E}}$ we have

$$\begin{aligned}
&\frac{\mathrm{d}}{\mathrm{d}r}\mathscr{F}(\boldsymbol{v}^0+\Theta_H(r,\boldsymbol{v}^{\pm})+\psi(\boldsymbol{v}^0))\\
&= -\mathrm{d}\mathscr{F}(\boldsymbol{v}^0+\Theta_H(r,\boldsymbol{v}^{\pm})+\psi(\boldsymbol{v}^0))\frac{H\Theta_H(r,\boldsymbol{v}^{\pm})}{\|H\Theta_H(r,\boldsymbol{v}^{\pm})\|_{\boldsymbol{E}}}\\
&= -\int_0^1 \mathrm{Hess}\mathscr{F}(\boldsymbol{v}^0+s\Theta_H(r,\boldsymbol{v}^{\pm})+\psi(\boldsymbol{v}^0))\left[\Theta_H(r,\boldsymbol{v}^{\pm}),\frac{H\Theta_H(r,\boldsymbol{v}^{\pm})}{\|H\Theta_H(r,\boldsymbol{v}^{\pm})\|_{\boldsymbol{E}}}\right]\mathrm{d}s\\
&\leq \tfrac{1}{2}\varepsilon\|\Theta_H(r,\boldsymbol{v}^{\pm})\|_{\boldsymbol{E}} - \int_0^1 \mathrm{Hess}\mathscr{F}(\boldsymbol{0})\left[\Theta_H(r,\boldsymbol{v}^{\pm}),\frac{H\Theta_H(r,\boldsymbol{v}^{\pm})}{\|H\Theta_H(r,\boldsymbol{v}^{\pm})\|_{\boldsymbol{E}}}\right]\mathrm{d}s\\
&= \tfrac{1}{2}\varepsilon\|\Theta_H(r,\boldsymbol{v}^{\pm})\|_{\boldsymbol{E}} - \|H\Theta_H(r,\boldsymbol{v}^{\pm})\|_{\boldsymbol{E}}\\
&\leq -\tfrac{1}{2}\varepsilon\|\Theta_H(r,\boldsymbol{v}^{\pm})\|_{\boldsymbol{E}}\\
&< 0.
\end{aligned}$$

Up to further shrinking the neighborhood $\mathscr{V}$ of $\boldsymbol{0}$, by the implicit function theorem there exists a C^1 function

$$\sigma:\mathscr{V}\setminus(\mathscr{V}^0\oplus\{\boldsymbol{0}\})\to\mathbb{R}$$

such that

$$\mathscr{F}^{\pm}(\boldsymbol{w}^{\pm}) = \mathscr{F}(\boldsymbol{w}^0+\Theta_H(\sigma(\boldsymbol{w}),\boldsymbol{w}^{\pm})+\psi(\boldsymbol{w}^0)) - \mathscr{F}(\boldsymbol{w}^0+\psi(\boldsymbol{w}^0)), \qquad \forall \boldsymbol{w}=\boldsymbol{w}^0+\boldsymbol{w}^{\pm}\in\mathscr{V}\setminus\{\boldsymbol{0}\}.$$

If we set $\boldsymbol{v}=\boldsymbol{v}^0+\boldsymbol{v}^{\pm}$, $\boldsymbol{v}^0=\boldsymbol{w}^0$ and $\boldsymbol{v}^{\pm}=\Theta_H(\sigma(\boldsymbol{w}),\boldsymbol{w}^{\pm})$, then

$$\boldsymbol{w}^{\pm}=\Theta_H(\tau(\boldsymbol{v}),\boldsymbol{v}^{\pm}), \tag{5.14}$$

and hence we can extend σ to a continuous function defined on all of $\mathscr{V}$ such that $\sigma(\boldsymbol{w}^0)=0$.

Finally, up to shrinking one last time the neighborhood $\mathscr{V}$ of $\boldsymbol{0}$, we can define the map $\phi:\mathscr{V}\to\mathscr{U}$ as in (5.8). This map is a homeomorphism onto its image with inverse

$$\phi^{-1}(\boldsymbol{w})=\boldsymbol{w}^0+\Theta_H(\sigma(\boldsymbol{w}),\boldsymbol{w}^{\pm})+\psi(\boldsymbol{w}^0), \qquad \forall \boldsymbol{w}=\boldsymbol{w}^0+\boldsymbol{w}^{\pm}\in\phi(\mathscr{V}).$$

Equation (5.9) is then a direct consequence of (5.13). □

5.3 Naturality of the Morse lemma

Let $H_\bullet$ be the bounded self-adjoint linear operator on $\boldsymbol{E}_\bullet \subset \boldsymbol{E}$ associated to the Hessian of the restricted function $\mathscr{F}_\bullet = \mathscr{F}|_{\boldsymbol{E}_\bullet}$ at $\mathbf{0}$. Then $H \circ J = J \circ H_\bullet$. This follows immediately from condition (5.2), for

$$\begin{aligned} H \circ J &= \mathrm{d}(\mathrm{Grad}\mathscr{F})(\mathbf{0}) \circ J \\ &= \mathrm{d}((\mathrm{Grad}\mathscr{F}) \circ J)(\mathbf{0}) \\ &= \mathrm{d}(J \circ \mathrm{Grad}(\mathscr{F}_\bullet))(\mathbf{0}) \\ &= J \circ \mathrm{d}(\mathrm{Grad}(\mathscr{F}_\bullet))(\mathbf{0}) \\ &= J \circ H_\bullet. \end{aligned}$$

Namely, the restriction of H to $\boldsymbol{E}_\bullet$ coincides with $H_\bullet$, i.e.,

$$H|_{\boldsymbol{E}_\bullet} = H_\bullet, \tag{5.15}$$

and in particular $H_\bullet$ is a self-adjoint Fredholm operator on $\boldsymbol{E}_\bullet$. If we denote by $\boldsymbol{E}_\bullet = \boldsymbol{E}_\bullet^0 \oplus \boldsymbol{E}_\bullet^+ \oplus \boldsymbol{E}_\bullet^-$ the orthogonal splitting associated to the operator $H_\bullet$ according to the spectral theorem, equation (5.15) readily implies that

$$\boldsymbol{E}_\bullet^0 \subseteq \boldsymbol{E}^0, \qquad \boldsymbol{E}_\bullet^+ \subseteq \boldsymbol{E}^+, \qquad \boldsymbol{E}_\bullet^- \subseteq \boldsymbol{E}^-. \tag{5.16}$$

Lemma 5.2.1 applies to both the functions $\mathscr{F}$ and $\mathscr{F}_\bullet$, and the following is a list of the various symbols involved in the corresponding statements. The top line contains the symbols used with $\mathscr{F}$, and the bottom line denotes the symbols used with $\mathscr{F}_\bullet$.

$$\begin{array}{ccccccccc} \boldsymbol{E}^\pm, & \boldsymbol{E}^0, & \mathscr{V}, & \Theta_H, & \phi, & \psi, & \tau, & \mathscr{F}^0, & \mathscr{F}^\pm, \\ \boldsymbol{E}_\bullet^\pm, & \boldsymbol{E}_\bullet^0, & \mathscr{V}_\bullet, & \Theta_{H_\bullet}, & \phi_\bullet, & \psi_\bullet, & \tau_\bullet, & \mathscr{F}_\bullet^0, & \mathscr{F}_\bullet^\pm. \end{array}$$

We want to show that, under the hypotheses of Theorem 5.1.1, the decomposition of $\mathscr{F}$ as $\mathscr{F}^\pm + \mathscr{F}^0$ restricts to the corresponding decomposition of $\mathscr{F}_\bullet$ as $\mathscr{F}_\bullet^\pm + \mathscr{F}_\bullet^0$.

Proposition 5.3.1. $\mathscr{F}^\pm|_{\boldsymbol{E}_\bullet^\pm} = \mathscr{F}_\bullet^\pm$.

Proof. This is a direct consequence of (5.15) and the definitions of the functions $\mathscr{F}^\pm$ and $\mathscr{F}_\bullet^\pm$. □

Lemma 5.3.2.

(i) *If* $\mathrm{ind}(\mathscr{F}, \mathbf{0}) = \mathrm{ind}(\mathscr{F}_\bullet, \mathbf{0})$, *then* $\boldsymbol{E}^- = \boldsymbol{E}_\bullet^-$.

(ii) *If* $\mathrm{nul}(\mathscr{F}, \mathbf{0}) = \mathrm{nul}(\mathscr{F}_\bullet, \mathbf{0})$, *then* $\boldsymbol{E}^0 = \boldsymbol{E}_\bullet^0$.

Proof. The claims follow at once from (5.16), since

$$\begin{aligned} \dim \boldsymbol{E}_\bullet^- &= \mathrm{ind}(\mathscr{F}_\bullet, \mathbf{0}) = \mathrm{ind}(\mathscr{F}, \mathbf{0}) = \dim \boldsymbol{E}^-, \\ \dim \boldsymbol{E}_\bullet^0 &= \mathrm{nul}(\mathscr{F}_\bullet, \mathbf{0}) = \mathrm{nul}(\mathscr{F}, \mathbf{0}) = \dim \boldsymbol{E}^0. \end{aligned}$$

□

Proposition 5.3.3. *If* $\mathrm{nul}(\mathscr{F},\mathbf{0}) = \mathrm{nul}(\mathscr{F}_\bullet,\mathbf{0})$, *the following equalities hold:*

(i) $\psi = \psi_\bullet$ *on a neighborhood of* $\mathbf{0}$ *in* $\boldsymbol{E}^0_\bullet = \boldsymbol{E}^0$,

(ii) $\mathscr{F}^0 = \mathscr{F}^0_\bullet$ *on a neighborhood of* $\mathbf{0}$ *in* $\boldsymbol{E}^0_\bullet = \boldsymbol{E}^0$,

(iii) $\phi = \phi_\bullet$ *on a neighborhood of* $\mathbf{0}$ *in* $\boldsymbol{E}_\bullet \subset \boldsymbol{E}$.

Proof. By Lemma 5.3.2(ii), the domains of the maps ψ and $\psi_\bullet$ are open neighborhoods of $\mathbf{0}$ in $\boldsymbol{E}^0 = \boldsymbol{E}^0_\bullet$. Up to shrinking these neighborhoods, we can assume that both ψ and $\psi_\bullet$ have common domain $\mathscr{V}^0 \subset \boldsymbol{E}^0$. Let $P^\pm : \boldsymbol{E} \to \boldsymbol{E}^\pm$ and $P^\pm_\bullet : \boldsymbol{E}_\bullet \to \boldsymbol{E}^\pm_\bullet$ be orthogonal projectors. Notice that

$$P^\pm|_{\boldsymbol{E}^\pm_\bullet} = P^\pm_\bullet.$$

The function $\psi_\bullet$ is implicitly defined as in point (i) of Lemma 5.2.1, or equivalently by the equation

$$P^\pm_\bullet\big(\mathrm{Grad}\mathscr{F}_\bullet(\boldsymbol{v}^0 + \psi_\bullet(\boldsymbol{v}^0))\big) = \mathbf{0}, \qquad \forall \boldsymbol{v}^0 \in \mathscr{V} \cap \boldsymbol{E}^0.$$

Assumption (5.2), together with the fact that $\boldsymbol{E}^0 = \boldsymbol{E}^0_\bullet$, implies that

$$\mathrm{Grad}\mathscr{F}(\boldsymbol{v}^0 + \psi_\bullet(\boldsymbol{v}^0)) = \mathrm{Grad}\mathscr{F}_\bullet(\boldsymbol{v}^0 + \psi_\bullet(\boldsymbol{v}^0)) \qquad \forall \boldsymbol{v}^0 \in \mathscr{V} \cap \boldsymbol{E}^0.$$

Hence

$$P^\pm\big(\mathrm{Grad}\mathscr{F}(\boldsymbol{v}^0 + \psi_\bullet(\boldsymbol{v}^0))\big) = \mathbf{0}, \qquad \forall \boldsymbol{v}^0 \in \mathscr{V} \cap \boldsymbol{E}^0.$$

This shows that $\psi_\bullet$ solves the equation that implicitly defines ψ, and therefore points (i) and (ii) follow.

As for (iii), up to shrinking the domains of ϕ and $\phi_\bullet$, we can assume that they are maps of the form $\phi : \mathscr{V} \to \mathscr{U}$ and $\phi_\bullet : \mathscr{V}_\bullet \to \mathscr{U}_\bullet$, where $\mathscr{V}_\bullet = \mathscr{V} \cap \boldsymbol{E}_\bullet$. Notice that (5.15) readily implies that the flow $\Theta_{H_\bullet}$ is the restriction of the flow Θ_H to $\boldsymbol{E}^\pm_\bullet \setminus \{\mathbf{0}\}$, i.e.,

$$\Theta_H(\cdot, \boldsymbol{v}^\pm) = \Theta_{H_\bullet}(\cdot, \boldsymbol{v}^\pm), \qquad \forall \boldsymbol{v}^\pm \in \boldsymbol{E}^\pm_\bullet \setminus \{\mathbf{0}\}.$$

By Lemma 5.2.1(ii) and since $\psi = \psi_\bullet$, for each $\boldsymbol{v} = \boldsymbol{v}^0 + \boldsymbol{v}^\pm \in \mathscr{V}_\bullet$ we have

$$\begin{aligned}\phi(\boldsymbol{v}) &= \boldsymbol{v}^0 + \Theta_H(\tau(\boldsymbol{v} - \psi(\boldsymbol{v}^0)), \boldsymbol{v}^\pm - \psi(\boldsymbol{v}^0)),\\ \phi_\bullet(\boldsymbol{v}) &= \boldsymbol{v}^0 + \Theta_H(\tau_\bullet(\boldsymbol{v} - \psi(\boldsymbol{v}^0)), \boldsymbol{v}^\pm - \psi(\boldsymbol{v}^0)).\end{aligned}$$

Hence in order to conclude the proof of (ii), we just need to show that, for each $\boldsymbol{v}$ in the domain of $\tau_\bullet$, we have $\tau(\boldsymbol{v}) = \tau_\bullet(\boldsymbol{v})$. This is easily verified since, by Lemma 5.2.1(ii) and Proposition 5.3.1, $\tau(\boldsymbol{v})$ and $\tau_\bullet(\boldsymbol{v})$ are implicitly defined by the same equation

$$\mathscr{F}^\pm(\Theta_H(\tau(\boldsymbol{v}), \boldsymbol{v}^\pm)) = \mathscr{F}(\boldsymbol{v} + \psi(\boldsymbol{v}^0)) - \mathscr{F}(\boldsymbol{v}^0 + \psi(\boldsymbol{v}^0)) = \mathscr{F}^\pm(\Theta_H(\tau_\bullet(\boldsymbol{v}), \boldsymbol{v}^\pm)). \qquad \square$$

5.4 Local homology

In this section we carry out the proof of Theorem 5.1.1. Before giving the proof, we need to establish another naturality property, this time for the isomorphism between the local homology of $\mathscr{F}$ at $\mathbf{0}$ and the homology of corresponding Gromoll-Meyer pairs (see Section A.5 in the Appendix).

Remark 5.4.1. Notice that our function $\mathscr{F}$ is only defined on an open set $\mathscr{U} \subset \boldsymbol{E}$, which is not a complete space unless $\mathscr{U} = \boldsymbol{E}$. Therefore in the following, we must implicitly require that all the Gromoll-Meyer pairs $(\mathscr{W}, \mathscr{W}_-)$ of $\mathscr{F}$ at $\mathbf{0}$ satisfy the following assumption (in addition to **(GM1)**,...,**(GM4)** from Section A.5): if $\Phi_{\mathrm{Grad}\mathscr{F}}$ denotes the anti-gradient flow of $\mathscr{F}$ as in Section A.3 of the Appendix, then for each $\boldsymbol{v} \in \mathscr{W}_-$ there exists $t = t(\boldsymbol{v}) > 0$ such that $\Phi_{\mathrm{Grad}\mathscr{F}}(t, \boldsymbol{v}) \in (\mathscr{F})_c$, where $c = \mathscr{F}(\mathbf{0})$. This can always be fulfilled by requiring $\mathscr{W}$ to be small enough. Indeed, one can even construct a Gromoll-Meyer pair $(\mathscr{W}, \mathscr{W}_-)$ in such a way that $\mathscr{W}_- \subset (\mathscr{F})_c$, see [GM69a, Section 2] or [Cha93, page 49]. □

Lemma 5.4.1. *Let $(\mathscr{W}, \mathscr{W}_-)$ be a Gromoll-Meyer pair for $\mathscr{F}$ at $\mathbf{0}$. Then the following holds.*

(i) *The pair $(\mathscr{W}_\bullet, \mathscr{W}_{\bullet -}) := (\mathscr{W} \cap \boldsymbol{E}_\bullet, \mathscr{W}_- \cap \boldsymbol{E}_\bullet) = (J^{-1}(\mathscr{W}), J^{-1}(\mathscr{W}_-))$ is a Gromoll-Meyer pair for $\mathscr{F}_\bullet = \mathscr{F}|_{\boldsymbol{E}_\bullet}$ at $\mathbf{0}$.*

(ii) *Consider the restrictions of $J : \boldsymbol{E}_\bullet \hookrightarrow \boldsymbol{E}$ given by*

$$J : ((\mathscr{F}_\bullet)_c \cup \{\mathbf{0}\}, (\mathscr{F}_\bullet)_c) \hookrightarrow ((\mathscr{F})_c \cup \{\mathbf{0}\}, (\mathscr{F})_c),$$
$$J : (\mathscr{W}_\bullet, \mathscr{W}_{\bullet -}) \hookrightarrow (\mathscr{W}, \mathscr{W}_-).$$

These restrictions induce the homology homomorphisms

$$\mathrm{H}_*(J) : \mathrm{C}_*(\mathscr{F}_\bullet, \mathbf{0}) \to \mathrm{C}_*(\mathscr{F}, \mathbf{0}),$$
$$\mathrm{H}_*(J) : \mathrm{H}_*(\mathscr{W}_\bullet, \mathscr{W}_{\bullet -}) \to \mathrm{H}_*(\mathscr{W}, \mathscr{W}_-).$$

Then there exist homology isomorphisms $\iota_{(\mathscr{W}_\bullet, \mathscr{W}_{\bullet -})}$ and $\iota_{(\mathscr{W}, \mathscr{W}_-)}$ such that the following diagram commutes.

$$\begin{array}{ccc}
\mathrm{C}_*(\mathscr{F}_\bullet, \mathbf{0}) & \xrightarrow{\mathrm{H}_*(J)} & \mathrm{C}_*(\mathscr{F}, \mathbf{0}) \\
{\scriptstyle \iota_{(\mathscr{W}_\bullet, \mathscr{W}_{\bullet -})}} \downarrow \simeq & & \simeq \downarrow {\scriptstyle \iota_{(\mathscr{W}, \mathscr{W}_-)}} \\
\mathrm{H}_*(\mathscr{W}_\bullet, \mathscr{W}_{\bullet -}) & \xrightarrow{\mathrm{H}_*(J)} & \mathrm{H}_*(\mathscr{W}, \mathscr{W}_-)
\end{array}$$

Proof. Point (i) is straightforward: it suffices to verify that the pair $(\mathscr{W}_\bullet, \mathscr{W}_{\bullet -})$ satisfies conditions **(GM1)**,...,**(GM4)** in the definition of Gromoll-Meyer pair (see

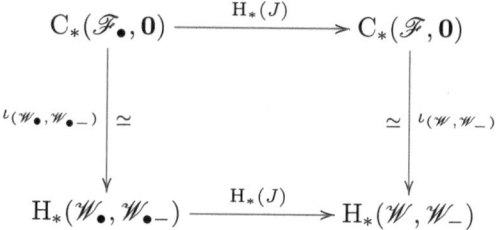

Section A.5). Point (ii) requires us to examine closely the proof of Theorem A.5.1 in the Appendix, and we refer the reader to [Cha93, page 48] for more details. The point, here, is to show that the isomorphism between the local homology and the homology of a Gromoll-Meyer pair is given by the composition of homology isomorphisms induced by maps, so that the assertion follows from the functoriality of singular homology.

We denote by $\Phi_{\mathrm{Grad}\mathscr{F}}$ the anti-gradient flow of $\mathscr{F}$. Notice that, by condition (5.2), $\Phi_{\mathrm{Grad}\mathscr{F}}$ restricts on $\boldsymbol{E}_\bullet$ to the anti-gradient flow $\Phi_{\mathrm{Grad}\mathscr{F}_\bullet}$ of the restricted function $\mathscr{F}_\bullet$. We choose a continuous function $T : \mathscr{W}_- \to [0,\infty)$ such that $\mathscr{F}(\Phi_{\mathrm{Grad}\mathscr{F}}(T(\boldsymbol{v}),\boldsymbol{v})) < c$ for each $\boldsymbol{v} \in \mathscr{W}_-$, and we introduce the closed sets

$$\begin{aligned}
\mathscr{Z} &:= \Big\{\Phi_{\mathrm{Grad}\mathscr{F}}(s,\boldsymbol{v})) \ \Big|\ \boldsymbol{v} \in \mathscr{W}_-,\ 0 \le s \le T(\boldsymbol{v})\Big\},\\
\mathscr{Y} &:= \mathscr{W} \cup \mathscr{Z},\\
\mathscr{Z}_\bullet &:= \Big\{\Phi_{\mathrm{Grad}\mathscr{F}}(s,\boldsymbol{v})) \ \Big|\ \boldsymbol{v} \in \mathscr{W}_{\bullet-},\ 0 \le s \le T(\boldsymbol{v})\Big\} = \mathscr{Z} \cap \boldsymbol{E}_\bullet,\\
\mathscr{Y}_\bullet &:= \mathscr{W}_\bullet \cup \mathscr{Z}_\bullet = \mathscr{Y} \cap \boldsymbol{E}_\bullet.
\end{aligned}$$

Consider the following diagram.

$$\begin{array}{ccc}
\mathrm{C}_*(\mathscr{F}_\bullet,\mathbf{0}) & \xrightarrow{\ \mathrm{H}_*(J)\ } & \mathrm{C}_*(\mathscr{F},\mathbf{0}) \\
\uparrow \simeq & & \uparrow \simeq \\
\mathrm{H}_*\left(\mathscr{Y}_\bullet \cap (\mathscr{F}_\bullet)_c \cup \{\mathbf{0}\}, \mathscr{Y}_\bullet \cap (\mathscr{F}_\bullet)_c\right) & \longrightarrow & \mathrm{H}_*\left(\mathscr{Y} \cap (\mathscr{F})_c \cup \{\mathbf{0}\}, \mathscr{Y} \cap (\mathscr{F})_c\right) \\
\uparrow \simeq & & \uparrow \simeq \\
\mathrm{H}_*\left(\mathscr{Y}_\bullet \cap (\mathscr{F}_\bullet)_c \cup \{\mathbf{0}\}, \mathscr{Z}_\bullet \cap (\mathscr{F}_\bullet)_c\right) & \longrightarrow & \mathrm{H}_*\left(\mathscr{Y} \cap (\mathscr{F})_c \cup \{\mathbf{0}\}, \mathscr{Z} \cap (\mathscr{F})_c\right) \\
\downarrow \simeq & & \downarrow \simeq \\
\mathrm{H}_*\left(\mathscr{Y}_\bullet, \mathscr{Z}_\bullet\right) & \longrightarrow & \mathrm{H}_*\left(\mathscr{Y}, \mathscr{Z}\right) \\
\uparrow \simeq & & \uparrow \simeq \\
\mathrm{H}_*(\mathscr{W}_\bullet, \mathscr{W}_{\bullet-}) & \xrightarrow{\ \mathrm{H}_*(J)\ } & \mathrm{H}_*(\mathscr{W}, \mathscr{W}_-)
\end{array}$$

In this diagram, all the arrows are homology homomorphisms induced by inclusions. Moreover, all the vertical arrows are isomorphisms (this can be easily proved by anti-gradient flow deformations and excisions), and we define the isomorphisms $\iota_{(\mathscr{W}_\bullet,\mathscr{W}_{\bullet -})}$ and $\iota_{(\mathscr{W},\mathscr{W}_-)}$ as the composition of the whole left vertical line and right vertical line respectively. By the functoriality of singular homology, this diagram is commutative, and the claim of point (ii) follows. □

After these preliminaries, let us go back to the proof of Theorem 5.1.1. If $\mathrm{nul}(\mathscr{F},\mathbf{0}) = \mathrm{nul}(\mathscr{F}_\bullet,\mathbf{0})$, then, by Proposition 5.3.3(ii), $\mathscr{F}^0 = \mathscr{F}^0_\bullet$ on a neighborhood $\mathscr{V}^0$ of the critical point $\mathbf{0}$. In particular

$$\mathrm{C}_*(\mathscr{F}^0_\bullet,\mathbf{0}) = \mathrm{C}_*(\mathscr{F}^0,\mathbf{0}). \tag{5.17}$$

As for the functions $\mathscr{F}^\pm$ and $\mathscr{F}^\pm_\bullet$, we have the following.

Lemma 5.4.2. *Assume that* $\mathrm{ind}(\mathscr{F},\mathbf{0}) = \mathrm{ind}(\mathscr{F}_\bullet,\mathbf{0})$. *Then the inclusion* J, *restricted as a map*

$$J : ((\mathscr{F}^\pm_\bullet)_0 \cup \{\mathbf{0}\}\,, (\mathscr{F}^\pm_\bullet)_0) \hookrightarrow ((\mathscr{F}^\pm)_0 \cup \{\mathbf{0}\}\,, (\mathscr{F}^\pm)_0), \tag{5.18}$$

induces the homology isomorphism

$$\mathrm{H}_*(J) : \mathrm{C}_*(\mathscr{F}^\pm_\bullet,\mathbf{0}) \xrightarrow{\simeq} \mathrm{C}_*(\mathscr{F}^\pm,\mathbf{0}).$$

Proof. The fact that J restricts to a map of the form (5.18) is guaranteed by Proposition 5.3.1. Since $\mathscr{F}^\pm$ and $\mathscr{F}^\pm_\bullet$ are quadratic forms with negative eigenspaces $\boldsymbol{E}^-$ and $\boldsymbol{E}^-_\bullet$ respectively, the inclusions

$$\begin{aligned} k &: (\boldsymbol{E}^-, \boldsymbol{E}^- \setminus \{\mathbf{0}\}) \hookrightarrow ((\mathscr{F}^\pm)_0 \cup \{\mathbf{0}\}\,, (\mathscr{F}^\pm)_0), \\ k_\bullet &: (\boldsymbol{E}^-_\bullet, \boldsymbol{E}^-_\bullet \setminus \{\mathbf{0}\}) \hookrightarrow ((\mathscr{F}^\pm_\bullet)_0 \cup \{\mathbf{0}\}\,, (\mathscr{F}^\pm_\bullet)_0) \end{aligned}$$

are homotopy equivalences, and in particular induce homology isomorphisms $\mathrm{H}_*(k)$ and $\mathrm{H}_*(k_\bullet)$. Moreover k and $k_\bullet$ are defined on the same domain, since $\boldsymbol{E}^-_\bullet = \boldsymbol{E}^-$ according to Lemma 5.3.2(i). Therefore $\mathrm{H}_*(J) = \mathrm{H}_*(k) \circ \mathrm{H}_*(k_\bullet)^{-1}$ is an isomorphism. □

Proof of Theorem 5.1.1. By Proposition 5.3.3(iii), the homeomorphisms ϕ and $\phi_\bullet$ provided by the generalized Morse lemma induce local homology isomorphisms $\mathrm{H}_*(\phi)$ and $\mathrm{H}_*(\phi_\bullet)$ such that the following diagram commutes.

$$\begin{array}{ccc} \mathrm{C}_*(\mathscr{F}_\bullet,\mathbf{0}) & \xrightarrow{\mathrm{H}_*(J)} & \mathrm{C}_*(\mathscr{F},\mathbf{0}) \\ {\scriptstyle \mathrm{H}_*(\phi_\bullet)}\uparrow{\scriptstyle \simeq} & & {\scriptstyle \simeq}\uparrow{\scriptstyle \mathrm{H}_*(\phi)} \\ \mathrm{C}_*(\mathscr{F}^0_\bullet + \mathscr{F}^\pm_\bullet,\mathbf{0}) & \xrightarrow{\mathrm{H}_*(J)} & \mathrm{C}_*(\mathscr{F}^0 + \mathscr{F}^\pm,\mathbf{0}) \end{array}$$

Hence, we only need to prove that the lower horizontal homomorphism $\mathrm{H}_*(J)$ is an isomorphism. We consider Gromoll-Meyer pairs $(\mathscr{W}^\pm, \mathscr{W}^\pm_-)$ and $(\mathscr{W}^0, \mathscr{W}^0_-)$ for $\mathscr{F}^\pm$ and $\mathscr{F}^0$ respectively at $\mathbf{0}$, so that the product of these pairs, that is

$$(\mathscr{W}, \mathscr{W}_-) := (\mathscr{W}^\pm \times \mathscr{W}^0, (\mathscr{W}^\pm_- \times \mathscr{W}^0) \cup (\mathscr{W}^\pm \times \mathscr{W}^0_-)),$$

is a Gromoll-Meyer pair for $\mathscr{F}^0 + \mathscr{F}^\pm$ at $\mathbf{0}$. By Lemma 5.4.1(i), we obtain Gromoll-Meyer pairs for the functions $\mathscr{F}^\pm_\bullet$, $\mathscr{F}^0_\bullet$ and $\mathscr{F}^0_\bullet + \mathscr{F}^\pm_\bullet$ at $\mathbf{0}$ respectively as

$$\begin{aligned}
(\mathscr{W}^\pm_\bullet, \mathscr{W}^\pm_{\bullet-}) &:= (\mathscr{W}^\pm \cap \boldsymbol{E}_\bullet, \mathscr{W}^\pm_- \cap \boldsymbol{E}_\bullet) = (J^{-1}(\mathscr{W}^\pm), J^{-1}(\mathscr{W}^\pm_-)),\\
(\mathscr{W}^0_\bullet, \mathscr{W}^0_{\bullet-}) &:= (\mathscr{W}^0 \cap \boldsymbol{E}_\bullet, \mathscr{W}^0_- \cap \boldsymbol{E}_\bullet) = (J^{-1}(\mathscr{W}^0), J^{-1}(\mathscr{W}^0_-)),\\
(\mathscr{W}_\bullet, \mathscr{W}_{\bullet-}) &:= (\mathscr{W} \cap \boldsymbol{E}_\bullet, \mathscr{W}_- \cap \boldsymbol{E}_\bullet) = (J^{-1}(\mathscr{W}), J^{-1}(\mathscr{W}_-)).
\end{aligned}$$

By Lemma 5.4.1(ii) and the Künneth formula we obtain the following commutative diagram.

$$\begin{array}{ccc}
\mathrm{C}_*(\mathscr{F}^0_\bullet + \mathscr{F}^\pm_\bullet, \mathbf{0}) & \xrightarrow{\ \mathrm{H}_*(J)\ } & \mathrm{C}_*(\mathscr{F}^0 + \mathscr{F}^\pm, \mathbf{0}) \\
\iota_{(\mathscr{W}_\bullet, \mathscr{W}_{\bullet-})} \downarrow \simeq & & \simeq \downarrow \iota_{(\mathscr{W}, \mathscr{W}_-)} \\
\mathrm{H}_*(\mathscr{W}_\bullet, \mathscr{W}_{\bullet-}) & \xrightarrow{\ \mathrm{H}_*(J)\ } & \mathrm{H}_*(\mathscr{W}, \mathscr{W}_-) \\
\text{Künneth} \downarrow \simeq & & \simeq \downarrow \text{Künneth} \\
\mathrm{H}_*(\mathscr{W}^\pm_\bullet, \mathscr{W}^\pm_{\bullet-}) \otimes \mathrm{H}_*(\mathscr{W}^0_\bullet, \mathscr{W}^0_{\bullet-}) & \xrightarrow{\ \mathrm{H}_*(J) \otimes \mathrm{H}_*(J)\ } & \mathrm{H}_*(\mathscr{W}^\pm, \mathscr{W}^\pm_-) \otimes \mathrm{H}_*(\mathscr{W}^0, \mathscr{W}^0_-) \\
\iota_{(\mathscr{W}^\pm_\bullet, \mathscr{W}^\pm_{\bullet-})} \otimes \iota_{(\mathscr{W}^0_\bullet, \mathscr{W}^0_{\bullet-})} \uparrow \simeq & & \simeq \uparrow \iota_{(\mathscr{W}^\pm, \mathscr{W}^\pm_-)} \otimes \iota_{(\mathscr{W}^0, \mathscr{W}^0_-)} \\
\mathrm{H}_*(\mathscr{F}^\pm_\bullet, \mathbf{0}) \otimes \mathrm{H}_*(\mathscr{F}^0_\bullet, \mathbf{0}) & \xrightarrow[\simeq]{\ \mathrm{H}_*(J) \otimes \mathrm{H}_*(J)\ } & \mathrm{H}_*(\mathscr{F}^\pm, \mathbf{0}) \otimes \mathrm{H}_*(\mathscr{F}^0, \mathbf{0})
\end{array}$$

The commutativity of the upper and lower squares follows from Lemma 5.4.1(ii), while the commutativity of the central square follows from the naturality of the Künneth formula (see for instance [Hat02, page 275]). By (5.17) and Lemma 5.4.2, the lower horizontal homomorphism $\mathrm{H}_*(J) \otimes \mathrm{H}_*(J)$ is an isomorphism, and therefore the other horizontal homomorphisms must be isomorphisms too. □

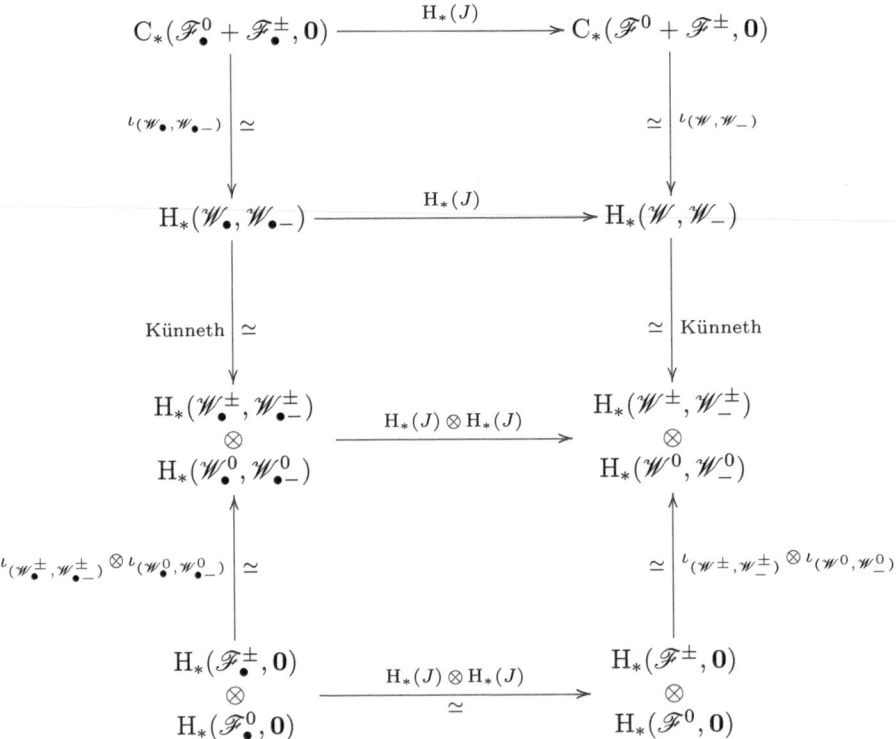

5.5 Application to the action functional

Let M be a smooth closed manifold of dimension $m \geq 1$, and $\mathscr{L} : \mathbb{T} \times \mathrm{T}M \to \mathbb{R}$ a 1-periodic convex quadratic-growth Lagrangian with associated action functional $\mathscr{A} : W^{1,2}(\mathbb{T}; M) \to \mathbb{R}$. We also consider, for each $n \in \mathbb{N}$, the mean action functional $\mathscr{A}^{[n]} : W^{1,2}(\mathbb{T}^{[n]}; M) \to \mathbb{R}$ (see Section 3.6) and, for each $k \in \mathbb{N}$ large enough, the discrete action functional $\mathscr{A}_k : \Delta_k \to \mathbb{R}$ (see Section 4.3) and the discrete mean action functional $\mathscr{A}_k^{[n]} : \Delta_{nk} \to \mathbb{R}$ (see Section 4.7).

We show that the abstract Theorem 5.1.1 can be applied to the case in which J is the discrete iteration map $\psi_k^{[n]} : \Delta_k \hookrightarrow \Delta_{nk}$ and $\mathscr{F}$ is the discrete mean action functional $\mathscr{A}_k^{[n]}$.

Theorem 5.5.1. *Let $\boldsymbol{q} \in \Delta_k$ be a critical point of the discrete action functional $\mathscr{A}_k$ such that $\mathscr{A}_k(\boldsymbol{q}) = c$. Assume that, for some $n \in \mathbb{N}$, $\boldsymbol{q}^{[n]}$ is an isolated critical point of $\mathscr{A}_k^{[n]}$ and*

$$\begin{aligned} \mathrm{ind}(\mathscr{A}_k, \boldsymbol{q}) &= \mathrm{ind}(\mathscr{A}_k^{[n]}, \boldsymbol{q}^{[n]}), \\ \mathrm{nul}(\mathscr{A}_k, \boldsymbol{q}) &= \mathrm{nul}(\mathscr{A}_k^{[n]}, \boldsymbol{q}^{[n]}). \end{aligned}$$

Then the discrete iteration map $\psi_k^{[n]}$, restricted as a map of the form

$$\psi_k^{[n]} : ((\mathscr{A}_k)_c \cup \{\boldsymbol{q}\}, (\mathscr{A}_k)_c) \hookrightarrow ((\mathscr{A}_k^{[n]})_c \cup \{\boldsymbol{q}^{[n]}\}, (\mathscr{A}_k^{[n]})_c),$$

induces the homology isomorphism

$$\mathrm{H}_*(\psi_k^{[n]}) : \mathrm{C}_*(\mathscr{A}_k, \boldsymbol{q}) \xrightarrow{\simeq} \mathrm{C}_*(\mathscr{A}_k^{[n]}, \boldsymbol{q}^{[n]}).$$

Proof. For notational convenience, let us assume that the pull-back vector bundle $\lambda_k(\boldsymbol{q})^*\mathrm{T}M$ is trivial (the general case can be handled analogously, and it is left to the reader). Applying the localization argument of Remark 4.2.1 we can assume that our Lagrangian function has the form $\mathscr{L} : \mathbb{T} \times \overline{U} \times \mathbb{R}^m \to \mathbb{R}$, where U is a bounded open subset of $\mathbb{R}^m$, and the corresponding action and mean action have the form $\mathscr{A} : W^{1,2}(\mathbb{T}; U) \to \mathbb{R}$ and $\mathscr{A}^{[n]} : W^{1,2}(\mathbb{T}^{[n]}; U) \to \mathbb{R}$. In this way, we consider $\boldsymbol{q}$ as lying in the open set $W_k = \lambda_k^{-1}(W^{1,2}(\mathbb{T}; U))$ of the k-fold product $\mathbb{R}^m \times \cdots \times \mathbb{R}^m$, and similarly

$$\boldsymbol{q}^{[n]} \in W_{nk} = (\lambda_k^{[n]})^{-1}(W^{1,2}(\mathbb{T}^{[n]}; U)) \subset \underbrace{\mathbb{R}^m \times \cdots \times \mathbb{R}^m}_{nk \text{ times}}.$$

The iteration map $\psi^{[n]} : W^{1,2}(\mathbb{T}; U) \hookrightarrow W^{1,2}(\mathbb{T}^{[n]}; U)$ is now the restriction of a bounded linear embedding $\psi^{[n]} : W^{1,2}(\mathbb{T}; \mathbb{R}^m) \hookrightarrow W^{1,2}(\mathbb{T}^{[n]}; \mathbb{R}^m)$, namely it is the restriction of the nth-iteration map on the Hilbert space $W^{1,2}(\mathbb{T}; \mathbb{R}^m)$. Analogously, the discrete iteration map $\psi_k^{[n]}$ is the restriction of the linear embedding

$\psi_k^{[n]} : \mathbb{R}^{km} \hookrightarrow \mathbb{R}^{nkm}$ given by

$$\psi_k^{[n]}(\boldsymbol{w}) = \boldsymbol{w}^{[n]} = \underbrace{(\boldsymbol{w}, \dots, \boldsymbol{w})}_{n \text{ times}}, \qquad \forall \boldsymbol{w} \in \mathbb{R}^{km}.$$

This latter embedding is an isometry with respect to the Euclidean inner product $\langle\langle \cdot, \cdot \rangle\rangle$ on $\mathbb{R}^{km}$ and the inner product $\langle\langle \cdot, \cdot \rangle\rangle^{[n]}$ on $\mathbb{R}^{nkm}$ obtained multiplying the Euclidean one by n^{-1}, i.e.,

$$\langle\langle \boldsymbol{w}, \boldsymbol{z} \rangle\rangle = \sum_{j=0}^{k-1} \langle w_j, z_j \rangle, \ \forall \boldsymbol{w} = (w_0, \dots, w_{k-1}), \boldsymbol{z} = (z_0, \dots, z_{k-1}) \in \mathbb{R}^{km},$$

$$\langle\langle \boldsymbol{w}', \boldsymbol{z}' \rangle\rangle^{[n]} = \frac{1}{n} \sum_{j=0}^{nk-1} \langle w'_j, z'_j \rangle, \ \forall \boldsymbol{w}' = (w'_0, \dots, w'_{nk-1}), \boldsymbol{z}' = (z'_0, \dots, z'_{nk-1}) \in \mathbb{R}^{nkm}.$$

Here we have denoted by $\langle \cdot, \cdot \rangle$ the Euclidean inner product on $\mathbb{R}^m$. Now, the functionals $\mathscr{A}_k$ and $\mathscr{A}_k^{[n]}$ are smooth (see Proposition 4.3.1), and therefore our claim follows from the abstract Theorem 5.1.1, provided we verify that condition (5.2) holds in our setting. Namely, we must verify that

$$\mathrm{Grad}\mathscr{A}_k^{[n]}(\boldsymbol{w}^{[n]}) = \psi^{[n]} \circ \mathrm{Grad}\mathscr{A}_k(\boldsymbol{w}), \qquad \forall \boldsymbol{w} \in W_k.$$

This follows easily from Proposition 4.4.1. In fact, if we fix an arbitrary $\boldsymbol{w} \in W_k$ and we define $\gamma_{\boldsymbol{w}} := \lambda_k(\boldsymbol{w})$ and $\boldsymbol{g} = (g_0, \dots, g_{k-1})$ with

$$g_j = \partial_v \mathscr{L}(\tfrac{j}{k}, \gamma_{\boldsymbol{w}}(\tfrac{j}{k}), \dot{\gamma}_{\boldsymbol{w}}(\tfrac{j^-}{k})) - \partial_v \mathscr{L}(\tfrac{j}{k}, \gamma_{\boldsymbol{w}}(\tfrac{j}{k}), \dot{\gamma}_{\boldsymbol{w}}(\tfrac{j^+}{k})) \in \mathbb{R}^m, \qquad \forall j \in \mathbb{Z}_k,$$

we have $\mathrm{Grad}\mathscr{A}_k(\boldsymbol{w}) = \boldsymbol{g}$ and $\mathrm{Grad}\mathscr{A}_k^{[n]}(\boldsymbol{w}^{[n]}) = \boldsymbol{g}^{[n]}$. □

Theorem 5.1.1 does not immediately apply to the mean action functional $\mathscr{A}^{[n]}$, since this latter functional is not C^2 in general (see Proposition 3.4.4). However, despite this lack of regularity, we can still obtain the assertion of the theorem as a consequence of the results of Chapter 4 and the above Theorem 5.5.1.

Corollary 5.5.2. *Let $\gamma \in W^{1,2}(\mathbb{T}; M)$ be a critical point of the action functional $\mathscr{A}$ such that $\mathscr{A}(\gamma) = c$. Assume that, for some $n \in \mathbb{N}$, $\gamma^{[n]}$ is an isolated critical point of $\mathscr{A}^{[n]}$ and*

$$\begin{aligned} \mathrm{ind}(\mathscr{A}, \gamma) &= \mathrm{ind}(\mathscr{A}^{[n]}, \gamma^{[n]}), \\ \mathrm{nul}(\mathscr{A}, \gamma) &= \mathrm{nul}(\mathscr{A}^{[n]}, \gamma^{[n]}). \end{aligned} \tag{5.19}$$

Then the iteration map $\psi^{[n]}$, restricted as a map of the form

$$\psi^{[n]} : ((\mathscr{A})_c \cup \{\gamma\}, (\mathscr{A})_c) \hookrightarrow ((\mathscr{A}^{[n]})_c \cup \{\gamma^{[n]}\}, (\mathscr{A}^{[n]})_c),$$

induces the homology isomorphism

$$\mathrm{H}_*(\psi^{[n]}) : \mathrm{C}_*(\mathscr{A}, \gamma) \xrightarrow{\simeq} \mathrm{C}_*(\mathscr{A}^{[n]}, \gamma^{[n]}).$$

Proof. As we have already remarked at the beginning of Section 4.4, for each $k \in \mathbb{N}$ sufficiently large there exists $\boldsymbol{q} \in \Delta_k$ such that

$$\gamma = \gamma_{\boldsymbol{q}} = \lambda_k(\boldsymbol{q}),$$

and $\boldsymbol{q}$ is a critical point of the discrete action functional $\mathscr{A}_k$. By Corollary 4.4.6, Lemma 4.4.9 and assumption (5.19), up to increasing k we have that

$$\begin{aligned}
\operatorname{ind}(\mathscr{A}_k, \boldsymbol{q}) &= \operatorname{ind}(\mathscr{A}, \gamma_{\boldsymbol{q}}) = \operatorname{ind}(\mathscr{A}^{[n]}, \gamma_{\boldsymbol{q}}^{[n]}) = \operatorname{ind}(\mathscr{A}_k^{[n]}, \boldsymbol{q}^{[n]}),\\
\operatorname{nul}(\mathscr{A}_k, \boldsymbol{q}) &= \operatorname{nul}(\mathscr{A}, \gamma_{\boldsymbol{q}}) = \operatorname{nul}(\mathscr{A}^{[n]}, \gamma_{\boldsymbol{q}}^{[n]}) = \operatorname{nul}(\mathscr{A}_k^{[n]}, \boldsymbol{q}^{[n]}).
\end{aligned}$$

By Corollary 4.5.3, up to further increasing k, the embeddings λ_k and $\lambda_k^{[n]}$ induce homology isomorphisms such that the following diagram commutes.

$$\begin{array}{ccc}
\mathrm{C}_*(\mathscr{A}, \gamma_{\boldsymbol{q}}) & \xrightarrow{\mathrm{H}_*(\psi^{[n]})} & \mathrm{C}_*(\mathscr{A}^{[n]}, \gamma_{\boldsymbol{q}}^{[n]}) \\
{\scriptstyle \mathrm{H}_*(\lambda_k)}\ \uparrow\ {\scriptstyle \simeq} & & {\scriptstyle \simeq}\ \uparrow\ {\scriptstyle \mathrm{H}_*(\lambda_k^{[n]})} \\
\mathrm{C}_*(\mathscr{A}_k, \boldsymbol{q}) & \xrightarrow[\simeq]{\mathrm{H}_*(\psi_k^{[n]})} & \mathrm{C}_*(\mathscr{A}_k^{[n]}, \boldsymbol{q}^{[n]})
\end{array}$$

By Theorem 5.5.1, the homomorphism $\mathrm{H}_*(\psi_k^{[n]})$ is an isomorphism, and hence $\mathrm{H}_*(\psi^{[n]})$ must also be an isomorphism. $\square$

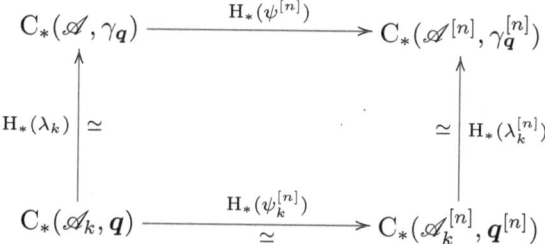

Chapter 6

Periodic Orbits of Tonelli Lagrangian Systems

This final chapter is devoted to the proof of multiplicity results for periodic orbits of Lagrangian systems given by Tonelli Lagrangians $\mathscr{L} : \mathbb{T} \times \mathrm{T}M \to \mathbb{R}$ with global Euler-Lagrange flow. As in the previous chapters, the approach that we follow is based on the critical point theory for the action functional. However, under the Tonelli assumptions, the machinery of abstract critical point theory is not directly applicable. The main problem is that a functional setting in which the Tonelli action is both regular and satisfies the Palais-Smale condition (the minimal requirements needed for critical point theory) is not known. The breakthrough required to overcome these difficulties was found by Abbondandolo and Figalli [AF07]. Their idea consists in modifying the Tonelli Lagrangian $\mathscr{L}$ outside a sufficiently large neighborhood of the zero section of $\mathrm{T}M$, making $\mathscr{L}$ fiberwise quadratic there, and then performing the critical point-theoretic analysis with the action functional of the modified Lagrangian. In a prescribed action sublevel, a suitable a priori estimate on the n-periodic orbits of the modified Lagrangian implies that these orbits must lie in the region where the Lagrangian is not modified. In particular, they are n-periodic solutions of the Euler-Lagrange system of the original Tonelli Lagrangian $\mathscr{L}$. This modification technique allows us to extend the validity of Benci's result on the multiplicity of periodic orbits with prescribed period (see Section 4.6) to the whole class of Tonelli Lagrangians with global Euler-Lagrange flow.

A harder class of results concerns the multiplicity of periodic orbits with unprescribed integer period. In 1984, Conley [Con84] conjectured that every 1-periodic Hamiltonian system on a standard symplectic m-torus admits contractible periodic orbits with arbitrarily high basic period, provided it admits only finitely many contractible 1-periodic orbits. This fact was soon established by Conley and Zehnder [CZ84] for generic Hamiltonians, and recently by Hingston [Hin09] in full generality. The conjecture was also extended to the wider class of closed symplectically aspherical manifolds by Salamon and Zehnder [SZ92] for generic Hamiltonians, and by Ginzburg [Gin10] in full generality. Further extensions are contained in [GG10, Hei09, Hei11]. Nowadays, people refer to the statement asserting the existence of infinitely many periodic orbits in a certain setting as "the Conley conjecture" in that setting. In this chapter we prove this result for Lagrangian systems associated to Tonelli Lagrangians with global Euler-Lagrange flow, or equivalently for Hamiltonian systems on cotangent bundles associated to Tonelli Hamiltonians with global Hamiltonian flow.

In the Lagrangian realm, the first important result on the multiplicity of periodic orbits with unprescribed period, which inspired the other ones mentioned here, was proved by Long [Lon00] and based on earlier work by Bangert and Klingenberg [BK83] on the closed geodesics problem. Long's result asserts the existence of infinitely many contractible periodic orbits in the case of fiberwise quadratic Lagrangians on the m-torus. Recently, Lu [Lu09] extended Long's result to the case of convex quadratic-growth Lagrangians on general closed configuration spaces. He also established the existence of infinitely many periodic orbits which are homotopic to some iteration of a given free loop, provided the connected components of the loop and of its iterations in the free loop space have nontrivial homology in some positive degree. Long and Lu [LL03] also proved Long's statement in the autonomous case. The extension to the whole class of Tonelli Lagrangians with global Euler-Lagrange flow is due to the author [Maz11].

Long's approach to the multiplicity of periodic orbits is Morse-theoretic, and can be roughly described as follows: assuming by contradiction that the considered Euler-Lagrange system admits only finitely many periodic solutions, it is possible to find a periodic orbit whose local homology persists under iteration, contradicting a homological vanishing principle. Here we extend this approach to the Tonelli case by combining the machinery of convex quadratic modifications with the discretization technique developed in Chapter 4.

In Section 6.1, following [AF07, Section 5], we recall the definition and the basic properties of convex quadratic modifications. In Section 6.2 we prove multiplicity results for 1-periodic orbits of Tonelli Lagrangian systems, extending Benci's results of Section 4.6. In Section 6.3 we introduce the discrete Tonelli action and we study the properties of its local homology. In Section 6.4 we recover Bangert and Klingenberg's homological vanishing principle in the Tonelli setting. Finally, in Section 6.5 we state and prove the Conley conjecture for Tonelli Lagrangians with global Euler-Lagrange flow.

6.1 Convex quadratic modifications

Let us fix, once for all, a smooth closed manifold M of dimension $m \geq 1$ with a Riemannian metric $\langle \cdot, \cdot \rangle_\cdot$, and a 1-periodic Tonelli Lagrangian $\mathscr{L} : \mathbb{T} \times \mathrm{T}M \to \mathbb{R}$.

As usual, we denote by $\mathscr{A}(\gamma) \in \mathbb{R} \cup \{+\infty\}$ the action of an absolutely continuous 1-periodic curve $\gamma : \mathbb{T} \to M$, i.e.,

$$\mathscr{A}(\gamma) = \int_0^1 \mathscr{L}(t, \gamma(t), \dot\gamma(t))\, \mathrm{d}t.$$

For each $n \in \mathbb{N}$, we denote by $\mathscr{A}^{[n]}(\zeta) \in \mathbb{R} \cup \{+\infty\}$ the mean action of an absolutely continuous n-periodic curve $\zeta : \mathbb{T}^{[n]} \to M$, i.e.,

$$\mathscr{A}^{[n]}(\zeta) = \frac{1}{n} \int_0^n \mathscr{L}(t, \zeta(t), \dot\zeta(t))\, \mathrm{d}t.$$

By the uniform fiberwise superlinearity of $\mathscr{L}$ (see Remark 1.2.1) there exists a real constant $C(\mathscr{L}) > 0$ such that

$$\mathscr{L}(t,q,v) \geq |v|_q - C(\mathscr{L}), \qquad \forall (t,q,v) \in \mathbb{T} \times \mathrm{T}M.$$

For each real $R > 0$, we say that $\mathscr{L}_R : \mathbb{T} \times \mathrm{T}M \to \mathbb{R}$ is a **convex quadratic R-modification** (or simply an **R-modification**) of a Tonelli Lagrangian $\mathscr{L} : \mathbb{T} \times \mathrm{T}M \to \mathbb{R}$ when:

(M1) $\mathscr{L}_R$ is a convex quadratic-growth Lagrangian,
(M2) $\mathscr{L}_R(t,q,v) = \mathscr{L}(t,q,v)$ for each $(t,q,v) \in \mathbb{T} \times \mathrm{T}M$ with $|v|_q \leq R$,
(M3) $\mathscr{L}_R(t,q,v) \geq |v|_q - C(\mathscr{L})$ for each $(t,q,v) \in \mathbb{T} \times \mathrm{T}M$.

Proposition 6.1.1 (Existence of modifications). *For every real $R > 0$, there exists a convex quadratic R-modification $\mathscr{L}_R$ of the Tonelli Lagrangian $\mathscr{L}$.*

Proof. We denote by ∂^2_{vv} the fiberwise Hessian operator on $\mathrm{T}M$, so that

$$\partial^2_{vv}\mathscr{L}(t,q,v) : \mathrm{T}_qM \times \mathrm{T}_qM \to \mathbb{R}$$

is the symmetric bilinear form given in local coordinates by

$$\partial^2_{vv}\mathscr{L}(t,q,v)[w,z] = \sum_{j,h=1}^{m} \frac{\partial^2 \mathscr{L}}{\partial v^j\, \partial v^h}(t,q,v) w^j z^h, \qquad \forall w,z \in \mathrm{T}_qM.$$

From now on, when we write inequalities (and equalities) between symmetric bilinear forms, by definition this means that the stated (in)equality holds between the associated quadratic forms. For instance, if $\ell \in \mathbb{R}$ and $\mathscr{Q} : \mathrm{T}M \to \mathbb{R}$, we will write $\partial^2_{vv}\mathscr{Q}(q,v) \geq \ell$ to mean

$$\partial^2_{vv}\mathscr{Q}(q,v)[w,w] \geq \ell\, |w|_q^2, \qquad \forall w \in \mathrm{T}_qM.$$

Let us fix any smooth convex function $\psi : \mathbb{R} \to \mathbb{R}$ such that

$$\begin{cases} \psi(r) = 0, & \forall r \leq R^2, \\ \psi(r) = r - 2R^2, & \forall r \geq 4R^2. \end{cases}$$

We define the autonomous Lagrangian $\mathscr{Q} : \mathrm{T}M \to \mathbb{R}$ by

$$\mathscr{Q}(q,v) := c_1\psi(|v|_q^2), \qquad \forall (t,q,v) \in \mathbb{T} \times \mathrm{T}M,$$

where c_1 is a positive real constant that we will fix later. Notice that $\mathscr{Q}$ is fiberwise convex everywhere and fiberwise quadratic outside a compact neighborhood of the zero section of $\mathrm{T}M$. Its fiberwise Hessian satisfies

$$\partial^2_{vv}\mathscr{Q}(q,v) = 2c_1, \qquad \forall (q,v) \in \mathrm{T}M \text{ with } |v|_q \geq 2R. \tag{6.1}$$

Next, let us fix any smooth function $\phi : \mathbb{R} \to \mathbb{R}$ such that

$$\begin{cases} \phi(r) = r, & \forall r \leq 1, \\ \phi(r) = 0, & \forall r \geq 2. \end{cases}$$

We define the Lagrangian $\mathscr{B} : \mathbb{T} \times \mathrm{T}M \to \mathbb{R}$ by

$$\mathscr{B}(t,q,v) := c_2\,\phi\biggl(\frac{\mathscr{L}(t,q,v)}{c_2}\biggr), \qquad \forall (t,q,v) \in \mathbb{T} \times \mathrm{T}M,$$

where

$$c_2 := \max\Bigl\{\mathscr{L}(t,q,v)\ \Big|\ (t,q,v) \in \mathbb{T} \times \mathrm{T}M,\ |v|_q \leq 2R\Bigr\}.$$

The Lagrangian $\mathscr{L}$ is coercive, meaning that $\mathscr{L}(t,q,v) \to \infty$ as $|v|_q \to \infty$. This implies that $\mathscr{B}$ is compactly supported. Moreover, by our choice of c_2, we have that $\mathscr{B}(t,q,v) = \mathscr{L}(t,q,v)$ whenever $|v|_q \leq 2R$. Up to fixing c_1, we define a smooth Lagrangian $\mathscr{L}_R : \mathbb{T} \times \mathrm{T}M \to \mathbb{R}$ by

$$\mathscr{L}_R(t,q,v) := \mathscr{B}(t,q,v) + \mathscr{Q}(q,v), \qquad \forall (t,q,v) \in \mathbb{T} \times \mathrm{T}M.$$

This Lagrangian clearly satisfies assumption **(M2)**. We want to show that it also satisfies assumptions **(M1)** and **(M3)** provided we choose $c_1 > 0$ large enough.

Note that

$$\mathscr{L}_R(t,q,v) = \mathscr{L}(t,q,v) + \mathscr{Q}(q,v), \qquad \forall (t,q,v) \in \mathbb{T} \times \mathrm{T}M \text{ with } |v|_q \leq 2R.$$

Since $\mathscr{Q}$ is fiberwise convex, we have

$$\partial^2_{vv}\mathscr{L}_R(t,q,v) \geq \partial^2_{vv}\mathscr{L}(t,q,v), \qquad \forall (t,q,v) \in \mathbb{T} \times \mathrm{T}M \text{ with } |v|_q \leq 2R. \tag{6.2}$$

Now, let us require that

$$c_1 \geq -\partial^2_{vv}\mathscr{B}(t,q,v), \qquad \forall (t,q,v) \in \mathbb{T} \times \mathrm{T}M. \tag{6.3}$$

By (6.1) and (6.3) we have

$$\partial^2_{vv}\mathscr{L}_R(t,q,v) \geq \partial^2_{vv}\mathscr{B}(t,q,v) + \partial^2_{vv}\mathscr{Q}(q,v) \geq c_1 > 0,$$
$$\forall (t,q,v) \in \mathbb{T} \times \mathrm{T}M \text{ with } |v|_q \geq 2R.$$

This estimate, together with (6.2), shows that $\mathscr{L}_R$ satisfies **(Q1)**. Now, for each $(t,q,v) \in \mathbb{T} \times \mathrm{T}M$ with $|v|_q$ large enough, we have

$$\mathscr{L}_R(t,q,v) = \mathscr{Q}(q,v) = c_1(|v|_q^2 - 2R^2).$$

Hence $\mathscr{L}_R$ also satisfies **(Q2)**, and therefore **(M1)**. It only remains to verify condition **(M3)**. On the one hand, we have

$$\mathscr{L}_R(t,q,v) \geq \mathscr{L}(t,q,v) \geq |v|_q - C(\mathscr{L}), \quad \forall (t,q,v) \in \mathbb{T} \times \mathrm{T}M \text{ with } |v|_q \leq 2R.$$

On the other hand, for c_1 large enough, we have

$$\mathscr{L}_R(t,q,v) \geq \min\{\mathscr{B}\} + c_1(|v|_q^2 - 2R^2) \geq |v|_q - C(\mathscr{L}), \quad \forall (t,q,v) \in \mathbb{T} \times \mathrm{T}M \text{ with } |v|_q \geq 2R. \qquad \square$$

From now on, we will always denote by $\mathscr{L}_R$ an arbitrary R-modification of $\mathscr{L}$, and by $\mathscr{A}_R : W^{1,2}(\mathbb{T};M) \to \mathbb{R}$ and $\mathscr{A}_R^{[n]} : W^{1,2}(\mathbb{T}^{[n]};M) \to \mathbb{R}$ (for each $n \in \mathbb{N}$) the associated action and mean action, i.e.,

$$\mathscr{A}_R(\gamma) = \int_0^1 \mathscr{L}_R(t,\gamma(t),\dot\gamma(t))\,\mathrm{d}t, \qquad \forall \gamma \in W^{1,2}(\mathbb{T};M),$$

$$\mathscr{A}_R^{[n]}(\zeta) = \frac{1}{n}\int_0^n \mathscr{L}_R(t,\zeta(t),\dot\zeta(t))\,\mathrm{d}t, \qquad \forall \zeta \in W^{1,2}(\mathbb{T}^{[n]};M).$$

If a curve $\gamma \in W^{1,2}(\mathbb{T};M)$ is a solution of the Euler-Lagrange system of $\mathscr{L}$, then, for each $R \geq \max\{|\dot\gamma(t)|_{\gamma(t)} \,|\, t \in \mathbb{T}\}$, γ is a critical point of $\mathscr{A}_R$ and the Gâteaux Hessian $\mathrm{Hess}\mathscr{A}_R(\gamma)$ coincides with the bilinear form $\mathscr{B}_\gamma$ associated to the Lagrangian $\mathscr{L}$ (see equation (2.1)). In particular, we have

$$\mathrm{ind}(\mathscr{A}^{[n]}, \gamma^{[n]}) = \mathrm{ind}(\mathscr{A}_R^{[n]}, \gamma^{[n]}),$$
$$\mathrm{nul}(\mathscr{A}^{[n]}, \gamma^{[n]}) = \mathrm{nul}(\mathscr{A}_R^{[n]}, \gamma^{[n]}).$$

One of the important features of convex quadratic modifications in the study of Tonelli Lagrangian systems is the following a priori estimate, due to Abbondandolo and Figalli [AF07, Lemma 5.2].

Lemma 6.1.2. *Assume that the Euler-Lagrange flow of the Tonelli Lagrangian $\mathscr{L}$ is global (see Section 1.1). Then for each $\tilde a > 0$ and $\tilde n \in \mathbb{N}$ there exists $\tilde R = \tilde R(\tilde a, \tilde n) > 0$ such that, for any R-modification $\mathscr{L}_R$ of $\mathscr{L}$ with $R > \tilde R$ and for any $n \in \{1, \ldots, \tilde n\}$, the following holds:*

(i) *if γ is a critical point of $\mathscr{A}_R^{[n]}$ such that $\mathscr{A}_R^{[n]}(\gamma) \leq \tilde a$, then*

$$|\dot\gamma(t)|_{\gamma(t)} \leq \tilde R, \qquad \forall t \in \mathbb{T}^{[n]},$$

and, in particular, γ is an n-periodic solution of the Euler-Lagrange system of $\mathscr{L}$ with $\mathscr{A}^{[n]}(\gamma) = \mathscr{A}_R^{[n]}(\gamma)$;

(ii) *if γ is an n-periodic solution of the Euler-Lagrange system of $\mathscr{L}$ such that $\mathscr{A}^{[n]}(\gamma) \leq \tilde{a}$, then*

$$|\dot\gamma(t)|_{\gamma(t)} \leq \tilde{R}, \qquad \forall t \in \mathbb{T}^{[n]},$$

and, in particular, γ is a critical point of $\mathscr{A}_R^{[n]}$ with $\mathscr{A}_R^{[n]}(\gamma) = \mathscr{A}^{[n]}(\gamma)$.

Proof. We introduce the compact subset of $\mathrm{T}M$ given by

$$K = K(\tilde{a}, \tilde{n}) = \left\{ \Phi_{\mathscr{L}}^{t_1,t_0}(q,v) \,|\, t_0, t_1 \in [-\tilde{n}, \tilde{n}],\ (q,v) \in \mathrm{T}M,\ |v|_q \leq \tilde{a} + C(\mathscr{L}) \right\},$$

where $\Phi_{\mathscr{L}}$ denotes the Euler-Lagrange flow associated to $\mathscr{L}$. We set

$$\tilde{R} = \tilde{R}(\tilde{a}, \tilde{n}) := \max\left\{ |v|_q \,|\, (q,v) \in K \right\}$$

and we fix $R > \tilde{R}$ and $n \in \{1, \ldots, \tilde{n}\}$.

Let γ be a critical point of $\mathscr{A}_R^{[n]}$ such that $\mathscr{A}_R^{[n]}(\gamma) \leq \tilde{a}$, as in point (i) of the statement. There exists $t_0 \in \mathbb{T}^{[n]}$ such that

$$\mathscr{L}_R(t_0, \gamma(t_0), \dot\gamma(t_0)) \leq \mathscr{A}_R^{[n]}(\gamma).$$

By **(M3)**, we have

$$|\dot\gamma(t_0)|_{\gamma(t_0)} \leq \mathscr{L}_R(t_0, \gamma(t_0), \dot\gamma(t_0)) + C(\mathscr{L}) \leq \mathscr{A}_R^{[n]}(\gamma) + C(\mathscr{L}) \leq \tilde{a} + C(\mathscr{L}),$$

and in particular $(\gamma(t_0), \dot\gamma(t_0)) \in K$. Let I be the closed subset of $\mathbb{T}^{[n]}$ given by

$$I = \left\{ t \in \mathbb{T}^{[n]} \,|\, (\gamma(t), \dot\gamma(t)) \in K \right\}.$$

If $t \in I$ we have $|\dot\gamma(t)|_{\gamma(t)} \leq \tilde{R} < R$, and hence the Lagrangian functions $\mathscr{L}$ and $\mathscr{L}_R$ coincide on a neighborhood of $(t, \gamma(t), \dot\gamma(t))$. This implies that there exists $\varepsilon > 0$ such that

$$(\gamma(s), \dot\gamma(s)) = \Phi_{\mathscr{L}_R}^{s,t}(\gamma(t), \dot\gamma(t)) = \Phi_{\mathscr{L}}^{s,t}(\gamma(t), \dot\gamma(t)), \qquad \forall s \in (t-\varepsilon, t+\varepsilon).$$

In particular $(t-\varepsilon, t+\varepsilon) \subset I$. Hence I is also open in $\mathbb{T}^{[n]}$, which forces $I = \mathbb{T}^{[n]}$ and thus $|\dot\gamma(t)|_{\gamma(t)} \leq \tilde{R}$ for each $t \in \mathbb{T}^{[n]}$. This proves point (i).

As for point (ii), let γ be an n-periodic solution of the Euler-Lagrange system of $\mathscr{L}$ such that $\mathscr{A}^{[n]}(\gamma) \leq \tilde{a}$. There exists $t_0 \in \mathbb{T}^{[n]}$ such that

$$|\dot\gamma(t_0)|_{\gamma(t_0)} \leq \mathscr{L}(t_0, \gamma(t_0), \dot\gamma(t_0)) + C(\mathscr{L}) \leq \mathscr{A}_R^{[n]}(\gamma) + C(\mathscr{L}) \leq \tilde{a} + C(\mathscr{L}).$$

By the definition of K and since γ is n-periodic, for each $t \in \mathbb{T}^{[n]}$ the point $(\gamma(t), \dot\gamma(t)) = \Phi_{\mathscr{L}}^{t,t_0}(\gamma(t_0), \dot\gamma(t_0))$ belongs to K, and point (ii) follows. □

6.2 Multiplicity of periodic orbits with prescribed period

In this section we are going to extend the multiplicity results for periodic orbits with prescribed period, discussed in Sections 3.5 and 4.6, to the case of Tonelli Lagrangians with global Euler-Lagrange flow. For syntactic convenience, we will work in period 1, but as usual everything can be carried out word by word in any integer period.

To begin with, we give a convex quadratic modifications-based proof of Theorem 3.5.5 (see [Mn91, Proposition 7.1] for a classical variational proof).

Theorem 6.2.1. *Let $\mathscr{L} : \mathbb{T} \times \mathrm{T}M \to \mathbb{R}$ be a 1-periodic Tonelli Lagrangian with global Euler-Lagrange flow. Then for each conjugacy class C of $\pi_1(M)$, the Euler-Lagrange system of $\mathscr{L}$ admits a 1-periodic orbit γ that is homotopic to some (and therefore all) $\zeta : \mathbb{T} \to M$ representing C. Moreover, γ minimizes the action among all the absolutely continuous 1-periodic curves that are homotopic to ζ.*

Proof. Consider a smooth $\zeta : \mathbb{T} \to M$ that represents C. We denote by $a \in \mathbb{R}$ its action with respect to the Lagrangian $\mathscr{L}$, i.e.,

$$a := \mathscr{A}(\zeta) = \int_0^1 \mathscr{L}(t, \zeta(t), \dot{\zeta}(t))\, \mathrm{d}t.$$

We fix a real $R > \tilde{R}(a,1)$, where $\tilde{R}(a,1)$ is the constant given by Lemma 6.1.2, and an R-modification $\mathscr{L}_R$ of $\mathscr{L}$ with associated action $\mathscr{A}_R$. By Theorem 3.5.5, there exists a critical point γ of $\mathscr{A}_R$ which is homotopic to ζ and minimizes the action $\mathscr{A}_R$ among all the absolutely continuous 1-periodic curves that are homotopic to ζ. By Lemma 6.1.2(i), γ is a 1-periodic orbit of the Euler-Lagrange system of $\mathscr{L}$ with $\mathscr{A}(\gamma) = \mathscr{A}_R(\gamma)$. By Lemma 6.1.2(ii), γ also minimizes the action $\mathscr{A}$ among all the absolutely continuous 1-periodic curves that are homotopic to ζ. □

The generalization of Theorem 4.6.1 to the Tonelli case is due to Abbondandolo and Figalli [AF07, Theorem 3.1].

Theorem 6.2.2. *Let M be a smooth closed manifold of positive dimension, and $\mathscr{L} : \mathbb{T} \times \mathrm{T}M \to \mathbb{R}$ a 1-periodic Tonelli Lagrangian with global Euler-Lagrange flow. Assume that the 1-periodic solutions of the Euler-Lagrange system of $\mathscr{L}$ are isolated. We denote by $\mathscr{A}$ the action functional associated to $\mathscr{L}$.*

(i) *If $\mathrm{H}_d(W^{1,2}(\mathbb{T}; M))$ is nontrivial, then the Euler-Lagrange system of $\mathscr{L}$ admits a 1-periodic solution γ such that*

$$\mathrm{ind}(\mathscr{A}, \gamma) \le d \le \mathrm{ind}(\mathscr{A}, \gamma) + \mathrm{nul}(\mathscr{A}, \gamma).$$

In particular, if $\mathrm{H}_d(W^{1,2}(\mathbb{T}; M))$ is nontrivial for infinitely many $d \in \mathbb{N}$, then the Euler-Lagrange system of $\mathscr{L}$ admits an infinite sequence $\{\gamma_n \,|\, n \in \mathbb{N}\}$ of

1-periodic solutions such that

$$\lim_{n\to\infty} \mathscr{A}(\gamma_n) = \infty, \qquad \lim_{n\to\infty} \mathrm{ind}(\mathscr{A}, \gamma_n) = \infty.$$

(ii) *The number of 1-periodic solutions of the Euler-Lagrange system of $\mathscr{L}$ is at least $\mathrm{CL}(W^{1,2}(\mathbb{T}; M)) + 1$, where $\mathrm{CL}(W^{1,2}(\mathbb{T}; M))$ is the cup-length of $W^{1,2}(\mathbb{T}; M)$.*

Theorems 6.2.1 and 6.2.2 readily allow us to extend Corollaries 4.6.2 and 4.6.3 to the class of Tonelli Lagrangians with global Euler-Lagrange flow.

Corollary 6.2.3. *Let M be a smooth closed manifold of positive dimension, and $\mathscr{L} : \mathbb{T} \times \mathrm{T}M \to \mathbb{R}$ a 1-periodic Tonelli Lagrangian with global Euler-Lagrange flow. We denote by $\mathscr{A}$ the action functional associated to $\mathscr{L}$.*

(i) *If $\pi_1(M)$ is finite, then the Euler-Lagrange system of $\mathscr{L}$ admits infinitely many contractible 1-periodic orbits. If they are all isolated, then there exists an infinite sequence of contractible 1-periodic orbits $\{\gamma_n \mid n \in \mathbb{N}\}$ such that*

$$\lim_{n\to\infty} \mathscr{A}(\gamma_n) = \infty, \qquad \lim_{n\to\infty} \mathrm{ind}(\mathscr{A}, \gamma_n) = \infty.$$

(ii) *If $\pi_1(M)$ is abelian, then the Euler-Lagrange system of $\mathscr{L}$ admits infinitely many 1-periodic orbits.* □

Proof of Theorem 6.2.2. Consider a cycle β representing a nonzero homology class $[\beta] \in \mathrm{H}_d(W^{1,2}(\mathbb{T}; M))$. Since the inclusion $C^\infty(\mathbb{T}; M) \hookrightarrow W^{1,2}(\mathbb{T}; M)$ is a homotopy equivalence (Proposition 3.2.1), we can assume that β is a cycle in $C^\infty(\mathbb{T}; M)$. We denote by $\mathscr{K}_\beta \subset C^\infty(\mathbb{T}; M)$ the compact support of β, and we set

$$R_\beta := \max\left\{ |\dot\zeta(t)|_{\zeta(t)} \;\middle|\; \zeta \in \mathscr{K}_\beta,\ t \in \mathbb{T} \right\}.$$

The Tonelli action functional is easily seen to be continuous on $C^\infty(\mathbb{T}; M)$, and therefore we can choose a real constant $a \in \mathbb{R}$ such that

$$a > \max_{\zeta \in \mathscr{K}_\beta} \mathscr{A}(\zeta).$$

Now, let us fix a real $R > \max\{\tilde R(a, 1), R_\beta\}$, where $\tilde R(a, 1)$ is the constant given by Lemma 6.1.2(i), and an R-modification $\mathscr{L}_R$ of $\mathscr{L}$ with associated action $\mathscr{A}_R$. Since the 1-periodic solutions of the Euler-Lagrange system of $\mathscr{L}$ are isolated, Lemma 6.1.2(i) implies that the critical points of $\mathscr{A}_R$ with critical value less than or equal to a are isolated. By Theorem 4.6.1(i), $\mathrm{minimax}_{\mathfrak{S}_{[\beta]}} \mathscr{A}_R$ is a critical value of $\mathscr{A}_R$. Note that, by our choice of R, we have

$$\underset{\mathfrak{S}_{[\beta]}}{\mathrm{minimax}}\, \mathscr{A}_R \le \max_{\zeta \in \mathscr{K}_\beta} \mathscr{A}_R(\zeta) = \max_{\zeta \in \mathscr{K}_\beta} \mathscr{A}(\zeta) < a.$$

Hence the critical points of $\mathscr{A}_R$ with critical value $\mathrm{minimax}_{\mathfrak{S}_{[\beta]}} \mathscr{A}_R$ are isolated. By Theorem 4.6.1(i), there exists a critical point γ of $\mathscr{A}_R$ such that

$$\mathscr{A}_R(\gamma) = \operatorname*{minimax}_{\mathfrak{S}_{[\beta]}} \mathscr{A}_R,$$
$$\mathrm{ind}(\mathscr{A}_R, \gamma) \leq d \leq \mathrm{ind}(\mathscr{A}_R, \gamma) + \mathrm{nul}(\mathscr{A}_R, \gamma).$$

By Lemma 6.1.2(i), $|\dot\gamma(t)|_{\gamma(t)} < R$ for all $t \in \mathbb{T}$, and therefore γ is also a 1-periodic solution of the Euler-Lagrange system of $\mathscr{L}$ with

$$\begin{aligned} \mathscr{A}(\gamma) &= \mathscr{A}_R(\gamma), \\ \mathrm{ind}(\mathscr{A}, \gamma) &= \mathrm{ind}(\mathscr{A}_R, \gamma), \\ \mathrm{nul}(\mathscr{A}, \gamma) &= \mathrm{nul}(\mathscr{A}_R, \gamma). \end{aligned}$$

This proves point (i). As for point (ii), consider a sequence of nonzero homology classes $[\beta_1] \prec \cdots \prec [\beta_r]$ in $\mathrm{H}_*(W^{1,2}(\mathbb{T}; M))$, where $r = \mathrm{CL}(W^{1,2}(\mathbb{T}; M)) + 1$ and, as before, β_j is a cycle in $C^\infty(\mathbb{T}; M)$ for each $j = 1, \ldots, r$. We choose a real constant $a \in \mathbb{R}$ such that

$$a > \max_{\zeta \in \mathscr{K}} \mathscr{A}(\zeta),$$

where $\mathscr{K} = \mathscr{K}_{\beta_1} \cup \cdots \cup \mathscr{K}_{\beta_r}$ and, for each $j = 1, \ldots, r$, the compact set $\mathscr{K}_{\beta_j} \subset C^\infty(\mathbb{T}; M)$ is the support of the cycle β_j. Assume that

$$R > \max\{\tilde R(a, 1), R_{\beta_1}, \ldots, R_{\beta_r}\}.$$

By Theorem 4.6.1(ii), there exist at least r many critical points of $\mathscr{A}_R$ with action less than or equal to $\mathrm{minimax}_{\mathfrak{S}_{[\beta_r]}} \mathscr{A}_R$, which in turn is less than a due to our choice of R. By Lemma 6.1.2(i), each of these critical points γ satisfies $|\dot\gamma(t)|_{\gamma(t)} < R$ for all $t \in \mathbb{T}$, and thus in particular is a 1-periodic solution of the Euler-Lagrange system of $\mathscr{L}$. □

6.3 Discrete Tonelli action

Let γ be a periodic solution of the Euler-Lagrange system associated to a Tonelli Lagrangian $\mathscr{L} : \mathbb{T} \times \mathrm{T}M \to \mathbb{R}$ with global Euler-Lagrange flow. In order to simplify the notation, we can assume that the period of γ is 1, so that it is a map of the form $\gamma : \mathbb{T} \to M$. We choose a constant $U \in \mathbb{R}$ such that

$$U > \max\left\{ |\dot\gamma(t)|_{\gamma(t)} \,\middle|\, t \in \mathbb{T} \right\}, \tag{6.4}$$

and we consider a convex quadratic-growth Lagrangian $\mathscr{L}_U : \mathbb{T} \times \mathrm{T}M \to \mathbb{R}$ that is a U-modification of $\mathscr{L}$. By our choice of the constant U, the curve γ is a critical point of the action $\mathscr{A}_U$, and $\mathscr{A}_U(\gamma) = \mathscr{A}(\gamma)$.

Let us apply the discretization technique of Chapter 4 to $\mathscr{L}_U$. For each integer $k \geq 1/\varepsilon_0(\mathscr{L}_U)$, where $\varepsilon_0(\mathscr{L}_U)$ is the constant given by Theorem 4.1.1, we obtain a map

$$\lambda_{k,\mathscr{L}_U} : \Delta_k \hookrightarrow C_k^\infty(\mathbb{T};M) \subset W^{1,2}(\mathbb{T};M),$$

as explained in Section 4.2. We define an open set $\mathscr{U}_k \subseteq C_k^\infty(\mathbb{T};M)$ by

$$\mathscr{U}_k := \left\{ \zeta \in C_k^\infty(\mathbb{T};M) \,\middle|\, \operatorname*{ess\,sup}_{t\in\mathbb{T}} \left\{ |\dot\zeta(t)|_{\zeta(t)} \right\} < U \right\}. \tag{6.5}$$

Since $\lambda_{k,\mathscr{L}_U}$ is smooth, we get an open set

$$U_k := \lambda_{k,\mathscr{L}_U}^{-1}(\mathscr{U}_k) \subseteq \Delta_k. \tag{6.6}$$

Notice that the action $\mathscr{A}_U$ coincides with the Tonelli action $\mathscr{A}$ on the open set $\mathscr{U}_k$. This allows us to define the **discrete Tonelli action** $\mathscr{A}_k : U_k \to \mathbb{R}$ as

$$\mathscr{A}_k := \mathscr{A} \circ \lambda_{k,\mathscr{L}_U}|_{U_k} = \mathscr{A}_U \circ \lambda_{k,\mathscr{L}_U}|_{U_k}.$$

This functional is smooth (Proposition 4.3.1) and $\boldsymbol{q} := \lambda_{k,\mathscr{L}_U}^{-1}(\gamma)$ is a critical point of it.

From now on we will just write $\lambda_k : U_k \hookrightarrow C_k^\infty(\mathbb{T};M)$ for $\lambda_{k,\mathscr{L}_U}|_{U_k}$. Even if this embedding does depend on the chosen U-modification $\mathscr{L}_U$, its germ at $\boldsymbol{q}$ does not, as stated by the following.

Lemma 6.3.1. *Let $\mathscr{L}_R$ be an R-modification of $\mathscr{L}$, where $R \geq U$, and consider an integer k large enough so that both $\lambda_{k,\mathscr{L}_U}$ and $\lambda_{k,\mathscr{L}_R}$ are defined. Then there exists an open neighborhood $V_k \subset U_k$ of $\boldsymbol{q}$ such that*

$$\lambda_k|_{V_k} = \lambda_{k,\mathscr{L}_U}|_{V_k} = \lambda_{k,\mathscr{L}_R}|_{V_k}.$$

Proof. The Lagrangian functions $\mathscr{L}_U$ and $\mathscr{L}_R$ coincide on a neighborhood of the support of the curve $(\gamma,\dot\gamma) : \mathbb{T} \to \mathrm{T}M$. Therefore, for each integer k larger than $\max\{1/\varepsilon_0(\mathscr{L}_U), 1/\varepsilon_0(\mathscr{L}_R)\}$, the k-broken Euler-Lagrange curves of $\mathscr{L}_U$ and $\mathscr{L}_R$ that are close to γ are the same and the claim follows. □

Let $c = \mathscr{A}_k(\boldsymbol{q}) = \mathscr{A}(\gamma)$. By Corollary 4.5.3 and the excision property, if $k \geq \bar{k}(\mathscr{L}_U, c)$, the embedding λ_k restricts to a map

$$\lambda_k : ((\mathscr{A}_k)_c \cup \{\boldsymbol{q}\}, (\mathscr{A}_k)_c) \hookrightarrow ((\mathscr{A}_U)_c \cup \{\gamma\}, (\mathscr{A}_U)_c)$$

that induces the homology isomorphism

$$\mathrm{H}_*(\lambda_k) : \mathrm{C}_*(\mathscr{A}_k, \boldsymbol{q}) \xrightarrow{\simeq} \mathrm{C}_*(\mathscr{A}_U, \gamma). \tag{6.7}$$

For each R-modification $\mathscr{L}_R$ of $\mathscr{L}$, with $R \geq U$, the action $\mathscr{A}_R$ coincides with $\mathscr{A}$ on the image of λ_k. Hence, the embedding λ_k also restricts to a map

$$\lambda_k : ((\mathscr{A}_k)_c \cup \{\boldsymbol{q}\}, (\mathscr{A}_k)_c) \hookrightarrow ((\mathscr{A}_R)_c \cup \{\gamma\}, (\mathscr{A}_R)_c). \tag{6.8}$$

A priori, the induced homomorphism on homology may not be an isomorphism. However, the next lemma guarantees that this cannot happen.

Lemma 6.3.2. *For each $R \geq U$, the map λ_k, restricted as in (6.8), induces a homology isomorphism*

$$\mathrm{H}_*(\lambda_k) : \mathrm{C}_*(\mathscr{A}_k, \boldsymbol{q}) \xrightarrow{\simeq} \mathrm{C}_*(\mathscr{A}_R, \gamma).$$

Proof. For each $h \in \mathbb{N}$ we define the embedding

$$j_h = \lambda_{hk}^{-1} \circ \lambda_k : U_k \hookrightarrow U_{hk}.$$

Since $hk \geq k \geq \bar{k}(\mathscr{L}_U, c)$, by Corollary 4.5.3 and the excision property we see that the map

$$\lambda_{hk} : ((\mathscr{A}_{hk})_c \cup \{j_h(\boldsymbol{q})\}, (\mathscr{A}_{jk})_c) \hookrightarrow ((\mathscr{A}_U)_c \cup \{\gamma\}, (\mathscr{A}_U)_c)$$

induces the homology isomorphism $\mathrm{H}_*(\lambda_{hk}) : \mathrm{C}_*(\mathscr{A}_{hk}, j_h(\boldsymbol{q})) \xrightarrow{\simeq} \mathrm{C}_*(\mathscr{A}_U, \gamma)$, as well as λ_k in (6.7). Hence, the map j_h, restricted as in the following commutative diagram

$$\begin{array}{ccc}
((\mathscr{A}_k)_c \cup \{\boldsymbol{q}\}, (\mathscr{A}_k)_c) & \xhookrightarrow{\quad\lambda_k\quad} & ((\mathscr{A}_U)_c \cup \{\gamma\}, (\mathscr{A}_U)_c) \\
{\scriptstyle j_h}\big\downarrow & \nearrow{\scriptstyle \lambda_{hk}} & \\
((\mathscr{A}_{hk})_c \cup \{j_h(\boldsymbol{q})\}, (\mathscr{A}_{hk})_c) & &
\end{array}$$

induces the homology isomorphism

$$\mathrm{H}_*(j_h) = \mathrm{H}_*(\lambda_{hk})^{-1} \circ \mathrm{H}_*(\lambda_k) : \mathrm{C}_*(\mathscr{A}_k, \boldsymbol{q}) \xrightarrow{\simeq} \mathrm{C}_*(\mathscr{A}_{hk}, j_h(\boldsymbol{q})).$$

Now, let us consider an R-modification $\mathscr{L}_R$ of $\mathscr{L}$ as in the statement, with associated maps

$$\lambda_{hk,\mathscr{L}_R} : \Delta_{hk} \hookrightarrow C_k^\infty(\mathbb{T}; M), \qquad \forall h \in \mathbb{N}.$$

By Lemma 6.3.1, for each $h \in \mathbb{N}$ large enough, there exists a neighborhood $V_{hk} \subset U_{hk}$ of $j_h(\boldsymbol{q})$ such that

$$\lambda_{hk}|_{V_{hk}} = \lambda_{hk,\mathscr{L}_R}|_{V_{hk}}.$$

If h is such that $hk \geq \bar{k}(\mathscr{L}_R, c)$, by Corollary 4.5.3 and the excision property we see that both λ_{hk} and $\lambda_{hk,\mathscr{L}_R}$ restrict to give the same map

$$\lambda'_{hk} : (V_{hk} \cap (\mathscr{A}_{hk})_c \cup \{j_h(\boldsymbol{q})\}, V_{hk} \cap (\mathscr{A}_{jk})_c) \hookrightarrow ((\mathscr{A}_R)_c \cup \{\gamma\}, (\mathscr{A}_R)_c)$$

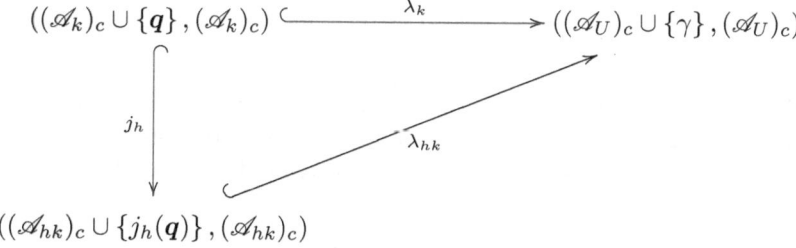

that induces an isomorphism in homology. Finally, consider the following commutative diagram of maps.

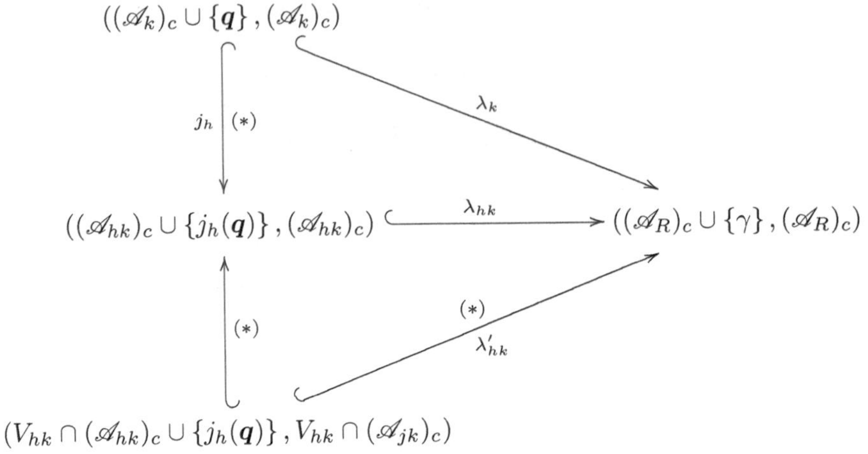

We already know that the maps marked with $(*)$ induce isomorphisms in homology. Therefore, the maps λ_{hk} and λ_k also induce isomorphisms in homology. □

Remark 6.3.1. As a consequence of the above lemma, we immediately obtain that the local homology groups $\mathrm{C}_*(\mathscr{A}_R, \gamma)$ do not depend (up to isomorphism) on the chosen real constant $R \geq U$ and on the chosen R-modification $\mathscr{L}_R$. □

6.4 Homological vanishing by iteration

All of the arguments of the last section can be carried out word by word if we work in an arbitrary period $n \in \mathbb{N}$. We introduce the open sets

$$\mathscr{U}_k^{[n]} := \left\{ \zeta \in C_k^\infty(\mathbb{T}^{[n]}; M) \,\middle|\, \operatorname*{ess\,sup}_{t \in \mathbb{T}^{[n]}} \left\{ |\dot\zeta(t)|_{\zeta(t)} \right\} < U \right\} \subset C_k^\infty(\mathbb{T}^{[n]}; M),$$

$$U_k^{[n]} := (\lambda_{k,\mathscr{L}_U}^{[n]})^{-1}(\mathscr{U}_k^{[n]}) \subset \Delta_{nk},$$

and we define the **discrete mean Tonelli action**

$$\mathscr{A}_k^{[n]} := \mathscr{A}^{[n]} \circ \lambda_k^{[n]} : U_k^{[n]} \to \mathbb{R},$$

where $\lambda_k^{[n]} := \lambda_{k,\mathscr{L}_U}^{[n]}$. Then Lemmas 6.3.1 and 6.3.2 go through without change.

We recall from Section 4.7 that we have a discrete iteration map

$$\psi_k^{[n]} : U_k \hookrightarrow U_k^{[n]}$$

such that the following diagram commutes.

$$
\begin{array}{ccc}
U_k & \xhookrightarrow{\ \psi_k^{[n]}\ } & U_{nk} \\
\Big\downarrow{\scriptstyle \lambda_k} & & \Big\downarrow{\scriptstyle \lambda_k^{[n]}} \\
W^{1,2}(\mathbb{T};M) & \xhookrightarrow{\ \psi^{[n]}\ } & W^{1,2}(\mathbb{T}^{[n]};M)
\end{array}
$$

Now, consider $\infty \geq c_2 > c_1 = c = \mathscr{A}(\gamma)$. For each $R \geq U$, the embeddings λ_k, $\lambda_k^{[n]}$ and $\psi^{[n]}$ restrict to give maps such that the following diagram commutes.

$$
\begin{array}{ccc}
((\mathscr{A}_k)_{c_1} \cup \{\boldsymbol{q}\}, (\mathscr{A}_k)_{c_1}) & \xhookrightarrow{\ \psi_k^{[n]}\ } & ((\mathscr{A}_k^{[n]})_{c_1} \cup \{\boldsymbol{q}^{[n]}\}, (\mathscr{A}_k^{[n]})_{c_1}) \\
\Big\downarrow{\scriptstyle \lambda_k} & & \Big\downarrow{\scriptstyle \lambda_k^{[n]}} \\
((\mathscr{A}_R)_{c_2}, (\mathscr{A}_R)_{c_1}) & \xhookrightarrow{\ \psi^{[n]}\ } & ((\mathscr{A}_R^{[n]})_{c_2}, (\mathscr{A}_R^{[n]})_{c_1})
\end{array}
$$

This latter diagram, in turn, induces the following commutative diagram of homology groups.

$$
\begin{array}{ccc}
\mathrm{C}_*(\mathscr{A}_k, \boldsymbol{q}) & \xrightarrow{\ \mathrm{H}_*(\psi_k^{[n]})\ } & \mathrm{C}_*(\mathscr{A}_k^{[n]}, \boldsymbol{q}^{[n]}) \\
\Big\downarrow{\scriptstyle \mathrm{H}_*(\lambda_k)} & & \Big\downarrow{\scriptstyle \mathrm{H}_*(\lambda_k^{[n]})} \\
\mathrm{H}_*((\mathscr{A}_R)_{c_2}, (\mathscr{A}_R)_{c_1}) & \xrightarrow{\ \mathrm{H}_*(\psi^{[n]})\ } & \mathrm{H}_*((\mathscr{A}_R^{[n]})_{c_2}, (\mathscr{A}_R^{[n]})_{c_1})
\end{array}
$$

The main result of this section is the following homological vanishing theorem, first established in the setting of closed geodesics by Bangert and Klingenberg [BK83], and later extended to more general Lagrangian systems by Long [Lon00], Lu [Lu09] and the author [Maz11].

Theorem 6.4.1 (Homological vanishing). *Let $[\eta]$ be an element in the local homology group $\mathrm{C}_*(\mathscr{A}_k, \boldsymbol{q})$, where the critical point $\boldsymbol{q}$ of $\mathscr{A}_k$ is not a local minimum. Then for each integer $j \geq 2$ there exist $\bar{R} = \bar{R}([\eta], j) \geq U$ and $\bar{n} = \bar{n}([\eta], j) \in \mathbb{N}$ that is a power of j such that, for each real $R \geq \bar{R}$, we have*

$$
\mathrm{H}_*(\psi^{[\bar{n}]}) \circ \mathrm{H}_*(\lambda_k)[\eta] = \mathrm{H}_*(\lambda_k^{[\bar{n}]}) \circ \mathrm{H}_*(\psi_k^{[\bar{n}]})[\eta] = 0 \qquad \text{in } \mathrm{H}_*((\mathscr{A}_R^{[\bar{n}]})_{c_2}, (\mathscr{A}_R^{[\bar{n}]})_{c_1}).
$$

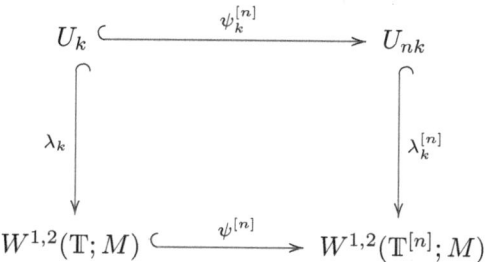

The proof of this theorem, which will take the remainder of this section, is based on a homotopical technique that is essentially due to Bangert (cf. [Ban80, Section 3]).

Lemma 6.4.2 (Bangert homotopy). *Let $\sigma : \Delta^p \to W^{1,2}(\mathbb{T}; M)$ be a continuous p-singular simplex such that*

$$\max_{\boldsymbol{z}\in\Delta^p} \{\mathscr{A}(\sigma(\boldsymbol{z}))\} < c_2,$$

$$\max_{\boldsymbol{z}\in\partial\Delta^p} \{\mathscr{A}(\sigma(\boldsymbol{z}))\} < c_1,$$

$$\sup_{\boldsymbol{z}\in\Delta^p} \operatorname*{ess\,sup}_{t\in\mathbb{T}} \left\{ \left| \frac{\mathrm{d}}{\mathrm{d}t}\sigma(\boldsymbol{z})(t) \right|_{\sigma(\boldsymbol{z})(t)} \right\} \leq \bar{r}(\sigma),$$

where $\bar{r}(\sigma)$ is a real constant. Then there exist a positive integer $\bar{n}(\sigma)$, a positive real $\bar{R}(\sigma) \geq \bar{r}(\sigma)$ and, for each $n \in \mathbb{N}$, a homotopy[1]

$$\mathrm{Ban}_\sigma^{[n]} : [0,1] \times \Delta^p \to W^{1,2}(\mathbb{T}; M) \qquad \textit{relative to } \partial\Delta^p$$

satisfying the following properties:

(i) $\mathrm{Ban}_\sigma^{[n]}(0,\cdot) = \psi^{[n]} \circ \sigma$, *for each $n \in \mathbb{N}$;*

(ii) *for each integer $n \geq \bar{n}(\sigma)$ we have*

$$\max_{(s,\boldsymbol{z})\in[0,1]\times\Delta^p} \left\{ \mathscr{A}^{[n]}(\mathrm{Ban}_\sigma^{[n]}(s,\boldsymbol{z})) \right\} < c_2,$$

$$\max_{(s,\boldsymbol{z})\in[0,1]\times\partial\Delta^p} \left\{ \mathscr{A}^{[n]}(\mathrm{Ban}_\sigma^{[n]}(s,\boldsymbol{z})) \right\} < c_1;$$

$$\max_{\boldsymbol{z}\in\Delta^p} \left\{ \mathscr{A}^{[n]}(\mathrm{Ban}_\sigma^{[n]}(1,\boldsymbol{z})) \right\} < c_1;$$

(iii) *for each $n \in \mathbb{N}$ we have*

$$\sup_{(s,\boldsymbol{z})\in[0,1]\times\Delta^p} \operatorname*{ess\,sup}_{t\in\mathbb{T}^{[n]}} \left\{ \left| \frac{\mathrm{d}}{\mathrm{d}t}\mathrm{Ban}_\sigma^{[n]}(s,\boldsymbol{z})(t) \right|_{\mathrm{Ban}_\sigma^{[n]}(s,\boldsymbol{z})(t)} \right\} \leq \bar{R}(\sigma).$$

Hereafter, the homotopies $\mathrm{Ban}_\sigma^{[n]}$ will be called **Bangert homotopies**. Before giving the proof of the above lemma, we will show that it readily gives the following homotopical vanishing result.

Theorem 6.4.3 (Homotopical vanishing). *Consider a continuous p-singular simplex $\sigma : (\Delta^p, \partial\Delta^p) \to ((\mathscr{A}_k)_{c_1} \cup \{\boldsymbol{q}\}, (\mathscr{A}_k)_{c_1})$. Then there exist a positive integer $\bar{n}(\sigma)$ and a real number $\bar{R}(\sigma)$ such that, for each real $R \geq \bar{R}(\sigma)$ and for each integer*

[1] See Section A.3 in the Appendix for a review of the classical terminology in homotopy theory.

$n \geq \bar{n}(\sigma)$, *we have*[2]

$$\pi_p(\psi^{[n]}) \circ \pi_p(\lambda_k)[\sigma] = \pi_p(\lambda_k^{[n]}) \circ \pi_p(\psi_k^{[n]})[\sigma] = 0 \qquad \text{in } \pi_p((\mathscr{A}_R^{[n]})_{c_2}, (\mathscr{A}_R^{[n]})_{c_1}).$$

Proof. Let $U > 0$ be the constant chosen in (6.4). The singular simplex $\tilde{\sigma} := \lambda_k \circ \sigma : \Delta^p \to W^{1,2}(\mathbb{T}; M)$ satisfies

$$\begin{aligned} \mathscr{A}(\tilde{\sigma}(\boldsymbol{z})) &= \mathscr{A}_k(\sigma(\boldsymbol{z})) \leq c_1 < c_2, & \forall \boldsymbol{z} \in \Delta^p, \\ \mathscr{A}(\tilde{\sigma}(\boldsymbol{z})) &= \mathscr{A}_k(\sigma(\boldsymbol{z})) < c_1, & \forall \boldsymbol{z} \in \partial\Delta^p. \end{aligned}$$

Moreover, since $\tilde{\sigma}(\Delta^p) \subset \lambda_k(U_k) \subset \mathscr{U}_k$ (see (6.5) and (6.6)), we further have

$$\sup_{\boldsymbol{z} \in \Delta^p} \ \operatorname*{ess\,sup}_{t \in \mathbb{T}} \left\{ \left| \frac{\mathrm{d}}{\mathrm{d}t} \tilde{\sigma}(\boldsymbol{z})(t) \right|_{\tilde{\sigma}(\boldsymbol{z})(t)} \right\} \leq U =: \bar{r}(\tilde{\sigma}).$$

By Lemma 6.4.2, we obtain $\bar{n}(\tilde{\sigma}) \in \mathbb{N}$, $\bar{R}(\tilde{\sigma}) \geq \bar{r}(\tilde{\sigma}) > 0$ and Bangert homotopies $\mathrm{Ban}_{\tilde{\sigma}}^{[n]}$ for each $n \in \mathbb{N}$. For each $R \geq \bar{R}(\tilde{\sigma})$ and $n \geq \bar{n}(\tilde{\sigma})$, $\mathrm{Ban}_{\tilde{\sigma}}^{[n]}$ is a homotopy of maps of pairs of the form

$$\mathrm{Ban}_{\tilde{\sigma}}^{[n]} : [0,1] \times (\Delta^p, \partial\Delta^p) \to ((\mathscr{A}_R^{[n]})_{c_2}, (\mathscr{A}_R^{[n]})_{c_1}),$$

with $\mathrm{Ban}_{\tilde{\sigma}}^{[n]}(0, \cdot) = \psi^{[n]} \circ \tilde{\sigma}$ and $\mathrm{Ban}_{\tilde{\sigma}}^{[n]}(\{1\} \times \Delta^p) \subset (\mathscr{A}_R^{[n]})_{c_1}$. Hence

$$0 = [\psi^{[n]} \circ \tilde{\sigma}] = \pi_p(\psi^{[n]})[\tilde{\sigma}] = \pi_p(\psi^{[n]}) \circ \pi_p(\lambda_k)[\sigma] \qquad \text{in } \pi_p((\mathscr{A}_R^{[n]})_{c_2}, (\mathscr{A}_R^{[n]})_{c_1}).$$

□

Proof of Lemma 6.4.2. First of all, let us introduce some notation. For a path $\alpha : [x_0, x_1] \to M$, we denote by $\overline{\alpha} : [x_0, x_1] \to M$ the inverse path

$$\overline{\alpha}(x) = \alpha(x_0 + x_1 - x), \qquad \forall x \in [x_0, x_1].$$

If we consider another path $\beta : [x_0', x_1'] \to M$ with $\alpha(x_1) = \beta(x_0')$, we denote by $\alpha \bullet \beta : [x_0, x_1 + x_1' - x_0'] \to M$ the concatenation of the paths α and β, namely

$$\alpha \bullet \beta(x) = \begin{cases} \alpha(x) & x \in [x_0, x_1], \\ \beta(x - x_1 + x_0') & x \in [x_1, x_1 + x_1' - x_0']. \end{cases}$$

Consider a continuous map $\vartheta : [x_0, x_1] \to W^{1,2}(\mathbb{T}; M)$ where $[x_0, x_1] \subset \mathbb{R}$. For each $n \in \mathbb{N}$, by composition with the iteration map we obtain a map $\vartheta^{[n]} := \psi^{[n]} \circ \vartheta : [x_0, x_1] \to W^{1,2}(\mathbb{T}^{[n]}; M)$. Now, we want to build a continuous map $\vartheta^{\langle n \rangle} : [x_0, x_1] \to W^{1,2}(\mathbb{T}^{[n]}; M)$ as explained in the following.

[2] Notice that $(\Delta^p, \partial\Delta^p)$ is homeomorphic to the pair (D^p, S^{p-1}), so that we can consider the singular simplex σ of the statement as an element of the relative homotopy group $\pi_p((\mathscr{A}_k)_{c_1} \cup \{\boldsymbol{q}\}, (\mathscr{A}_k)_{c_1})$.

To begin with, let us denote by $\mathrm{ev} : W^{1,2}(\mathbb{T};M) \to M$ the **evaluation map**, given by

$$\mathrm{ev}(\zeta) = \zeta(0), \qquad \forall \zeta \in W^{1,2}(\mathbb{T};M).$$

This map is smooth (see the proof of Proposition 3.2.3), and hence the initial point curve $\mathrm{ev} \circ \vartheta : [x_0, x_1] \to M$ is a uniformly continuous path. In particular there exists a constant $\rho(\vartheta) > 0$ such that, for each $x, x' \in [x_0, x_1]$ with $|x - x'| \leq \rho(\vartheta)$, we have

$$\mathrm{dist}(\mathrm{ev} \circ \vartheta(x), \mathrm{ev} \circ \vartheta(x')) < \mathrm{injrad}(M).$$

In this inequality we have denoted by $\mathrm{dist}(\cdot, \cdot)$ the Riemannian distance on M, and by $\mathrm{injrad}(M)$ the injectivity radius of M. Now, for each $x, x' \in [x_0, x_1]$ with $0 \leq x' - x \leq \rho(\vartheta)$, we define the **horizontal geodesic** $\vartheta_x^{x'} : [x, x'] \to M$ as the unique shortest geodesic[3] that connects the points $\mathrm{ev} \circ \vartheta(x)$ and $\mathrm{ev} \circ \vartheta(x')$. Let $J \in \mathbb{N}$ be such that $x_0 + J\rho \leq x_1 \leq x_0 + (J+1)\rho$. For each $x \in [x_0, x_1]$ we further choose $j \in \mathbb{N}$ such that $x_0 + j\rho \leq x \leq x_0 + (j+1)\rho$, and we define the **horizontal broken geodesics** $\vartheta_{x_0}^{x} : [x_0, x] \to M$ and $\vartheta_x^{x_1} : [x, x_1] \to M$ by

$$\begin{aligned}
\vartheta_{x_0}^{x} &:= \vartheta_{x_0}^{x_0+\rho} \bullet \vartheta_{x_0+\rho}^{x_0+2\rho} \bullet \cdots \bullet \vartheta_{x_0+j\rho}^{x},\\
\vartheta_x^{x_1} &:= \vartheta_x^{x_0+(j+1)\rho} \bullet \vartheta_{x_0+(j+1)\rho}^{x_0+(j+2)\rho} \bullet \cdots \bullet \vartheta_{x_0+J\rho}^{x_1}.
\end{aligned}$$

We define a preliminary map $\tilde{\vartheta}^{\langle n \rangle} : [x_0, x_1] \to W^{1,2}(\mathbb{T}^{[n]};M)$ in the following way. For each $j \in \{1, \ldots, n-2\}$ and $y \in [0, \frac{x_1 - x_0}{n}]$ we set

$$\begin{aligned}
\tilde{\vartheta}^{\langle n \rangle}(x_0 + y) &= \vartheta^{[n-1]}(x_0) \bullet \vartheta_{x_0}^{x_0+ny} \bullet \vartheta(x_0 + ny) \bullet \overline{\vartheta_{x_0}^{x_0+ny}},\\
\tilde{\vartheta}^{\langle n \rangle}\big(x_0 + \tfrac{j}{n}(x_1 - x_0) + y\big) &= \vartheta^{[n-j-1]}(x_0) \bullet \vartheta_{x_0}^{x_0+ny} \bullet \vartheta(x_0 + ny) \bullet \vartheta_{x_0+ny}^{x_1}\\
&\quad \bullet \vartheta^{[j]}(x_1) \bullet \overline{\vartheta_{x_0}^{x_1}},\\
\tilde{\vartheta}^{\langle n \rangle}\big(x_0 + \tfrac{n-1}{n}(x_1 - x_0) + y\big) &= \vartheta(x_0 + ny) \bullet \vartheta_{x_0+ny}^{x_1} \bullet \vartheta^{[n-1]}(x_1) \bullet \overline{\vartheta_{x_0+ny}^{x_1}}.
\end{aligned}$$

For each $x \in [x_0, x_1]$, we reparametrize the loop $\tilde{\vartheta}^{\langle n \rangle}(x)$ as follows: in the above formulas, each fixed part $\vartheta(x_0)$ and $\vartheta(x_1)$ spends the original time 1, while the moving parts $\vartheta(x_0 + ny)$ and the pieces of horizontal broken geodesics share the remaining time 1 proportionally to their original parametrizations. We define the map $\vartheta^{\langle n \rangle} : [x_0, x_1] \to W^{1,2}(\mathbb{T}^{[n]};M)$ as the continuous path in the loop space obtained in this manner (see the example in Figure 6.1(a)).

For each $x \in [x_0, x_1]$, we define the **pulling loop** $\widehat{\vartheta}(x) : \mathbb{T} \to M$ as the loop obtained erasing from the formula of $\tilde{\vartheta}^{\langle n \rangle}(x)$ the fixed parts $\vartheta(x_0)$ and $\vartheta(x_1)$

[3]In the following, we will implicitly use the fact that the shortest geodesics lying in $C^\infty([t_0,t_1];M)$ depend smoothly on their endpoints (cf. Theorem 4.1.2).

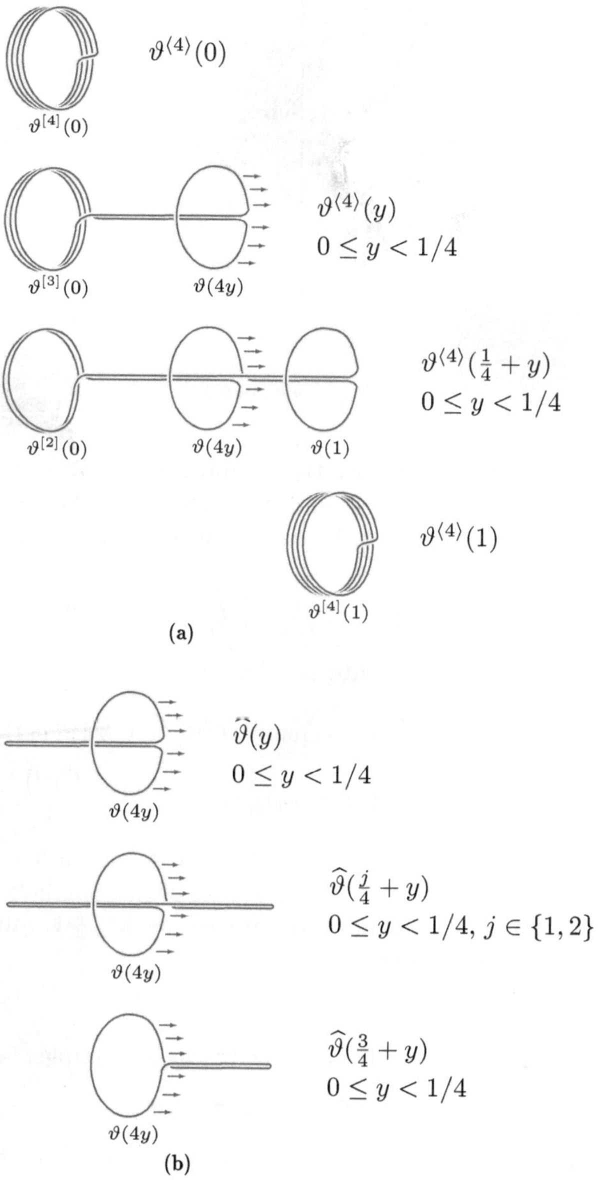

Figure 6.1. **(a)** Description of $\vartheta^{\langle 4\rangle} : [0,1] \to W^{1,2}(\mathbb{T}^{[4]}; M)$, obtained from a continuous map $\vartheta : [0,1] \to W^{1,2}(\mathbb{T}; M)$. Here, for simplicity, we are assuming that the diameter of $\vartheta([x_0, x_1])$ is less than the injectivity radius of M, so that the horizontal geodesics are not broken. The arrows show the direction in which the loop $\vartheta(4y)$ is pulled as y grows. **(b)** Description of the map of pulling loops $\widehat{\vartheta} : [0,1] \to W^{1,2}(\mathbb{T}; M)$.

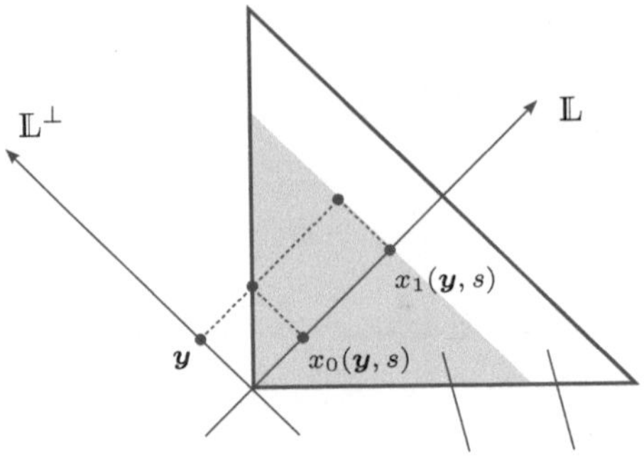

Figure 6.2.

and reparametrizing on $[0,1]$ (see the example in Figure 6.1(b)). Notice that $\widehat{\vartheta}$ is independent of the integer $n \in \mathbb{N}$ and, for each $x \in \mathbb{N}$, the action $\mathscr{A}(\widehat{\vartheta}(x))$ is finite and depends continuously on x. In particular we obtain a finite constant

$$C(\vartheta) := \max_{x\in[x_0,x_1]} \left\{\mathscr{A}(\widehat{\vartheta}(x))\right\} = \max_{x\in[x_0,x_1]} \left\{\int_0^1 \mathscr{L}\left(t, \widehat{\vartheta}(x)(t), \frac{\mathrm{d}}{\mathrm{d}t}\widehat{\vartheta}(x)(t)\right) \mathrm{d}t\right\} < \infty,$$

and, for each $n \subset \mathbb{N}$, the estimate

$$\begin{aligned}\mathscr{A}^{[n]}(\vartheta^{\langle n\rangle}(x)) &\leq \frac{1}{n}\left[(n-1)\max\{\mathscr{A}(\vartheta(x_0)), \mathscr{A}(\vartheta(x_1))\} + \mathscr{A}(\widehat{\vartheta}(x))\right] \\ &\leq \max\{\mathscr{A}(\vartheta(x_0)), \mathscr{A}(\vartheta(x_1))\} + \frac{C(\vartheta)}{n}.\end{aligned} \tag{6.9}$$

Now, following Long [Lon00, page 461], we denote by $\mathbb{L} \subseteq \mathbb{R}^p$ the straight line passing through the origin and the barycenter of the standard p-simplex $\Delta^p \subset \mathbb{R}^p$. We have an orthogonal decomposition of $\mathbb{R}^p$ as $\mathbb{L}^\perp \oplus \mathbb{L}$, and according to this decomposition we can write every $\boldsymbol{z} \in \mathbb{R}^p$ as

$$\boldsymbol{z} = (\boldsymbol{y}, x) \in \mathbb{L}^\perp \oplus \mathbb{L}.$$

For each $s \in [0,1]$ we denote by $s\Delta^p$ the rescaled p-simplex

$$s\Delta^p = \{s\boldsymbol{z} \mid \boldsymbol{z} \in \Delta^p\}.$$

Varying s from 1 to 0 we obtain a deformation retraction of Δ^p onto the origin of $\mathbb{R}^p$. For each $(\boldsymbol{y}, x) \in s\Delta^p$, we denote by $[x_0(\boldsymbol{y},s), x_1(\boldsymbol{y},s)] \subseteq \mathbb{L}$ the maximal interval such that $(\boldsymbol{y}, x')$ belongs to $s\Delta^p$ for all $x' \in [x_0(\boldsymbol{y},s), x_1(\boldsymbol{y},s)]$ (see Figure 6.2).

Consider the p-singular simplex σ from the statement of the lemma. For each $n \in \mathbb{N}$, we define the **Bangert homotopy**

$$\mathrm{Ban}_\sigma^{[n]} : [0,1] \times \Delta^p \to W^{1,2}(\mathbb{T}^{[n]}; M)$$

by

$$\mathrm{Ban}_\sigma^{[n]}(s,\boldsymbol{z}) := \begin{cases} (\sigma|_{[x_0(\boldsymbol{y},s),x_1(\boldsymbol{y},s)]})^{\langle n\rangle}(x) & \boldsymbol{z}=(\boldsymbol{y},x)\in s\Delta^p, \\ \psi^{[n]}\circ\sigma(\boldsymbol{z}) & \boldsymbol{z}\notin s\Delta^p, \end{cases}$$

for each $(s,\boldsymbol{z})\in[0,1]\times\Delta^p$. This homotopy $\mathrm{Ban}_\sigma^{[n]}$ is relative to $\partial\Delta^p$, for

$$\mathrm{Ban}_\sigma^{[n]}(s,\boldsymbol{z}) = \psi^{[n]}\circ\sigma(\boldsymbol{z}), \qquad \forall(s,\boldsymbol{z})\in[0,1]\times\partial\Delta^p,$$

and clearly $\mathrm{Ban}_\sigma^{[n]}(0,\cdot) = \psi^{[n]}\circ\sigma$, proving assertion (i) of the lemma. By the assumptions on σ, there exists $\varepsilon>0$ such that

$$\max_{\boldsymbol{z}\in\Delta^p}\mathscr{A}(\sigma(\boldsymbol{z}))\leq c_2-\varepsilon, \qquad \max_{\boldsymbol{z}\in\partial\Delta^p}\mathscr{A}(\sigma(\boldsymbol{z}))\leq c_1-\varepsilon.$$

For each $s\in[0,1]$, $n\in\mathbb{N}$ and $\boldsymbol{z}=(\boldsymbol{y},x)\in s\Delta^p$, by the estimate in (6.9) we have

$$\mathscr{A}^{[n]}(\mathrm{Ban}_\sigma^{[n]}(s,\boldsymbol{z}))\leq\max\left\{\mathscr{A}(\sigma(x_0(\boldsymbol{y},s))),\mathscr{A}(\sigma(x_1(\boldsymbol{y},s)))\right\}+\frac{C(\sigma|_{[x_0(\boldsymbol{y},s),x_1(\boldsymbol{y},s)]})}{n},$$

while, for each $\boldsymbol{z}\in\Delta^p\setminus s\Delta^p$, we have

$$\mathscr{A}^{[n]}(\mathrm{Ban}_\sigma^{[n]}(s,\boldsymbol{z}))=\mathscr{A}(\sigma(\boldsymbol{z})).$$

In particular, there exists a finite constant

$$C(\sigma):=\max\left\{C(\sigma|_{[x_0(\boldsymbol{y},s),x_1(\boldsymbol{y},s)]})\;\middle|\;s\in[0,1],\ (\boldsymbol{y},x)\in s\Delta^p\right\}$$

such that, for each $n\in\mathbb{N}$ and $(s,\boldsymbol{z})\in[0,1]\times\Delta^p$, we have

$$\begin{aligned}\mathscr{A}^{[n]}(\mathrm{Ban}_\sigma^{[n]}(s,\boldsymbol{z}))&\leq\max_{\boldsymbol{w}\in\Delta^p}\left\{\mathscr{A}(\sigma(\boldsymbol{w}))\right\}+\frac{C(\sigma)}{n}\leq c_2-\varepsilon+\frac{C(\sigma)}{n},\\ \mathscr{A}^{[n]}(\mathrm{Ban}_\sigma^{[n]}(1,\boldsymbol{z}))&\leq\max_{\boldsymbol{w}\in\partial\Delta^p}\left\{\mathscr{A}(\sigma(\boldsymbol{w}))\right\}+\frac{C(\sigma)}{n}\leq c_1-\varepsilon+\frac{C(\sigma)}{n},\end{aligned}$$

proving assertion (ii). Finally, for all $s\in[0,1]$ and $\boldsymbol{z}=(\boldsymbol{y},x)\in s\Delta^p$, there is a uniform bound $\bar{r}'(\sigma)>0$ for the derivative of the pulling loops associated to the paths $\sigma|_{[x_0(\boldsymbol{y},s),x_1(\boldsymbol{y},s)]}$. Therefore, for $\bar{R}(\sigma):=\max\{\bar{r}(\sigma),\bar{r}'(\sigma)\}$, assertion (iii) follows. □

The idea of the proof of the homological vanishing theorem (Theorem 6.4.1) is to apply the Bangert homotopy lemma successively to all the faces of the singular simplices that compose the relative cycle η. In order to do this, we will make use of the following elementary result from algebraic topology, that is stated in [BK83, Lemma 1]. For the sake of completeness, we include a detailed proof here (basically, the argument is a variation of the one that proves that singular homology is a homotopy invariant, see for instance [Hat02, page 112]).

Lemma 6.4.4. *Let (X,Y) be a pair of topological spaces, μ a relative p-cycle*[4] *in (X,Y) and $\Sigma(\mu)$ the set of singular simplices that make up μ together with their faces. Suppose that, for each singular simplex $\sigma : \Delta^q \to X$ that belongs to $\Sigma(\mu)$, where $0 \le q \le p$, there exists a homotopy*

$$P_\sigma : \Delta^q \times [0,1] \to X$$

such that

(i) $P_\sigma(\cdot, 0) = \sigma$;

(ii) $P_\sigma(\cdot, s) = \sigma$ *for each* $s \in [0,1]$, *if* $\sigma(\Delta^q) \subset Y$;

(iii) $P_\sigma(\Delta^q \times \{1\}) \subset Y$;

(iv) $P_\sigma(F_j(\cdot), \cdot) = P_{\sigma \circ F_j}(\cdot, \cdot)$ *for each* $j = 0, \dots, p$, *where* $F_j : \Delta^{q-1} \to \Delta^q$ *is the standard affine map onto the* j^{th}*-face of* Δ^q.

Then $[\mu] = 0$ *in* $\mathrm{H}_p(X,Y)$.

Proof. If $\boldsymbol{v}_0, \dots, \boldsymbol{v}_h$ are points in $\mathbb{R}^q$, we will denote by $\langle \boldsymbol{v}_0, \dots, \boldsymbol{v}_h \rangle$ their convex hull, that is, the minimal convex closed subset of $\mathbb{R}^q$ containing these points. We will denote by ∂ the usual boundary operator from algebraic topology, i.e., $\partial \langle \boldsymbol{v}_0, \dots, \boldsymbol{v}_h \rangle$ is the following formal sum of $(h-1)$-simplices

$$\partial \langle \boldsymbol{v}_0, \dots, \boldsymbol{v}_h \rangle := \sum_{j=0}^{h} (-1)^j \langle \boldsymbol{v}_0, \dots, \widehat{\boldsymbol{v}}_j, \dots, \boldsymbol{v}_h \rangle ,$$

where, as usual, we add a hat over $\boldsymbol{v}_j$ to indicate that it is missing in the corresponding term, that is

$$\langle \boldsymbol{v}_0, \dots, \widehat{\boldsymbol{v}}_j, \dots, \boldsymbol{v}_h \rangle := \langle \boldsymbol{v}_0, \dots, \boldsymbol{v}_{j-1}, \boldsymbol{v}_{j+1}, \dots, \boldsymbol{v}_h \rangle .$$

If $\boldsymbol{e}_1, \dots, \boldsymbol{e}_q$ is the standard basis of $\mathbb{R}^q$ and $\boldsymbol{e}_0$ is the origin, then $\langle \boldsymbol{e}_0, \dots, \boldsymbol{e}_q \rangle$ is the standard q-simplex Δ^q. We identify $\mathbb{R}^q$ with $\mathbb{R}^q \times \{0\} \subset \mathbb{R}^{q+1}$ and we define $\boldsymbol{f}_j = (\boldsymbol{e}_j, 1)$ for each $j \in \{0, \dots, q\}$. The product $\Delta^q \times [0,1]$ can be decomposed as the union of $(q+1)$-simplices as follows:

$$\Delta^q \times [0,1] = \bigcup_{j=0}^{q} \langle \boldsymbol{e}_0, \dots, \boldsymbol{e}_j, \boldsymbol{f}_j, \dots, \boldsymbol{f}_q \rangle .$$

For each $j \in \{0, \dots, q-1\}$, $\langle \boldsymbol{e}_0, \dots, \boldsymbol{e}_j, \boldsymbol{f}_j, \dots, \boldsymbol{f}_q \rangle$ intersects $\langle \boldsymbol{e}_0, \dots, \boldsymbol{e}_{j+1}, \boldsymbol{f}_{j+1}, \dots, \boldsymbol{f}_q \rangle$ in the q-simplex face $\langle \boldsymbol{e}_0, \dots, \boldsymbol{e}_j, \boldsymbol{f}_{j+1}, \dots, \boldsymbol{f}_q \rangle$ (see the example in Figure 6.3).

[4]Namely $[\mu] \in \mathrm{H}_p(X,Y)$.

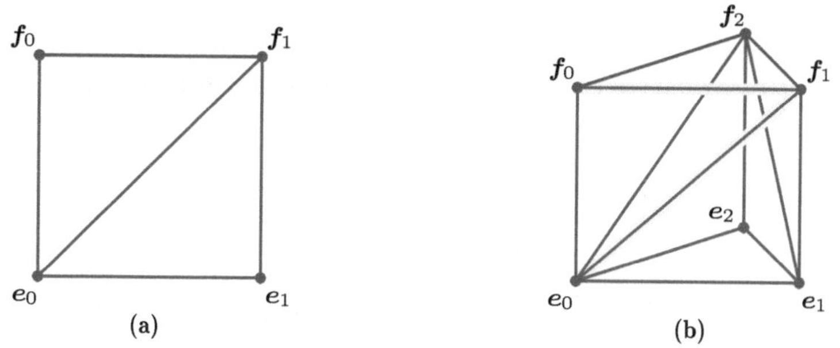

Figure 6.3. (a) Decomposition of $\Delta^1 \times [0,1] = [0,1] \times [0,1]$ into 2-simplices. (b) Decomposition of $\Delta^2 \times [0,1]$ into 3-simplices.

After these preliminaries, consider the relative cycle μ of the lemma. For each q-simplex $\sigma : \Delta^q \to X$ that belongs to $\Sigma(\mu)$, we define an associated $q+1$-chain p_σ by

$$p_\sigma = \sum_{j=0}^{q} (-1)^j P_\sigma|_{\langle \boldsymbol{e}_0,\ldots,\boldsymbol{e}_j,\boldsymbol{f}_j,\ldots,\boldsymbol{f}_q\rangle}.$$

We extend this definition for chains η that are made up of simplices that belong to $\Sigma(\mu)$, in the following way: if we write η as a formal sum of q-simplices

$$\eta = \sum_{v=1}^{V} \eta_v,$$

then we set

$$p_\eta := \sum_{v=1}^{V} p_{\eta_v}.$$

Notice that p_η is well defined: if η_v and $\eta_{v'}$ have a common face, say $\eta_v \circ F_j = \eta_{v'} \circ F_{j'}$, then assumption (iv) guarantees

$$p_{\eta_v} \circ F_j = p_{\eta_v \circ F_j} = p_{\eta_{v'} \circ F_{j'}} = p_{\eta_{v'}} \circ F_{j'}.$$

We define a chain $\tilde{\mu}$ that is "homotopic" to the relative cycle μ (by the homotopies given in the lemma) in the following way: if we write μ as a formal sum of q-simplices

$$\mu = \sum_{w=1}^{W} \mu_w,$$

then we set

$$\tilde{\mu} := \sum_{w=1}^{W} \tilde{\mu}_w,$$

where $\tilde{\mu}_w = P_{\mu_w}(\cdot,1) : \Delta^p \to Y$ for each $w \in \{1,\dots,W\}$. Here, the fact that $\tilde{\mu}_w(\Delta^p) \subset Y$ for each $w \in \{1,\dots,W\}$ is guaranteed by assumption (iii), and it implies that $\tilde{\mu}$ is a relative cycle whose homology class in $\mathrm{H}_p(X,Y)$ is zero, i.e.,

$$[\tilde{\mu}] = 0 \qquad \text{in } \mathrm{H}_p(X,Y). \tag{6.10}$$

Now,

$$\partial p_\mu = \sum_{w=1}^{W} \partial p_{\mu_w} = \sum_{w=1}^{W} \Bigg[\sum_{j\leq i} (-1)^{i+j} P_{\mu_w}|_{\langle e_0,\dots,\widehat{e}_j,\dots,e_i,\boldsymbol{f}_i,\dots,\boldsymbol{f}_q\rangle} + \sum_{j\geq i} (-1)^{i+j+1} P_{\mu_w}|_{\langle e_0,\dots,e_i,\boldsymbol{f}_i,\dots,\widehat{\boldsymbol{f}}_j,\dots,\boldsymbol{f}_q\rangle} \Bigg].$$

The terms with $i=j$ in the two inner sums cancel, except for $P_{\mu_w}|_{\langle \widehat{e}_0,\boldsymbol{f}_0,\dots,\boldsymbol{f}_q\rangle} = \tilde{\mu}_w$ and $-P_{\mu_w}|_{\langle e_0,\dots,e_q,\widehat{\boldsymbol{f}}_q\rangle} = -\mu_w$. The terms with $i \neq j$ are precisely $p_{\partial\mu_w}$, for

$$p_{\partial\mu_w} = \sum_{i<j} (-1)^{i+j} P_{\mu_w}|_{\langle e_0,\dots,e_i,\boldsymbol{f}_i,\dots,\widehat{\boldsymbol{f}}_j,\dots,\boldsymbol{f}_q\rangle} + \sum_{i>j} (-1)^{i-1+j} P_{\mu_w}|_{\langle e_0,\dots,\widehat{e}_j,\dots,e_i,\boldsymbol{f}_i,\dots,\boldsymbol{f}_q\rangle}.$$

Thus we have shown

$$\tilde{\mu} - \mu = \partial p_\mu - p_{\partial\mu}. \tag{6.11}$$

The above equality must be understood in the pth-relative chain group of (X,Y). Since μ is a relative cycle, assumption (ii) implies that the singular simplices that make up $p_{\partial\mu}$ have image inside Y. Hence $p_{\partial\mu}$ is a relative cycle whose homology class in $\mathrm{H}_p(X,Y)$ is zero, i.e.,

$$[p_{\partial\mu}] = 0 \qquad \text{in } \mathrm{H}_p(X,Y). \tag{6.12}$$

By (6.10), (6.11) and (6.12), we conclude

$$[\mu] = [\tilde{\mu}] - [\partial p_\mu] + [p_{\partial\mu}] = 0 \qquad \text{in } \mathrm{H}_p(X,Y). \qquad \square$$

Proof of Theorem 6.4.1. Let $U > 0$ be the constant chosen in (6.4). We denote by $\Sigma(\eta)$ the set of singular simplices in η together with all their faces, and by $\mathbb{K} \subset \mathbb{N}$ the set of nonnegative integer powers of j, i.e., $\mathbb{K} = \{j^n \mid n \in \mathbb{N} \cup \{0\}\}$. For each singular simplex $\sigma : \Delta^p \to (\mathscr{A}_k)_{c_1} \cup \{\boldsymbol{q}\}$ that belongs to $\Sigma(\eta)$ we will find an integer $\bar{n} = \bar{n}(\sigma,j) \in \mathbb{K}$, a positive real $\bar{R} = \bar{R}(\sigma,j) \geq U$ and a homotopy

$$P_\sigma^{[\bar{n}]} : [0,1] \times \Delta^p \to W^{1,2}(\mathbb{T}^{[\bar{n}]};M)$$

such that

(i) $P_\sigma^{[\bar{n}]}(0,\cdot) = \psi^{[\bar{n}]} \circ \lambda_k \circ \sigma$;

(ii) if $\sigma(\Delta^p) \subset (\mathscr{A}_k)_{c_1}$, then $P_\sigma^{[\bar{n}]}(s,\cdot) = \psi^{[\bar{n}]} \circ \lambda_k \circ \sigma$ for each $s \in [0,1]$;

(iii) $\mathscr{A}^{[\bar{n}]}(P_\sigma^{[\bar{n}]}(s,\boldsymbol{z})) < c_2$ and $\mathscr{A}^{[\bar{n}]}(P_\sigma^{[\bar{n}]}(1,\boldsymbol{z})) < c_1$ for each $(s,\boldsymbol{z}) \in [0,1] \times \Delta^p$;

(iv) $P_\sigma^{[\bar{n}]}(\cdot, F_i(\cdot)) = P_{\sigma\circ F_i}^{[\bar{n}]}(\cdot,\cdot)$ for each $i = 0,\dots,p$, where $F_i : \Delta^{p-1} \to \Delta^p$ is the standard affine map onto the ith-face of Δ^p.

(v) $$\sup_{(s,\boldsymbol{z})\in[0,1]\times\Delta^p} \ \operatorname*{ess\,sup}_{t\in\mathbb{T}^{[\bar{n}]}} \left\{ \left| \frac{\mathrm{d}}{\mathrm{d}t} P_\sigma^{[\bar{n}]}(s,\boldsymbol{z})(t) \right|_{P_\sigma^{[\bar{n}]}(s,\boldsymbol{z})(t)} \right\} < \bar{R}.$$

For each $n \in \mathbb{K}$ greater than $\bar{n}$, we define a homotopy

$$P_\sigma^{[n]} : [0,1] \times \Delta^q \to W^{1,2}(\mathbb{T}^{[n]}; M)$$

by $P_\sigma^{[n]} := \psi^{[n/\bar{n}]} \circ P_\sigma^{[\bar{n}]}$. This homotopy satisfies the analogous properties (i),...,(v) in period n. Notice that property (iv) implicitly requires that $\bar{n}(\sigma,j) \geq \bar{n}(\sigma\circ F_i, j)$ for each $i = 0,\dots,p$.

Now, assume that such homotopies exist and set

$$\begin{aligned} \bar{R} = \bar{R}([\eta],j) &:= \max\{\bar{R}(\sigma,j) \mid \sigma \in \Sigma(\eta)\}, \\ \bar{n} = \bar{n}([\eta],j) &:= \max\{\bar{n}(\sigma,j) \mid \sigma \in \Sigma(\eta)\}. \end{aligned}$$

For each $R \geq \bar{R}$, the set of homotopies $\{P_\sigma^{[\bar{n}]} \mid \sigma \in \Sigma(\eta)\}$ satisfies the hypotheses of Lemma 6.4.4 with respect to the relative cycle $\psi^{[\bar{n}]} \circ \lambda_k \circ \eta$ in $((\mathscr{A}_R^{[\bar{n}]})_{c_2}, (\mathscr{A}_R^{[\bar{n}]})_{c_1})$, and we conclude

$$\mathrm{H}_*(\psi^{[\bar{n}]}) \circ \mathrm{H}_*(\lambda_k)[\eta] = [\psi^{[\bar{n}]} \circ \lambda_k \circ \eta] = 0 \qquad \text{in } \mathrm{H}_*((\mathscr{A}_R^{[\bar{n}]})_{c_2}, (\mathscr{A}_R^{[\bar{n}]})_{c_1}).$$

In order to conclude the proof, it remains to build the above homotopies. We do it inductively on the dimension of the relative cycle η. If η is a 0-relative cycle, $\Sigma(\eta)$ is a finite set of points $\{\boldsymbol{w}_1,\dots,\boldsymbol{w}_h\}$ that is contained in $(\mathscr{A}_k)_{c_1} \cup \{\boldsymbol{q}\}$. Since we are assuming that $\boldsymbol{q}$ is not a minimum, the sublevel $(\mathscr{A}_k)_{c_1}$ is not empty. Hence, for each $\boldsymbol{w} \in \Sigma(\eta)$, we can find a path $\Gamma_{\boldsymbol{w}} : [0,1] \to (\mathscr{A}_k)_{c_1} \cup \{\boldsymbol{q}\}$ such that $\Gamma_{\boldsymbol{w}}(0) = \boldsymbol{w}$ and $\Gamma_{\boldsymbol{w}}(s) \in (\mathscr{A}_k)_{c_1}$ for each $s \in (0,1]$ (if $\boldsymbol{w} \in (\mathscr{A}_k)_{c_1}$, we simply choose $\Gamma_{\boldsymbol{w}}(s) := \boldsymbol{w}$ for each $s \in [0,1]$). We set $\bar{R}(\boldsymbol{w},j) := U$, $\bar{n}(\boldsymbol{w},j) := 1$ and

$$P_{\boldsymbol{w}}^{[1]} := \lambda_k \circ \Gamma_{\boldsymbol{w}} : [0,1] \times \{\boldsymbol{0}\} \to W^{1,2}(\mathbb{T}; M).$$

If η is a p-relative cycle, with $p \geq 1$, we can apply the inductive hypothesis: for each nonnegative integer $i < p$ and for each i-singular simplex $\nu \in \Sigma(\eta)$, we obtain $\bar{n}(\nu,j) \in \mathbb{K}$, $\bar{R}(\nu,j) \geq U$ and, for each $n \in \mathbb{K}$ greater than or equal to $\bar{n}(\nu,j)$, a homotopy $P_\nu^{[n]}$ satisfying the above properties (i),...,(v). Now, consider a p-singular simplex $\sigma : \Delta^p \to (\mathscr{A}_k)_{c_1} \cup \{\boldsymbol{q}\}$ that belongs to $\Sigma(\eta)$. If $\sigma(\Delta^p) \subset (\mathscr{A}_k)_{c_1}$ we simply set $\bar{R}(\sigma,j) := U$, $\bar{n}(\sigma,j) := 1$ and $P_\sigma^{[1]}(s,\cdot) := \lambda_k \circ \sigma$ for each $s \in [0,1]$. In the other case, $\sigma(\Delta^p) \not\subset (\mathscr{A}_k)_{c_1}$, we proceed as follows. We denote by $\bar{R}' = \bar{R}'(\sigma,j)$

and $\bar{n}' = \bar{n}'(\sigma, j)$ respectively the maximum of the $\bar{R}(\nu, j)$'s and $\bar{n}(\nu, j)$'s for all the proper faces ν of σ. Thus, for each $n \in \mathbb{K}$ greater than or equal to $\bar{n}'(\sigma, j)$, every proper face ν of σ has an associated homotopy

$$P_\nu^{[n]} : [0,1] \times \Delta^{p-1} \to W^{1,2}(\mathbb{T}^{[n]}; M).$$

For technical reasons which will become clear shortly we assume that

$$P_\nu^{[n]}(s,\cdot) = P_\nu^{[n]}(\tfrac{1}{2},\cdot), \qquad s \in [\tfrac{1}{2}, 1].$$

Patching together the homotopies of the proper faces of σ, we obtain

$$P_\sigma^{[n]} : ([0,\tfrac{1}{2}] \times \partial\Delta^p) \cup (0 \times \Delta^p) \to W^{1,2}(\mathbb{T}^{[n]}; M), \qquad \forall n \in \mathbb{K},\ n \geq \bar{n}',$$

such that $P_\sigma^{[n]}(0,\cdot) = \psi^{[n]} \circ \lambda_k \circ \sigma$ and $P_\sigma^{[n]}(\cdot, F_i(\cdot)) = P_{\sigma\circ F_i}^{[n]}$ for all $i = 1, \ldots, p$. By retracting $([0,\frac{1}{2}]\times\Delta^p)$ onto $([0,\frac{1}{2}]\times\partial\Delta^p)\cup(0\times\Delta^p)$, we can extend the homotopies $P_\sigma^{[n]}$ to the whole of $([0,\frac{1}{2}] \times \Delta^p)$, obtaining

$$P_\sigma^{[n]} : [0,\tfrac{1}{2}] \times \Delta^p \to W^{1,2}(\mathbb{T}^{[n]}; M), \qquad \forall n \in \mathbb{K},\ n \geq \bar{n}'. \tag{6.13}$$

Let us denote the singular simplex $P_\sigma^{[\bar{n}']}(\frac{1}{2},\cdot) : \Delta^p \to W^{1,2}(\mathbb{T}^{[\bar{n}']}; M)$ simply by $\tilde{\sigma}$. Note that

$$\max_{\boldsymbol{z}\in\Delta^p} \left\{ \mathscr{A}^{[\bar{n}']}(\tilde{\sigma}(\boldsymbol{z})) \right\} < c_2,$$

$$\max_{\boldsymbol{z}\in\partial\Delta^p} \left\{ \mathscr{A}^{[\bar{n}']}(\tilde{\sigma}(\boldsymbol{z})) \right\} < c_1,$$

$$\sup_{\boldsymbol{z}\in\Delta^p} \ \operatorname*{ess\,sup}_{t\in\mathbb{T}^{[\bar{n}']}} \left\{ \left| \frac{\mathrm{d}}{\mathrm{d}t}\tilde{\sigma}(\boldsymbol{z})(t) \right|_{\tilde{\sigma}(\boldsymbol{z})(t)} \right\} \leq \bar{R}'.$$

Hence we can apply Lemma 6.4.2, obtaining an integer $\bar{n}(\tilde{\sigma})$, a positive real $\bar{R}(\tilde{\sigma}) \geq \bar{R}'(\sigma, j)$ and, if we choose the smallest $\bar{n}'' \in \mathbb{K}$ greater than or equal to $\bar{n}(\tilde{\sigma})$, a Bangert homotopy

$$\mathrm{Ban}_{\tilde{\sigma}}^{[\bar{n}'']} : [0,1] \times \Delta^p \to W^{1,2}(\mathbb{T}^{[\bar{n}'\bar{n}'']}; M) \qquad \text{relative to } \partial\Delta^p$$

such that $\mathrm{Ban}_{\tilde{\sigma}}^{[\bar{n}'']}(0,\cdot) = \psi^{[\bar{n}'']} \circ \tilde{\sigma}$ and

$$\max_{(s,\boldsymbol{z})\in[0,1]\times\Delta^p} \left\{ \mathscr{A}^{[\bar{n}'\bar{n}'']}(\mathrm{Ban}_{\tilde{\sigma}}^{[\bar{n}'']}(s,\boldsymbol{z})) \right\} < c_2,$$

$$\max_{(s,\boldsymbol{z})\in[0,1]\times\partial\Delta^p} \left\{ \mathscr{A}^{[\bar{n}'\bar{n}'']}(\mathrm{Ban}_{\tilde{\sigma}}^{[\bar{n}'']}(s,\boldsymbol{z})) \right\} < c_1,$$

$$\max_{\boldsymbol{z}\in\Delta^p} \left\{ \mathscr{A}^{[\bar{n}'\bar{n}'']}(\mathrm{Ban}_{\tilde{\sigma}}^{[\bar{n}'']}(1,\boldsymbol{z})) \right\} < c_1,$$

$$\sup_{(s,\boldsymbol{z})\in[0,1]\times\Delta^p} \ \operatorname*{ess\,sup}_{t\in\mathbb{T}^{[\bar{n}'\bar{n}'']}} \left\{ \left| \frac{\mathrm{d}}{\mathrm{d}t}\mathrm{Ban}_{\tilde{\sigma}}^{[\bar{n}'']}(s,\boldsymbol{z})(t) \right|_{\mathrm{Ban}_{\tilde{\sigma}}^{[\bar{n}'']}(s,\boldsymbol{z})(t)} \right\} \leq \bar{R}(\sigma).$$

Finally, we set $\bar{n} = \bar{n}(\sigma, j) := \bar{n}'\bar{n}''$, $\bar{R}(\sigma, j) := \bar{R}(\tilde{\sigma})$ and we build the homotopy $P_\sigma^{[\bar{n}]} : [0,1] \times \Delta^p \to W^{1,2}(\mathbb{T}^{[\bar{n}]}; M)$ extending the one in (6.13) by

$$P_\sigma^{[\bar{n}]}(s, \cdot) := \mathrm{Ban}_{\tilde{\sigma}}^{[\bar{n}'']}(2s - 1, \cdot), \qquad \forall s \in [\tfrac{1}{2}, 1]. \qquad \square$$

6.5 The Conley conjecture

In this final section we prove a theorem about the multiplicity of periodic orbits with unprescribed period that confirms the **Conley conjecture** for Lagrangian systems associated to Tonelli Lagrangians with global Euler-Lagrange flow. This theorem was first established, in the case of fiberwise quadratic Lagrangians on the m-torus, by Long [Lon00, Theorem 1.1]. Later, it was extended by Lu [Lu09, Corollary 1.2] to the class of convex quadratic-growth Lagrangians on general closed configuration spaces. The version that we give here is due to the author [Maz11, Theorem 1.1].

Theorem 6.5.1. *Let M be a smooth closed manifold of positive dimension, and $\mathscr{L} : \mathbb{T} \times \mathrm{T}M \to \mathbb{R}$ a 1-periodic Tonelli Lagrangian with global Euler-Lagrange flow. Fix a constant $a \in \mathbb{R}$ greater than*

$$\max_{q \in M} \left\{ \int_0^1 \mathscr{L}(t, q, 0)\, \mathrm{d}t \right\}. \tag{6.14}$$

Assume that only finitely many contractible 1-periodic solutions of the Euler-Lagrange system of $\mathscr{L}$ have action less than a. Then for each prime number p the Euler-Lagrange system of $\mathscr{L}$ admits infinitely many contractible periodic solutions with period that is a non-negative integer power of p and mean action less than a.

Here, it is worthwhile to stress that the infinitely many periodic orbits that are found in the above statement are **geometrically distinct**, i.e., none of them is an iteration of another one.

If we consider the Tonelli Hamiltonian $\mathscr{H}$ that is Legendre-dual to the Tonelli Lagrangian $\mathscr{L}$ (see Sections 1.1 and 1.2), Theorem 6.5.1 can be rephrased in the Hamiltonian formulation. The two statements are completely equivalent.

Theorem 6.5.2 (Hamiltonian formulation). *Let M be a smooth closed manifold of positive dimension, and $\mathscr{H} : \mathbb{T} \times \mathrm{T}^*M \to \mathbb{R}$ a 1-periodic Tonelli Hamiltonian with global Hamiltonian flow. Fix a constant $a \in \mathbb{R}$ greater than*

$$-\min_{q \in M} \left\{ \int_0^1 \min_{p \in \mathrm{T}_q^*M} \{\mathscr{H}(t, q, p)\}\, \mathrm{d}t \right\}.$$

Assume that only finitely many contractible 1-periodic solutions of the Hamilton system of $\mathscr{H}$ have action less than a. Then for each prime number p the Hamilton system of $\mathscr{H}$ admits infinitely many contractible periodic solutions with period that is a non-negative integer power of p and mean action less than a.

In this statement, the mean action of a periodic orbit $\Gamma : \mathbb{T}^{[n]} \to \mathrm{T}^*M$ is its **Hamiltonian mean action**, defined by

$$\frac{1}{n}\int_0^n \Big(\Gamma^*\lambda - \mathscr{H}(t,\Gamma(t))\Big)\,\mathrm{d}t,$$

where λ is the Liouville form on T^*M (see Section 1.1). This quantity coincides with the usual Lagrangian mean action of the associated Lagrangian periodic orbit, i.e., if $\mathscr{L}$ is Legendre-dual to $\mathscr{H}$ and we write Γ as (γ,ρ), where $\gamma : \mathbb{T}^{[n]} \to M$ is a periodic solution of the Euler-Lagrange system of $\mathscr{L}$, we have

$$\begin{aligned}\frac{1}{n}\int_0^n \Big(\Gamma^*\lambda - \mathscr{H}(t,\Gamma(t))\Big)\,\mathrm{d}t &= \frac{1}{n}\int_0^n \Big(\rho(t)[\dot\gamma(t)] - \mathscr{H}(t,\gamma(t),\rho(t))\Big)\,\mathrm{d}t\\ &= \frac{1}{n}\int_0^n \mathscr{L}(t,\gamma(t),\dot\gamma(t))\,\mathrm{d}t.\end{aligned}$$

Before proving Theorem 6.5.1, we will introduce a notion that will be crucial for the proof. Let $\mathscr{L} : \mathbb{T}\times \mathrm{T}M \to \mathbb{R}$ be a Tonelli Lagrangian with global Euler-Lagrange flow, and let γ be a periodic solution of the Euler-Lagrange system of $\mathscr{L}$, say of period 1 for simplicity. Let $\mathscr{A}_k = \mathscr{A}\circ\lambda_k$ be the discrete Tonelli action associated to $\mathscr{L}$ (see Section 6.3), and $\boldsymbol{q}$ the critical point of $\mathscr{A}_k$ corresponding to γ, i.e., $\gamma = \lambda_k(\boldsymbol{q})$. We say that γ is **homologically visible** in degree d when the local homology group $\mathrm{C}_d(\mathscr{A}_k,\boldsymbol{q})$ is nontrivial. By Lemma 6.3.2, this is equivalent to the local homology group $\mathrm{C}_d(\mathscr{A}_R,\gamma)$ being nontrivial for all sufficiently large $R>0$. Hereafter, all the homology groups that appear are assumed to have coefficients in the finite field $\mathbb{Z}_2$. Let p be a prime number, and let us denote by $\mathbb{K}_p$ the set of non-negative integer powers of p, i.e.,

$$\mathbb{K}_p = \big\{p^j \;\big|\; j\in\mathbb{N}\cup\{0\}\big\}.$$

We say that γ is **strongly homologically visible** in degree d and periods $\mathbb{K}_p$ when, for infinitely many periods $n\in\mathbb{K}_p$, its iteration $\gamma^{[n]}$ is homologically visible in degree d. Strongly homologically visible periodic orbits have a very intriguing property: their existence automatically implies the existence of infinitely many other periodic orbits with close mean action (however, at the time in which this monograph is being written, it is not known whether the other orbits accumulate on the strongly homologically visible one). In the setting of closed geodesics, this important result is due to Bangert and Klingenberg [BK83, Theorem 3], and it is implicitly contained in the work of Long [Lon00] as well as in the subsequent ones [Lu09, Maz11].

Theorem 6.5.3. *Let p be a prime number and γ a periodic orbit with period $p' \in \mathbb{K}_p$. Assume that γ is strongly homologically visible in some degree $d\geq 1$ and periods $\mathbb{K}_p$. Then there exists a sequence $\{\gamma_v \,|\, v\in\mathbb{N}\}$ of geometrically distinct periodic orbits such that:*

- *γ_v has period $p_v \in \mathbb{K}_p$,*
- *the iterated periodic orbits $\gamma_v^{[p']}$ and $\gamma^{[p_v]}$ are homotopic,*
- *$\mathscr{A}^{[p_v]}(\gamma_v) \to \mathscr{A}^{[p']}(\gamma)$ as $v \to \infty$.*

Proof. Without loss of generality, we can assume that γ has period $p' = 1$. This can be easily achieved by time-rescaling the Lagrangian $\mathscr{L}$ in the following way. Let $\widetilde{\mathscr{L}} : \mathbb{T} \times \mathrm{T}M \to \mathbb{R}$ be the Tonelli Lagrangian given by

$$\widetilde{\mathscr{L}}(t, q, v) := \mathscr{L}(p't, q, \tfrac{1}{p'}v), \qquad \forall (t, q, v) \in \mathbb{T} \times \mathrm{T}M.$$

For each $n \in \mathbb{N}$, a curve $\widetilde{\zeta} : \mathbb{R} \to M$ is an n-periodic solution of the Euler-Lagrange system of $\widetilde{\mathscr{L}}$ if and only if the reparametrized curve $\zeta : \mathbb{R} \to M$, given by

$$\zeta(t) := \widetilde{\zeta}(\tfrac{1}{p'}t), \qquad \forall t \in \mathbb{R},$$

is a $p'n$-periodic solution of the Euler-Lagrange system of $\mathscr{L}$. Moreover, $\widetilde{\zeta}$ and ζ have the same mean action with respect to the Lagrangians $\widetilde{\mathscr{L}}$ and $\mathscr{L}$ respectively.

Let $\mathbb{K}'$ be the infinite subset of those $n \in \mathbb{K}_p$ such that the iterated orbit $\gamma^{[n]}$ is homologically visible in degree d. Consider the constant $U > 0$ associated to γ as in (6.4). By Lemma 6.3.2,

$$\mathrm{C}_d(\mathscr{A}_R^{[n]}, \gamma^{[n]}) \neq 0, \qquad \forall R \geq U,\ n \in \mathbb{K}'.$$

We recall that the Morse index and nullity of $\mathscr{A}_R^{[n]}$ at $\gamma^{[n]}$ are the same as the Morse index and nullity of the Tonelli mean action $\mathscr{A}^{[n]}$ at $\gamma^{[n]}$ (see the paragraph before Lemma 6.1.2), and we denote them simply by $\mathrm{ind}(\gamma^{[n]})$ and $\mathrm{nul}(\gamma^{[n]})$ respectively. From now on we can assume that $\gamma^{[n]}$ is an isolated critical point of $\mathscr{A}_R^{[n]}$ (indeed, if $\gamma^{[n]}$ is a non-isolated critical point the assertion of the theorem readily follows). By Corollary 4.5.4 we infer that $\mathrm{ind}(\gamma^{[n]}) \leq d$ for each $n \in \mathbb{K}'$. Hence, Proposition 2.2.4(iii) implies that $\overline{\mathrm{ind}}(\gamma) = 0$, and Proposition 2.2.4(ii) further gives

$$\mathrm{ind}(\gamma^{[n]}) = 0, \qquad \forall n \in \mathbb{N}. \tag{6.15}$$

By Proposition 2.2.4(i) we know that the nullity of a periodic orbit is always less than or equal to $2\dim(M)$. Hence, by the pigeonhole principle we can find an infinite subset $\mathbb{K}'' \subseteq \mathbb{K}'$ such that

$$\mathrm{nul}(\gamma^{[n_1]}) = \mathrm{nul}(\gamma^{[n_2]}), \qquad \forall n_1, n_2 \in \mathbb{K}''. \tag{6.16}$$

Up to time-rescaling the Lagrangian $\mathscr{L}$ as before, we can assume that $1 \in \mathbb{K}''$. By (6.15), (6.16) and Corollary 5.5.2, the iteration map induces the homology isomorphism

$$\mathrm{H}_*(\psi^{[n]}) : \mathrm{C}_*(\mathscr{A}_R, \gamma) \xrightarrow{\simeq} \mathrm{C}_*(\mathscr{A}_R^{[n]}, \gamma^{[n]}), \qquad \forall n \in \mathbb{K}'',\ R \geq U. \tag{6.17}$$

Let $\lambda_k^{[n]} : U_k^{[n]} \hookrightarrow W^{1,2}(\mathbb{T}^{[n]}; M)$ be the embedding defined in Section 6.3 (see also the first paragraph of Section 6.4), and set $\boldsymbol{q} := \lambda_k^{-1}(\gamma)$, so that $\lambda_k^{[n]}(\boldsymbol{q}^{[n]}) = \gamma^{[n]}$. For each $n \in \mathbb{K}''$ and $R \geq U$, the homology isomorphism in (6.17) fits into the following commutative diagram.

$$\begin{array}{ccc} \mathrm{C}_*(\mathscr{A}_k, \boldsymbol{q}) & \xrightarrow{\mathrm{H}_*(\psi_k^{[n]})} & \mathrm{C}_*(\mathscr{A}_k^{[n]}, \boldsymbol{q}^{[n]}) \\ {\scriptstyle \mathrm{H}_*(\lambda_k)} \Big\downarrow \simeq & & \simeq \Big\downarrow {\scriptstyle \mathrm{H}_*(\lambda_k^{[n]})} \\ \mathrm{C}_*(\mathscr{A}_R, \gamma) & \underset{\simeq}{\xrightarrow{\mathrm{H}_*(\psi^{[n]})}} & \mathrm{C}_*(\mathscr{A}_R^{[n]}, \gamma^{[n]}) \end{array}$$

In particular, we infer that the discrete iteration map induces the homology isomorphism

$$\mathrm{H}_*(\psi_k^{[n]}) : \mathrm{C}_*(\mathscr{A}_k, \boldsymbol{q}) \xrightarrow{\simeq} \mathrm{C}_*(\mathscr{A}_k^{[n]}, \boldsymbol{q}^{[n]}), \qquad \forall n \in \mathbb{K}''. \tag{6.18}$$

Set $c := \mathscr{A}(\gamma)$. We will prove the theorem by contradiction: we assume that, for some $\varepsilon > 0$, the only periodic solutions of the Euler-Lagrange system of $\mathscr{L}$ with mean action in $[c - \varepsilon, c + \varepsilon]$, period in $\mathbb{K}_p$, and that are homotopic to some iteration of γ are

$$\gamma_1, \ldots, \gamma_r,$$

where $\gamma_1 = \gamma$ and $r \geq 1$. Up to choosing a smaller $\varepsilon > 0$, we can further assume that all the periodic orbits $\gamma_1, \ldots, \gamma_r$ have mean action equal to c.

Now, for each $n \in \mathbb{K}''$, consider the constant $\tilde{R}(c+\varepsilon, n)$ given by Lemma 6.1.2 and denote by $\mathscr{C}^{[n]}$ the connected component of $\gamma^{[n]}$ in $W^{1,2}(\mathbb{T}^{[n]}; M)$. By our choice of $\varepsilon > 0$, for each $R > \tilde{R}(c + \varepsilon, n)$ the action functional $\mathscr{A}_R^{[n]}|_{\mathscr{C}^{[n]}}$ does not have any critical values in the interval $(c, c + \varepsilon)$.

By Theorem A.4.2(i), the inclusion

$$((\mathscr{A}_R^{[n]})_c \cup \{\gamma^{[n]}\}, (\mathscr{A}_R^{[n]})_c) \hookrightarrow ((\mathscr{A}_R^{[n]})_{c+\varepsilon}, (\mathscr{A}_R^{[n]})_c)$$

induces a monomorphism in homology. Hence, the embedding $\lambda_k^{[n]}$, seen as a map

$$\lambda_k^{[n]} : ((\mathscr{A}_k^{[n]})_c \cup \{\boldsymbol{q}^{[n]}\}, (\mathscr{A}_k^{[n]})_c) \hookrightarrow ((\mathscr{A}_R^{[n]})_{c+\varepsilon}, (\mathscr{A}_R^{[n]})_c),$$

induces a monomorphism in homology as well. By combining this with (6.18) we obtain the following commutative diagram.

$$\begin{array}{ccc}
0 \neq \mathrm{C}_d(\mathscr{A}_k, \boldsymbol{q}) & \xrightarrow[\simeq]{\mathrm{H}_d(\psi_k^{[n]})} & \mathrm{C}_d(\mathscr{A}_k^{[n]}, \boldsymbol{q}^{[n]}) \\
\Big\downarrow {\scriptstyle \mathrm{H}_d(\lambda_k)} & & \Big\downarrow {\scriptstyle \mathrm{H}_d(\lambda_k^{[n]})} \\
\mathrm{H}_d((\mathscr{A}_R)_{c+\varepsilon}, (\mathscr{A}_R)_c) & \xrightarrow{\mathrm{H}_d(\psi^{[n]})} & \mathrm{H}_d((\mathscr{A}_R^{[n]})_{c+\varepsilon}, (\mathscr{A}_R^{[n]})_c)
\end{array}$$

This diagram contradicts the homological vanishing principle of Theorem 6.4.1. In fact, since the local homology group $\mathrm{C}_d(\mathscr{A}_k, \boldsymbol{q})$ is nontrivial and $d > 0$, the point $\boldsymbol{q}$ is not a local minimum of $\mathscr{A}_k$. For each nonzero $[\eta] \in \mathrm{C}_d(\mathscr{A}_k, \boldsymbol{q})$, there exist $\bar{R} = \bar{R}([\eta], p) \geq U$ and $\bar{n} = \bar{n}([\eta], p) \in \mathbb{K}_p$ such that, for each real $R \geq \bar{R}$ and for each $n \in \mathbb{K}$ greater than or equal to $\bar{n}$, we have

$$\mathrm{H}_d(\psi^{[n]}) \circ \mathrm{H}_d(\lambda_k)[\eta] = \mathrm{H}_d(\psi^{[n/\bar{n}]}) \circ \underbrace{\mathrm{H}_d(\psi^{[\bar{n}]}) \circ \mathrm{H}_d(\lambda_k)[\eta]}_{=0} = 0. \qquad \square$$

Proof of Theorem 6.5.1. Fix a period $n \in \mathbb{N}$ and a real $R > \tilde{R}(a, n)$, where a is as in the statement and $\tilde{R}(a, n)$ is the constant given by Lemma 6.1.2. Choose any R-modification $\mathscr{L}_R$ of $\mathscr{L}$. From now on, we implicitly restrict the mean action functional $\mathscr{A}_R^{[n]}$ of $\mathscr{L}_R$ to the connected component of $W^{1,2}(\mathbb{T}^{[n]}; M)$ containing the contractible loops. In particular, the action sublevel $(\mathscr{A}_R^{[n]})_a$ is understood to be contained in this connected component.

Let $m \geq 1$ be the dimension of the manifold M. We want to show that the homology of the sublevel $(\mathscr{A}_R^{[n]})_a$ is nontrivial in dimension m, i.e.,

$$\mathrm{H}_m((\mathscr{A}_R^{[n]})_a) \neq 0.$$

This is a refinement of Corollary 3.2.4, and can be proved in the following way. To begin with, note that the quantity in (6.14) is finite (due to the compactness of M) and indeed it is equal to the maximum of the function $\mathscr{A}^{[n]} \circ \iota^{[n]} : M \to \mathbb{R}$, where $\iota^{[n]} : M \hookrightarrow W^{1,2}(\mathbb{T}^{[n]}; M)$ denotes the embedding that maps a point to the constant loop at that point. By our choice of the constant a we have

$$\mathscr{A}_R^{[n]} \circ \iota^{[n]}(q) = \mathscr{A}^{[n]} \circ \iota^{[n]}(q) < a, \qquad \forall q \in M.$$

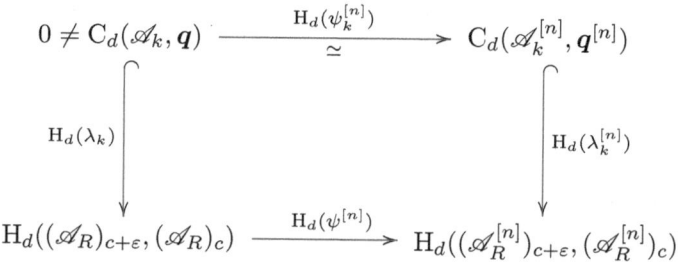

Therefore $\iota^{[n]}$ can be seen as a map of the form $\iota^{[n]} : M \hookrightarrow (\mathscr{A}_R^{[n]})_a$. Since M is an m-dimensional closed manifold and we consider homology groups with $\mathbb{Z}_2$ coefficients, $\mathrm{H}_m(M)$ is nontrivial. The following commutative diagram readily implies that $\mathrm{H}_m(\iota^{[n]})$ is a monomorphism, and the claim follows.

$$
\begin{array}{ccc}
 & \mathrm{H}_m((\mathscr{A}_R^{[n]})_a) & \\
 {\scriptstyle \mathrm{H}_m(\iota^{[n]})} \nearrow & & \searrow {\scriptstyle \mathrm{H}_m(\mathrm{ev})} \\
0 \neq \mathrm{H}_m(M) & \xrightarrow[\simeq]{\mathrm{H}_m(\mathrm{id}_M)} & \mathrm{H}_m(M)
\end{array}
$$

Now, the Morse inequality (Corollary A.4.4) for the sublevel $(\mathscr{A}_R^{[n]})_a$ implies that the mean action functional $\mathscr{A}_R^{[n]}$ has a critical point γ such that $\mathscr{A}_R^{[n]}(\gamma) < a$ and $\mathrm{H}_m(\mathscr{A}_R^{[n]}, \gamma) \neq 0$. By Lemma 6.1.2(i), γ is a contractible n-periodic solution of the Euler-Lagrange system of $\mathscr{L}$, and $\mathscr{A}^{[n]}(\gamma) = \mathscr{A}_R^{[n]}(\gamma) < a$.

Summing up, we have proved that for each period $n \in \mathbb{N}$ there exists a contractible n-periodic solution γ_n of the Euler-Lagrange system of $\mathscr{L}$ that is homologically visible in degree m and has mean action less than a. Now, fix a prime number p and consider the family of contractible periodic orbits $\mathscr{O}_p = \{\gamma_n \mid n \in \mathbb{K}_p\}$. If this family contains infinitely many geometrically distinct periodic orbits, we are done. Otherwise, there exists a contractible periodic orbit $\gamma \in \mathscr{O}_p$ that has mean action less than a and is strongly homologically visible in degree m and periods $\mathbb{K}_p$. Hence, Theorem 6.5.3 completes the proof. □

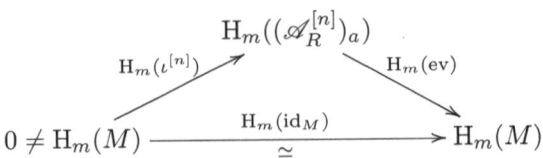

Appendix

An Overview of Morse Theory

Morse theory is a beautiful subject that sits between differential geometry, topology and calculus of variations. It was first developed by Morse [Mor25] in the middle of the 1920s and further extended, among many others, by Bott, Milnor, Palais, Smale, Gromoll and Meyer. The general philosophy of the theory is that the topology of a smooth manifold is intimately related to the number and "type" of critical points that a smooth function defined on it can have. In this brief appendix we would like to give an overview of the topic, from the classical point of view of Morse, but with the more recent extensions that allow the theory to deal with so-called degenerate functions on infinite-dimensional manifolds. A comprehensive treatment of the subject can be found in the first chapter of the book of Chang [Cha93].

There is also another, more recent, approach to the theory that we are not going to touch on in this brief note. It is based on the so-called Morse complex. This approach was pioneered by Thom [Tho49] and, later, by Smale [Sma61] in his proof of the generalized Poincaré conjecture in dimensions greater than 4 (see the beautiful book of Milnor [Mil56] for an account of that stage of the theory). The definition of Morse complex appeared in 1982 in a paper by Witten [Wit82]. See the book of Schwarz [Sch93], the one of Banyaga and Hurtubise [BH04] or the survey of Abbondandolo and Majer [AM06] for a modern treatment.

We will try to keep our exposition as elementary as possible. In order to do this and avoid subtle technicalities, we will often not give the results in their maximal generality. Nevertheless, keeping in mind the application of Morse theory to the study of the critical points of the Lagrangian action functional, we will stress the regularity that the function under consideration must have, which will be mostly C^1 and occasionally C^2.

A.1 Preliminaries

Throughout this appendix we will denote by $\mathscr{M}$ a Hilbert manifold, i.e., a Hausdorff topological space that is locally homeomorphic to a real separable Hilbert space $\boldsymbol{E}$ with smooth change of charts. If $\boldsymbol{E}$ is finite-dimensional, i.e., $\boldsymbol{E} = \mathbb{R}^m$ for some $m \in \mathbb{N}$, then $\mathscr{M}$ is just an ordinary m-dimensional smooth manifold. In the context of Morse theory, the most relevant difference between the finite-dimensional and the infinite-dimensional situations is that in this latter case the manifold $\mathscr{M}$ is not locally compact. We will come back to this point later on.

Throughout this appendix, $\mathscr{F} : \mathscr{M} \to \mathbb{R}$ will be a C^1 function, unless more regularity will be explicitly required. A point $p \in \mathscr{M}$ is called a **critical point** of $\mathscr{F}$ when the differential $\mathrm{d}\mathscr{F}(p)$ vanishes. In this case, the corresponding image $\mathscr{F}(p)$ is called a **critical value**. We denote by $\mathrm{Crit}\mathscr{F}$ the set of critical points of $\mathscr{F}$. In what follows, we will only deal with functions $\mathscr{F}$ having isolated critical points. We wish to investigate the relationship between the critical points of $\mathscr{F}$ and the topological properties (or, more precisely, the homological properties) of the underlying manifold $\mathscr{M}$.

Consider an open neighborhood $\mathscr{U}$ of a critical point p such that there exists a chart $\phi : \mathscr{U} \to \boldsymbol{E}$ of $\mathscr{M}$. We denote by $\mathscr{F}_\phi$ the function $\mathscr{F} \circ \phi^{-1} : \boldsymbol{E} \to \mathbb{R}$. By differentiating, we have

$$\mathrm{d}\mathscr{F}_\phi(\phi(p)) = \mathrm{d}\mathscr{F}(p) \circ \mathrm{d}\phi^{-1}(\phi(p)).$$

Therefore $\phi(p)$ is a critical point of the function $\mathscr{F}_\phi$. If $\mathscr{F}$ is C^2 or at least twice Gâteaux-differentiable, then so is $\mathscr{F}_\phi$. We recall that, in these cases, the second Gâteaux derivative of $\mathscr{F}_\phi$ at $\boldsymbol{x}$ can be seen as a symmetric bounded bilinear form $\mathrm{Hess}\mathscr{F}_\phi(\boldsymbol{x}) : \boldsymbol{E} \times \boldsymbol{E} \to \mathbb{R}$ given by

$$\mathrm{Hess}\mathscr{F}_\phi(\boldsymbol{x})[\boldsymbol{v}, \boldsymbol{w}] = (\mathrm{d}(\mathrm{d}\mathscr{F}_\phi)(\boldsymbol{x})\boldsymbol{v})\, \boldsymbol{w}, \qquad \forall \boldsymbol{v}, \boldsymbol{w} \in \boldsymbol{E}.$$

In the above formula, we have denoted by $\mathrm{d}(\mathrm{d}\mathscr{F}_\phi)$ the Gâteaux derivative of $\mathrm{d}\mathscr{F}_\phi$, i.e.,

$$\mathrm{d}(\mathrm{d}\mathscr{F}_\phi)(\boldsymbol{x})\boldsymbol{v} = \left.\frac{\mathrm{d}}{\mathrm{d}\varepsilon}\right|_{\varepsilon=0} \mathrm{d}\mathscr{F}_\phi(\boldsymbol{x} + \varepsilon \boldsymbol{v}) \qquad \forall \boldsymbol{x}, \boldsymbol{v} \in \boldsymbol{E}.$$

This latter expression coincides with the Fréchet derivative of $\mathrm{d}\mathscr{F}_\phi$ in case $\mathscr{F}$ is C^2. We define the **Hessian** of $\mathscr{F}$ at a critical point p as the symmetric bilinear form

$$\mathrm{Hess}\mathscr{F}(p) : \mathrm{T}_p\mathscr{M} \times \mathrm{T}_p\mathscr{M} \to \mathbb{R}$$

given by

$$\mathrm{Hess}\mathscr{F}(p)[v, w] = \mathrm{Hess}\mathscr{F}_\phi(\boldsymbol{x})[\boldsymbol{v}_\phi, \boldsymbol{w}_\phi], \quad \forall v, w \in \mathrm{T}_p\mathscr{M},$$

where $\boldsymbol{x} = \phi(p)$, $\boldsymbol{v}_\phi = \mathrm{d}\phi(p)v$ and $\boldsymbol{w}_\phi = \mathrm{d}\phi(p)w$ are in $\boldsymbol{E}$. It is easy to see that $\mathrm{Hess}\mathscr{F}(p)$ is intrinsically defined, i.e., its definition is independent of the chosen chart ϕ as long as p is a critical point of $\mathscr{F}$.

Remark A.1.1. If $\mathscr{M}$ is finite-dimensional and $(x^1, \dots, x^m)$ is a system of local coordinates around p, the Hessian of $\mathscr{F}$ at the critical point p is given by

$$\mathrm{Hess}\mathscr{F}(p) = \sum_{i,j=1}^{m} \frac{\partial^2 \mathscr{F}}{\partial x^i \, \partial x^j}(p) \, \mathrm{d}x^i \otimes \mathrm{d}x^j .$$

Moreover, if ∇ is any linear connection on $\mathscr{M}$, the Hessian of $\mathscr{F}$ at p is given by

$$\mathrm{Hess}\mathscr{F}(p) = \nabla(\mathrm{d}\mathscr{F})(p). \qquad \square$$

We denote by $\mathrm{ind}(\mathscr{F}, p)$ the supremum of the dimensions of those subspaces of $\mathrm{T}_p\mathscr{M}$ on which $\mathrm{Hess}\mathscr{F}(p)$ is negative-definite, and by $\mathrm{nul}(\mathscr{F}, p)$ the dimension of the nullspace of $\mathrm{Hess}\mathscr{F}(p)$, i.e., the Hilbert space consisting of all $v \in \mathrm{T}_p\mathscr{M}$ such that $\mathrm{Hess}\mathscr{F}(p)[v, w] = 0$ for all $w \in \mathrm{T}_p\mathscr{M}$. We call $\mathrm{ind}(\mathscr{F}, p)$ and $\mathrm{nul}(\mathscr{F}, p)$ respectively **Morse index** and **nullity** of the function $\mathscr{F}$ at p. Notice that both may be infinite. Their sum is sometimes called the **large Morse index**. Morse Theory was initially developed for the so-called **Morse functions**, which are functions all of whose critical points have nullity equal to zero. Such critical points are called **non-degenerate**. Nowadays we are able to deal with functions having possibly degenerate critical points.

Since the inner product $\langle\!\langle \cdot, \cdot \rangle\!\rangle_{\boldsymbol{E}}$ of $\boldsymbol{E}$ is a non-degenerate bilinear form, there exists a bounded self-adjoint linear operator $H_\phi = H_\phi(p) : \boldsymbol{E} \to \boldsymbol{E}$ such that

$$\mathrm{Hess}\mathscr{F}(p)[v, w] = \langle\!\langle H_\phi \boldsymbol{v}_\phi, \boldsymbol{w}_\phi \rangle\!\rangle_{\boldsymbol{E}}, \qquad \forall v, w \in \mathrm{T}_p\mathscr{M}.$$

By the spectral theorem, this operator induces an orthogonal splitting

$$\boldsymbol{E} = \boldsymbol{E}_\phi^0 \oplus \boldsymbol{E}_\phi^- \oplus \boldsymbol{E}_\phi^+ ,$$

where $\boldsymbol{E}_\phi^0$ is the kernel of H_ϕ, and $\boldsymbol{E}_\phi^-$ [resp. $\boldsymbol{E}_\phi^+$] is a subspace of $\boldsymbol{E}$ on which H_ϕ is negative-definite [resp. positive-definite], i.e.,

$$\begin{aligned} &\boldsymbol{E}_\phi^0 = \{ \boldsymbol{v} \in \boldsymbol{E} \mid H_\phi \boldsymbol{v} = 0 \} , \\ &\langle\!\langle H_\phi \boldsymbol{v}, \boldsymbol{v} \rangle\!\rangle_{\boldsymbol{E}} < 0, && \forall \boldsymbol{v} \in \boldsymbol{E}_\phi^- \setminus \{\boldsymbol{0}\} , \\ &\langle\!\langle H_\phi \boldsymbol{w}, \boldsymbol{w} \rangle\!\rangle_{\boldsymbol{E}} > 0, && \forall \boldsymbol{w} \in \boldsymbol{E}_\phi^+ \setminus \{\boldsymbol{0}\} . \end{aligned}$$

Notice that $\mathrm{ind}(\mathscr{F}, p) = \dim \boldsymbol{E}_\phi^-$ and $\mathrm{nul}(\mathscr{F}, p) = \dim \boldsymbol{E}_\phi^0$.

A.2 The generalized Morse Lemma

A starting point for Morse Theory might be the so-called **Morse Lemma**, originally due to Morse and generalized to infinite-dimensional Hilbert manifolds by Palais [Pal63]. The version that we give here, valid for possibly degenerate functions, is due to Gromoll and Meyer [GM69a] and is sometimes referred to as the **splitting**

lemma. Since it is a local result, we can assume that our Hilbert manifold $\mathscr{M}$ is just the Hilbert space $\boldsymbol{E}$. Let $\mathbf{0}$ be an isolated critical point of a C^2 function $\mathscr{F} : \mathscr{U} \to \mathbb{R}$, where $\mathscr{U}$ is an open subset of $\boldsymbol{E}$. We denote by $H : \boldsymbol{E} \to \boldsymbol{E}$ the bounded self-adjoint linear operator associated to the Hessian of $\mathscr{F}$ at $\mathbf{0}$, and by $\boldsymbol{E}^0 \oplus \boldsymbol{E}^- \oplus \boldsymbol{E}^+$ of $\boldsymbol{E}$ the orthogonal splitting induced by the spectral theorem as in the previous section. According to this splitting, we will write every vector $\boldsymbol{v} \in \boldsymbol{E}$ as

$$\boldsymbol{v} = \boldsymbol{v}^0 + \boldsymbol{v}^\pm,$$

where $\boldsymbol{v}^\pm \in \boldsymbol{E}^\pm := \boldsymbol{E}^- \oplus \boldsymbol{E}^+$ and $\boldsymbol{v}^0 \in \boldsymbol{E}^0$. We assume that H is **Fredholm**, meaning that its image $H(\boldsymbol{E})$ is closed and that its kernel and cokernel have both finite dimension. In particular $\mathrm{nul}(\mathscr{F}, \mathbf{0})$ is finite.

Lemma A.2.1 (Generalized Morse Lemma). *If $\mathscr{V} \subseteq \mathscr{U}$ is a sufficiently small open neighborhood of $\mathbf{0}$, there exist*

- *a map $\phi : (\mathscr{V}, \mathbf{0}) \to (\mathscr{U}, \mathbf{0})$ that is a homeomorphism onto its image,*
- *a C^2 function $\mathscr{F}^0 : \mathscr{V} \cap \boldsymbol{E}^0 \to \mathbb{R}$ whose Hessian vanishes at the origin,*

such that

$$\mathscr{F} \circ \phi(\boldsymbol{v}) = \mathscr{F}^0(\boldsymbol{v}^0) + \mathscr{F}^\pm(\boldsymbol{v}^\pm), \qquad \forall \boldsymbol{v} = \boldsymbol{v}^0 + \boldsymbol{v}^\pm \in \mathscr{V}, \tag{A.1}$$

where $\mathscr{F}^\pm : \boldsymbol{E}^\pm \to \mathbb{R}$ is the quadratic form defined by

$$\mathscr{F}^\pm(\boldsymbol{v}^\pm) = \tfrac{1}{2}\mathrm{Hess}\mathscr{F}(\mathbf{0})[\boldsymbol{v}^\pm, \boldsymbol{v}^\pm] = \tfrac{1}{2}\langle\!\langle H\boldsymbol{v}^\pm, \boldsymbol{v}^\pm \rangle\!\rangle_{\boldsymbol{E}}, \qquad \forall \boldsymbol{v}^\pm \in \boldsymbol{E}^\pm. \qquad \square$$

Notice that $\mathbf{0}$ is a critical point of $\mathscr{F}^\pm$, and therefore of $\mathscr{F}^0$ as well. If $\mathbf{0}$ is a non-degenerate critical point of $\mathscr{F}$, namely $\mathrm{nul}(\mathscr{F}, \mathbf{0}) = 0$, equation (A.1) reduces to

$$\mathscr{F} \circ \phi(\boldsymbol{v}) = \mathscr{F}(\mathbf{0}) + \mathscr{F}^\pm(\boldsymbol{v}) = \mathscr{F}(\mathbf{0}) + \frac{1}{2}\langle\!\langle H\boldsymbol{v}, \boldsymbol{v} \rangle\!\rangle_{\boldsymbol{E}}.$$

This is a fundamental rigidity result for functions around a non-degenerate critical point: up to a local reparametrization and to an additive constant, they are all non-degenerate quadratic forms. In the general case, equation (A.1) gives us a local representation of a function $\mathscr{F}$ as the sum of a non-degenerate quadratic form $\mathscr{F}^\pm$ and of a function $\mathscr{F}^0$ with a **totally degenerate** critical point at the origin.

A.3 Deformation of sublevels

At this point, let us recall some standard terminology from topology. Two continuous maps f_0 and f_1 from a topological space X to another topological space Y are **homotopic**, which we write as $f_0 \sim f_1$, when there exists a continuous map $f : [0,1] \times X \to Y$, called **homotopy**, such that $f_0 = f(0, \cdot)$ and $f_1 = f(1, \cdot)$. If we consider the two maps as maps of topological pairs $f_0, f_1 : (X, A) \to (Y, B)$, i.e., $A \subseteq X$, $B \subseteq Y$ and $f_0(A) \cup f_1(A) \subseteq B$, then they are homotopic (as maps of pairs) when there exists a homotopy f as before that further satisfies $f(t, A) \subseteq B$ for all $t \in [0, 1]$. We write this as $f : [0, 1] \times (X, A) \to (Y, B)$ and still $f_0 \sim f_1$. A homotopy f is said to be **relative** to C when C is a subspace of X and $f(t, x) = f(0, x)$ for all $(t, x) \in [0, 1] \times C$.

A map $j : X \to Y$ is a **homotopy equivalence**, which we write as $j : X \xrightarrow{\sim} Y$ or $X \sim Y$ if j is implicit from the context, when there exists a map $l : Y \to X$, called **homotopy inverse**, such that $l \circ j \sim \mathrm{id}_X$ and $j \circ l \sim \mathrm{id}_Y$. A map of topological pairs $j : (X, A) \to (Y, B)$ is a homotopy equivalence (of pairs) when it has a homotopy inverse of the form $l : (Y, B) \to (X, A)$ and the homotopies of $j \circ l$ and $l \circ j$ to the identities are homotopies of maps of pairs.

Now, consider an inclusion $\iota : X \hookrightarrow Y$ and a **retraction** $r : Y \to X$, i.e., a surjective map r such that the restriction $r|_X$ is equal to the identity id_X. If there exists a homotopy $R : [0, 1] \times (Y, X) \to (Y, X)$ such that $R(0, \cdot) = \mathrm{id}_Y$ and $R(1, \cdot) = r$, then we say that Y **deformation retracts** onto X and we call the homotopy R a **deformation retraction**. Notice that R is assumed to be a homotopy of maps of pairs, therefore $R(t, X) \subseteq X$ for all $t \in [0, 1]$. This implies that the inclusion ι is a homotopy equivalence. The deformation retraction R is called **strong** if we further assume that it is relative to X, i.e., $R(t, x) = x$ for all $(t, x) \in [0, 1] \times X$.

After this excursion, let us go back to Morse Theory. So far, we have just discussed local aspects, but Morse Theory allows us to say something about the global properties of $\mathscr{M}$ and $\mathscr{F} : \mathscr{M} \to \mathbb{R}$. To start with, notice that the function $\mathscr{F}$ defines a **filtration** of the manifold $\mathscr{M}$. In fact, if $c \in \mathbb{R}$, let us denote by $(\mathscr{F})_c$ the open subspace $\mathscr{F}^{-1}(-\infty, c)$, called c-**sublevel** of $\mathscr{F}$. If $\{c_n \mid n \in \mathbb{N}\}$ is a monotone increasing sequence of real numbers tending to infinity, we have the sequence of inclusions

$$(\mathscr{F})_{c_1} \subseteq (\mathscr{F})_{c_2} \subseteq (\mathscr{F})_{c_3} \subseteq \cdots \subseteq \mathscr{M}. \tag{A.2}$$

We also define $(\mathscr{F})_{-\infty} = \varnothing$ and $(\mathscr{F})_{+\infty} = \mathscr{M}$. We wish to investigate the relation between the homology of pairs of sublevels $((\mathscr{F})_b, (\mathscr{F})_a)$ and the critical points of $\mathscr{F}$ contained in the region $\mathscr{F}^{-1}[a, b)$, at first in the simple case in which the interval $[a, b)$ contains a single critical value. Then, by standard algebraic-topological manipulations, we will derive some information about the homology of pairs $((\mathscr{F})_b, (\mathscr{F})_a)$, where the interval $[a, b)$ contains an arbitrary number of critical values.

Example A.1. From now on, it will be useful to keep in mind a classical finite-dimensional example: a torus $\mathscr{M} = \mathbb{T}^2$ in $\mathbb{R}^3$ sitting on a plane as shown in Figure A.1. On this torus we consider the height function $\mathscr{F} : \mathscr{M} \to \mathbb{R}$, i.e., $\mathscr{F}(p)$ is the height of $p \in \mathbb{T}^2$ above the plane. For simplicity, we will assume that $\mathscr{F}$ is a Morse function. Its critical points p_1, p_2, p_3 and p_4 have indices 0, 1, 1 and 2 respectively. □

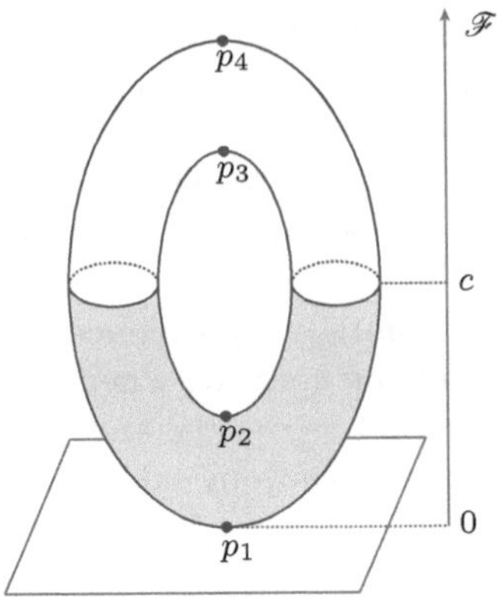

Figure A.1. Height function on the 2-torus in $\mathbb{R}^3$. The shaded region is the sublevel $(\mathcal{F})_c$, for $\mathcal{F}(p_2) < c < \mathcal{F}(p_3)$.

At this point, let us consider a **Hilbert-Riemannian metric** $\langle\langle \cdot, \cdot \rangle\rangle_\cdot$ on $\mathcal{M}$, i.e., a bounded bilinear form on $\mathrm{T}\mathcal{M}$ that is symmetric and positive-definite, meaning

$$\langle\langle v, w \rangle\rangle_q = \langle\langle w, v \rangle\rangle_q, \qquad \langle\langle z, z \rangle\rangle_q > 0 \quad \forall q \in \mathcal{M},\ v, w, z \in \mathrm{T}_q\mathcal{M},\ z \neq 0.$$

In local coordinates given by a chart $\phi : \mathcal{U} \to \boldsymbol{E}$, this metric can be expressed in terms of the inner product of $\boldsymbol{E}$ as

$$\langle\langle v, w \rangle\rangle_q = \langle\langle G_\phi(q)\boldsymbol{v}_\phi, \boldsymbol{w}_\phi \rangle\rangle_{\boldsymbol{E}}, \qquad \forall q \in M,\ v, w \in \mathrm{T}_q\mathcal{M},$$

where G_ϕ is the unique map from $\mathcal{U}$ to the space of bounded self-adjoint operators on $\boldsymbol{E}$ that realizes this equality. If $\psi : \mathcal{V} \to \boldsymbol{E}$ is another chart of $\mathcal{M}$, then

$$G_\psi(q) = \left(\mathrm{d}(\phi \circ \psi^{-1})(\psi(q))\right)^* \circ G_\phi(q) \circ \mathrm{d}(\phi \circ \psi^{-1})(\psi(q)) \qquad \forall q \in \mathcal{U} \cap \mathcal{V}.$$

We denote by $\|\cdot\|_\cdot$ the associated **Hilbert-Finsler metric** given by

$$\|v\|_q = \sqrt{\langle\langle v, v \rangle\rangle_q}, \qquad \forall q \in \mathcal{M},\ v \in \mathrm{T}_q\mathcal{M}.$$

This metric induces, as usual, a corresponding Hilbert-Finsler metric (still denoted by $\|\cdot\|_\cdot$) on the cotangent bundle $\mathrm{T}^*\mathcal{M}$ as

$$\|\nu\|_q = \max\left\{\nu(v) \mid v \in \mathrm{T}_q\mathcal{M},\ \|v\|_q = 1\right\}, \qquad \forall q \in \mathcal{M},\ \nu \in \mathrm{T}^*_q\mathcal{M}.$$

For each C^1 curve $\sigma : [a, b] \to \mathcal{M}$, we define its **length** (with respect to the Hilbert-Finsler metric $\|\cdot\|_\cdot$) as

$$\int_a^b \|\dot\sigma(t)\|_{\sigma(t)}\,\mathrm{d}t.$$

If q and q' are two points that belong to the same connected component of $\mathscr{M}$, we can define their **distance** (again, with respect to the Hilbert-Finsler metric $\|\cdot\|$.) as the infimum of the lengths of all C^1 curves joining q and q'. If each component of $\mathscr{M}$ is a complete metric space with respect to this distance, we say that $\mathscr{M}$ is a **complete** Hilbert-Riemannian manifold.

A C^1 vector field $\mathscr{X}$ on $\mathscr{M} \setminus \mathrm{Crit}(\mathscr{F})$ is called a **pseudo-gradient** of $\mathscr{F}$ when, for each $q \in \mathscr{M}$, it satisfies

$$\begin{aligned} \|\mathscr{X}(q)\|_q &\leq 2\|\mathrm{d}\mathscr{F}(q)\|_q, \\ \mathrm{d}\mathscr{F}(q)\,\mathscr{X}(q) &\geq \|\mathrm{d}\mathscr{F}(q)\|_q^2. \end{aligned}$$

By means of a partition of unity, one can show that pseudo-gradient vector fields always exist on Hilbert manifolds. Integrating $\mathscr{X}$ we obtain its **(anti) pseudo-gradient flow**, which is a map

$$\Phi_{\mathscr{X}} : \mathscr{W} \to \mathscr{M},$$

where $\mathscr{W} \subseteq \mathbb{R} \times \mathscr{M} \setminus \mathrm{Crit}(\mathscr{F})$ is an open neighborhood of $\{0\} \times \mathscr{M} \setminus \mathrm{Crit}(\mathscr{F})$, satisfying the following Cauchy problem:

$$\begin{aligned} \frac{\partial \Phi_{\mathscr{X}}}{\partial t}(t,q) &= -\mathscr{X}(\Phi_{\mathscr{X}}(t,q)), \\ \Phi_{\mathscr{X}}(0,\cdot) &= \mathrm{id}_{\mathscr{M}}. \end{aligned}$$

It is easy to verify that the function $\mathscr{F}$ is decreasing along pseudo-gradient flow lines. In fact, for every $(t,q) \in \mathscr{W}$, we have

$$\begin{aligned} \mathscr{F}(\Phi_{\mathscr{X}}(t,q)) - \mathscr{F}(q) &= -\int_0^t \mathrm{d}\mathscr{F}(\Phi_{\mathscr{X}}(s,q))\,\mathscr{X}(\Phi_{\mathscr{X}}(s,q))\,\mathrm{d}s \\ &\leq -\int_0^t \underbrace{\|\mathrm{d}\mathscr{F}(\Phi_{\mathscr{X}}(s,q))\|^2_{\Phi_{\mathscr{X}}(s,q)}}_{>0}\,\mathrm{d}s \\ &< 0. \end{aligned}$$

Example A.2. If $\mathscr{F}$ is C^2, a pseudo-gradient is given by the **gradient** of $\mathscr{F}$, that is the vector field $\mathrm{Grad}\mathscr{F}$ defined by

$$\langle\!\langle \mathrm{Grad}\mathscr{F}(q), v \rangle\!\rangle_q = \mathrm{d}\mathscr{F}(q)v, \qquad \forall q \in \mathscr{M},\ v \in \mathrm{T}_q\mathscr{M}.$$ □

We would like to use the pseudo-gradient flow to deform a certain sublevel $(\mathscr{F})_{c_2}$ to a lower one, say $(\mathscr{F})_{c_1}$, for some $c_1 < c_2$ such that the interval $[c_1, c_2)$ does not contain critical values (see Figure A.2). In case $\mathscr{M}$ is finite-dimensional and compact, there are no obstacles for performing such an operation. However, if we deal with a non-compact manifold (but still complete), some assumption is needed in order to replace the lack of compactness. The "right" assumption on

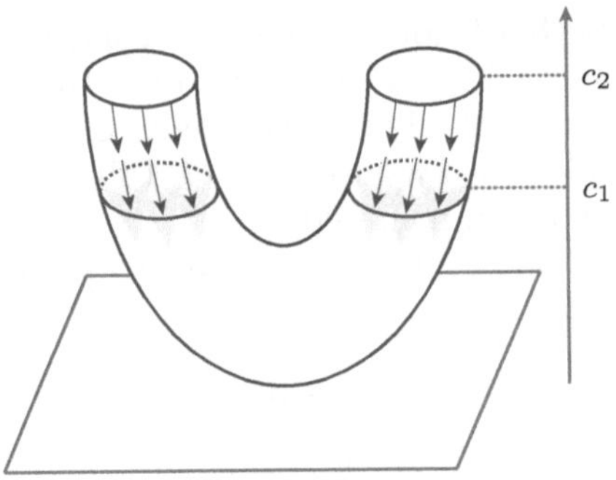

Figure A.2. Deformation of $(\mathscr{F})_{c_2}$ over $(\mathscr{F})_{c_1}$ along gradient flow lines in the torus Example A.1.

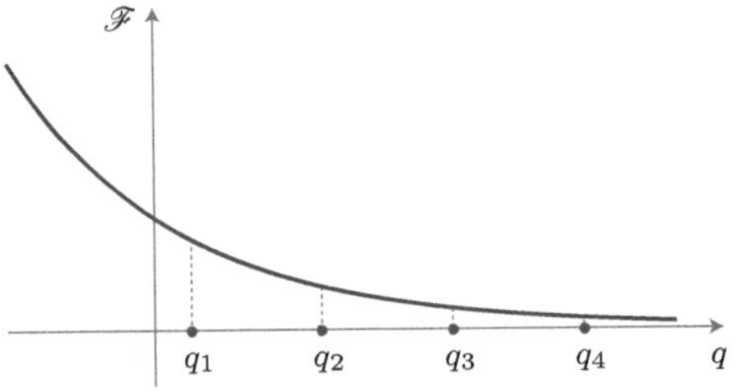

Figure A.3.

$\mathscr{F}$ was found in the 1960s by Palais and Smale [PS64], and it now carries their name: we say that $\mathscr{F}$ satisfies the **Palais-Smale condition** at level c when, for each sequence $\{q_n \mid n \in \mathbb{N}\} \subseteq \mathscr{M}$ such that

$$\lim_{n\to\infty} \mathscr{F}(q_n) = c, \\ \lim_{n\to\infty} \|\mathrm{d}\mathscr{F}(q_n)\|_{q_n} = 0, \tag{A.3}$$

there exists a subsequence converging to a point $q \in \mathscr{M}$. By (A.3), the limit point q must be a critical point of $\mathscr{F}$. We say that $\mathscr{F}$ satisfies the Palais-Smale condition if it satisfies it at every level $c \in \mathbb{R}$.

Example A.3. The following are two slightly different examples of situations that the Palais-Smale condition wants to avoid.

- Consider the function $\mathscr{F}(q) = \exp(-q)$ on $\mathscr{M} = \mathbb{R}$. For any diverging sequence $q_n \uparrow \infty$ we have that $\mathscr{F}(q_n) \to 0$ and $\mathscr{F}'(q_n) \to 0$, however $\{q_n\}$ does not admit any converging subsequence (see Figure A.3).
- Consider a function $\mathscr{F} : \mathscr{M} \to \mathbb{R}$ such that, for a certain level $c \in \mathbb{R}$, the set $\mathrm{Crit}\mathscr{F} \cap \mathscr{F}^{-1}(c)$ is not compact (e.g., $\mathscr{F} : \mathbb{R} \to \mathbb{R}$ with $\mathscr{F}(q) = \sin(q)$ and

$c = 1$). Then there is a sequence of critical points $\{q_n\} \subset \mathscr{F}^{-1}(c)$ that does not admit any converging subsequence. □

As promised, here is the important consequence of the Palais-Smale condition.

Lemma A.3.1 (Deformation Lemma). *Let $\mathscr{M}$ be a complete Hilbert-Riemannian manifold and assume that $\mathscr{F} : \mathscr{M} \to \mathbb{R}$ satisfies the Palais-Smale condition at every level $c \in [a, b]$ and does not have any critical value in the interval $[a, b)$. Then the inclusion $(\mathscr{F})_a \hookrightarrow (\mathscr{F})_b$ is a homotopy equivalence.* □

Notice that, by this lemma, the inclusion $(\mathscr{F})_a \hookrightarrow (\mathscr{F})_b$ induces the homology isomorphism $\mathrm{H}_*((\mathscr{F})_a) \xrightarrow{\simeq} \mathrm{H}_*((\mathscr{F})_b)$, and $\mathrm{H}_*((\mathscr{F})_b, (\mathscr{F})_a) = 0$. The same is true if we substitute the singular homology with any other homotopy invariant functor: singular cohomology, K-theory, and so forth.

A.4 Passing a critical level

We now want to study the changes that occur, in term of homology, whenever we pass a critical level. In order to do this, we first introduce a fundamental invariant of isolated critical points p of a C^1 function $\mathscr{F} : \mathscr{M} \to \mathbb{R}$. The **local homology** of $\mathscr{F}$ at p is defined as the relative homology group

$$\mathrm{C}_*(\mathscr{F}, p) := \mathrm{H}_*((\mathscr{F})_c \cup \{p\}, (\mathscr{F})_c),$$

where $c = \mathscr{F}(p)$. This is a local invariant, in the sense that it depends only on the germ of $\mathscr{F}$ at p. In fact, if $\mathscr{U} \subseteq \mathscr{M}$ is an open neighborhood of p, by excision we obtain that the inclusion

$$(\mathscr{U} \cap (\mathscr{F})_c \cup \{p\}, \mathscr{U} \cap (\mathscr{F})_c) \hookrightarrow ((\mathscr{F})_c \cup \{p\}, (\mathscr{F})_c)$$

induces an isomorphism in homology, and therefore the local homology of $\mathscr{F}$ at p coincides with $\mathrm{H}_*(\mathscr{U} \cap (\mathscr{F})_c \cup \{p\}, \mathscr{U} \cap (\mathscr{F})_c)$.

Remark A.4.1. It is easy to see that the local homology of $\mathscr{F}$ at p can be equivalently defined as $\mathrm{H}_*\big(\overline{(\mathscr{F})_c}, \overline{(\mathscr{F})_c} \setminus \{p\}\big)$. □

The role that local homology plays in Morse Theory is illustrated by the following statements. We recall that a **λ-cell**, for $\lambda \in \mathbb{N}$, is simply a λ-dimensional closed disk D^λ. A topological space Y is obtained by the attachment of a λ-cell to a topological manifold X when there exists a continuous map $f : \partial D^\lambda \to \partial X$ such that

$$Y = X \cup D^\lambda / \sim,$$

where $\sim$ is the identification given by $z \sim f(z)$ for each $z \in \partial D^\lambda$.

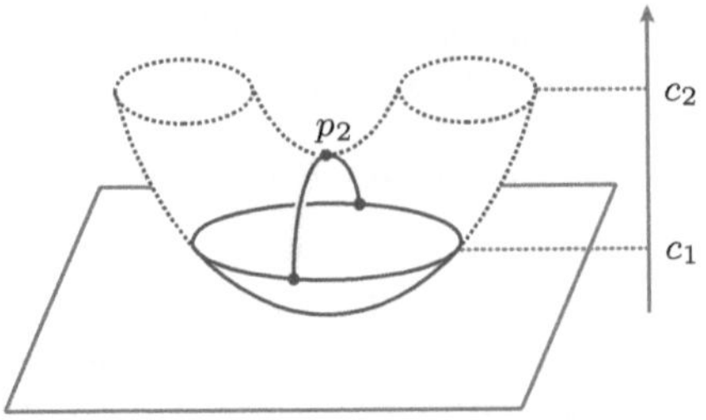

Figure A.4. Attachment of a cell of dimension $1 = \mathrm{ind}(\mathscr{F}, p_2)$ to the closed sublevel $\overline{(\mathscr{F})_{c_1}}$ of the height function in Example A.1. Notice that the result of the attachment is homotopically equivalent to $\overline{(\mathscr{F})_{c_2}}$.

Theorem A.4.1. *Consider a complete Hilbert-Riemannian manifold $\mathscr{M}$, a C^2 function $\mathscr{F} : \mathscr{M} \to \mathbb{R}$ that satisfies the Palais-Smale condition, and an interval $(c_1, c_2) \subset \mathbb{R}$. Assume that p is the only critical point of $\mathscr{F}$ with critical value inside (c_1, c_2), and that it is non-degenerate. Then $\overline{(\mathscr{F})_{c_2}}$ is homotopically equivalent to $\overline{(\mathscr{F})_{c_1}}$ with an $\mathrm{ind}(\mathscr{F}, p)$-cell attached. In particular*

$$\mathrm{C}_n(\mathscr{F}, p) = \begin{cases} \mathbb{F} & n = \mathrm{ind}(\mathscr{F}, p), \\ 0 & n \neq \mathrm{ind}(\mathscr{F}, p), \end{cases}$$

where $\mathbb{F}$ is the coefficient group of the homology. □

In Figure A.4 the assertions are illustrated for the case of the height function on the torus of Example A.1. Notice that Theorem A.4.1 implies that critical points p with infinite Morse index do not produce any change in the homotopy type as we cross their sublevel (in particular their local homology $\mathrm{C}_*(\mathscr{F}, p)$ is trivial). This follows from the fact that the unit sphere of an infinite-dimensional Hilbert space is contractible, as one can easily prove by constructing an explicit deformation retraction to a point (a stronger and much harder theorem due to Bessaga [Bes66] asserts that every infinite-dimensional Hilbert space is even diffeomorphic to its unit sphere).

Theorem A.4.1 tells us that, for a non-degenerate critical point of a C^2 function, the knowledge of the local homology at it coincides with the knowledge of its Morse index. This is no longer true in the degenerate case, in which the Morse index and nullity of a critical point do not completely determine its local homology. An easy example on $\mathscr{M} = \mathbb{R}^2$ is the following.

Example A.4. Consider the functions $\mathscr{F}, \mathscr{G} : \mathbb{R}^2 \to \mathbb{R}$ given by

$$\mathscr{F}(x, y) = (y - 2x^2)(y - x^2),$$
$$\mathscr{G}(x, y) = x^4 + y^2.$$

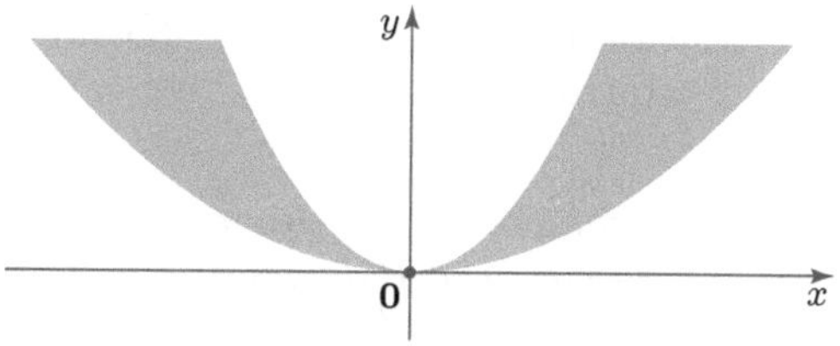

Figure A.5. Behaviour of $\mathscr{F}(x, y) = (y - 2x^2)(y - x^2)$ around the critical point $\mathbf{0}$. The shaded region corresponds to the sublevel $(\mathscr{F})_0 = \mathscr{F}^{-1}(-\infty, 0)$.

Both functions have $\mathbf{0} = (0, 0)$ as isolated critical point with

$$\begin{aligned} \operatorname{ind}(\mathscr{F}, \mathbf{0}) &= \operatorname{ind}(\mathscr{G}, \mathbf{0}) = 0, \\ \operatorname{nul}(\mathscr{F}, \mathbf{0}) &= \operatorname{nul}(\mathscr{G}, \mathbf{0}) = 1. \end{aligned}$$

However $\mathbf{0}$ is a saddle point for $\mathscr{F}$ (see Figure A.5) and a global minimum for $\mathscr{G}$, therefore

$$\mathrm{C}_n(\mathscr{F}, \mathbf{0}) = \begin{cases} \mathbb{F} & n = 1, \\ 0 & n \neq 1, \end{cases} \qquad \mathrm{C}_n(\mathscr{G}, \mathbf{0}) = \begin{cases} \mathbb{F} & n = 0, \\ 0 & n \neq 0. \end{cases}$$ □

If the function $\mathscr{F}$ is merely of class C^1 and has possibly degenerate critical points, we can still describe the homological change that occurs by crossing a critical level in terms of local homology.

Theorem A.4.2. *Consider a complete Hilbert-Riemannian manifold $\mathscr{M}$ and a C^1 function $\mathscr{F} : \mathscr{M} \to \mathbb{R}$ satisfying the Palais-Smale condition. Assume that the critical points of $\mathscr{F}$ having critical value c are isolated (hence there are only finitely many), and choose $\varepsilon > 0$ small enough such that $\mathscr{F}$ does not have critical values in $(c, c + \varepsilon)$. The following claims hold:*

(i) *For each $p \in \mathrm{Crit}(\mathscr{F})$ with $\mathscr{F}(p) = c$, the inclusion*

$$j_p : ((\mathscr{F})_c \cup \{p\}, (\mathscr{F})_c) \hookrightarrow ((\mathscr{F})_{c+\varepsilon}, (\mathscr{F})_c)$$

induces the homology monomorphism

$$\mathrm{H}_*(j_p) : \mathrm{C}_*(\mathscr{F}, p) \hookrightarrow \mathrm{H}_*((\mathscr{F})_{c+\varepsilon}, (\mathscr{F})_c).$$

(ii) *If $\{p_1, \dots, p_s\} \subset \mathscr{M}$ is the set of critical point of $\mathscr{F}$ with critical value c, we have an isomorphism*

$$J_c : \mathrm{C}_*(\mathscr{F}, p_1) \oplus \cdots \oplus \mathrm{C}_*(\mathscr{F}, p_s) \xrightarrow{\simeq} \mathrm{H}_*((\mathscr{F})_{c+\varepsilon}, (\mathscr{F})_c),$$

where $J_c = \mathrm{H}_*(j_{p_1}) \oplus \cdots \oplus \mathrm{H}_*(j_{p_s})$. □

By playing with the filtration (A.2) given by the sublevels of $\mathscr{F}$ and with the associated long exact sequences in homology we can deduce the following statement.

Theorem A.4.3. *Consider a complete Hilbert-Riemannian manifold $\mathscr{M}$ and a C^1 function $\mathscr{F} : \mathscr{M} \to \mathbb{R}$ that satisfies the Palais-Smale condition and has isolated critical points. Fix a bounded interval $[a,b] \subset \mathbb{R}$, and let $\{c_1, \dots, c_t\}$ be the set of critical values of $\mathscr{F}$ inside $[a,b)$. Then for each ε such that*

$$0 < \varepsilon < \min\{c_h - c_k \mid h,k = 1,\dots,t,\ h \neq k\})$$

we have

$$\sum_{n=0}^{N} (-1)^{N-n} \operatorname{rank} \mathrm{H}_n((\mathscr{F})_b, (\mathscr{F})_a) \leq \sum_{n=0}^{N} (-1)^{N-n} \sum_{h=1}^{t} \operatorname{rank} \mathrm{H}_n((\mathscr{F})_{c_h+\varepsilon}, (\mathscr{F})_{c_h}), \qquad \forall N \in \mathbb{N}.$$

Moreover

$$\sum_{n=0}^{\infty} (-1)^{n} \operatorname{rank} \mathrm{H}_n((\mathscr{F})_b, (\mathscr{F})_a) = \sum_{n=0}^{\infty} (-1)^{n} \sum_{h=1}^{t} \operatorname{rank} \mathrm{H}_n((\mathscr{F})_{c_h+\varepsilon}, (\mathscr{F})_{c_h}),$$

provided the above sums are finite. □

As a consequence of this result and of Theorem A.4.2, we obtain the celebrated Morse inequalities.

Corollary A.4.4 (Morse inequalities). *Assume that $\mathscr{F}$ satisfies the hypotheses of Theorem A.4.3. For each bounded interval $[a,b] \subset \mathbb{R}$, if $\{p_1, \dots, p_u\} \subset \mathscr{M}$ is the set of critical points of $\mathscr{F}$ with critical values inside $[a,b)$, we have*

$$\operatorname{rank} \mathrm{H}_n((\mathscr{F})_b, (\mathscr{F})_a) \leq \sum_{h=1}^{u} \operatorname{rank} \mathrm{C}_n(\mathscr{F}, p_h). \qquad \forall n \in \mathbb{N} \tag{A.4}$$

□

A.5 Local homology and Gromoll-Meyer pairs

In their work on degenerate Morse theory [GM69a], Gromoll and Meyer showed that the local homology of an isolated critical point can also be computed as the homology of a suitable closed neighborhood of the critical point relative to a part of its boundary. The homotopy type of the pair

(closed neighborhood, part of boundary)

is what in the 1980s, after the seminal work of Conley [Con78], would be called **Conley index** of the critical point. In Conley theory, the critical point is viewed as an isolated invariant set for the dynamical system defined by a pseudo-gradient flow. We refer the interested reader to Chang and Ghoussoub [CG96] for a detailed investigation of the relation between Gromoll-Meyer theory and Conley theory.

For this section, let us assume that our function $\mathscr{F} : \mathscr{M} \to \mathbb{R}$ is C^2 with Fredholm Hessian, so that we can choose as a pseudo-gradient $\mathscr{X}$ the gradient of $\mathscr{F}$, see Example A.2. Some, but not all, of the statements that we will give still hold under the C^1 assumption on $\mathscr{F}$ (up to choosing a C^1 pseudo-gradient).

Let $p \in \mathscr{M}$ be a critical point of $\mathscr{F}$ with $\mathscr{F}(p) = c$. A pair of topological spaces $(\mathscr{W}, \mathscr{W}_-)$ is called a **Gromoll-Meyer pair** for $\mathscr{F}$ at p when

(GM1) $\mathscr{W} \subset \mathscr{M}$ is a closed neighborhood of p that does not contain other critical points of $\mathscr{F}$,

(GM2) there exists $\varepsilon > 0$ such that $[c-\varepsilon, c)$ does not contain critical values of $\mathscr{F}$, and $\mathscr{W} \cap (\mathscr{F})_{c-\varepsilon} = \varnothing$,

(GM3) if $t_1 < t_2$ are such that $\Phi_{\mathscr{X}}(t_1, q) \in \mathscr{W}$ and $\Phi_{\mathscr{X}}(t_2, q) \in \mathscr{W}$ for some $q \in \mathscr{M}$, then $\Phi_{\mathscr{X}}(t, q) \in \mathscr{W}$ for all $t \in [t_1, t_2]$,

(GM4) $\mathscr{W}_-$ is defined as

$$\mathscr{W}_- = \left\{ q \in \mathscr{W} \;\middle|\; \max\{t \in \mathbb{R} \,|\, \Phi_{\mathscr{X}}(t, q) \in \mathscr{W}\} = 0 \right\}$$

and it is a piecewise submanifold of $\mathscr{M}$ transversal to the flow $\Phi_{\mathscr{X}}$.

Given an open neighborhood $\mathscr{U}$ of the isolated critical point p, it is always possible to build a Gromoll-Meyer pair $(\mathscr{W}, \mathscr{W}_-)$ such that $\mathscr{W} \subset \mathscr{U}$. Moreover, all the Gromoll-Meyer pairs of a critical point have the same homology type, and in fact they provide an alternative definition for the local homology.

Theorem A.5.1. *If $(\mathscr{W}, \mathscr{W}_-)$ is a Gromoll-Meyer pair for $\mathscr{F}$ at an isolated critical point $p \in \mathscr{M}$, we have*

$$\mathrm{H}_*(\mathscr{W}, \mathscr{W}_-) \simeq \mathrm{C}_*(\mathscr{F}, p). \qquad \square$$

Now, let us apply the generalized Morse Lemma (Lemma A.2.1) and the notation adopted therein: without loss of generality, we can identify an isolated critical point p of $\mathscr{F}$ with the origin $\mathbf{0}$ in the Hilbert space $\boldsymbol{E}$, and we can assume that $\mathscr{F}$ is defined on a neighborhood $\mathscr{V} \subset \boldsymbol{E}$ of $\mathbf{0}$ and has the form

$$\mathscr{F}(\boldsymbol{v}) = \mathscr{F}^0(\boldsymbol{v}^0) + \mathscr{F}^\pm(\boldsymbol{v}^\pm), \qquad \forall \boldsymbol{v} = \boldsymbol{v}^0 + \boldsymbol{v}^\pm \in \mathscr{V} \subset \boldsymbol{E} = \boldsymbol{E}^0 \oplus \boldsymbol{E}^\pm.$$

The origin is a non-degenerate critical point of $\mathscr{F}^\pm : \mathscr{V} \cap \boldsymbol{E}^\pm \to \mathbb{R}$ and a totally degenerate critical point of $\mathscr{F}^0 : \mathscr{V} \cap \boldsymbol{E}^0 \to \mathbb{R}$. If we consider Gromoll-Meyer pairs $(\mathscr{W}^\pm, \mathscr{W}^\pm_-)$ and $(\mathscr{W}^0, \mathscr{W}^0_-)$ for $\mathscr{F}^\pm$ and $\mathscr{F}^0$ respectively at $\mathbf{0}$, it is easy to verify that the product of these pairs, that is

$$(\mathscr{W}^\pm, \mathscr{W}^\pm_-) \times (\mathscr{W}^0, \mathscr{W}^0_-) = (\mathscr{W}^\pm \times \mathscr{W}^0, (\mathscr{W}^\pm_- \times \mathscr{W}^0) \cup (\mathscr{W}^\pm \times \mathscr{W}^0_-)),$$

is a Gromoll-Meyer pair for $\mathscr{F}$ at $\mathbf{0}$. Now, assume that the coefficient group $\mathbb{F}$ of the homology is a field. By the Künneth formula we get an isomorphism of graded vector spaces

$$\mathrm{H}_*((\mathscr{W}^{\pm},\mathscr{W}^{\pm}_{-})\times(\mathscr{W}^{0},\mathscr{W}^{0}_{-}))\simeq \mathrm{H}_*(\mathscr{W}^{\pm},\mathscr{W}^{\pm}_{-})\otimes \mathrm{H}_*(\mathscr{W}^{0},\mathscr{W}^{0}_{-}),$$

and by the above Theorem A.5.1 we obtain the following.

Theorem A.5.2. $\mathrm{C}_*(\mathscr{F},\mathbf{0})\simeq \mathrm{C}_*(\mathscr{F}^{\pm},\mathbf{0})\otimes \mathrm{C}_*(\mathscr{F}^{0},\mathbf{0})$. □

By Theorem A.4.1, the local homology of the non-degenerate function $\mathscr{F}^{\pm}$ at $\mathbf{0}$ is nontrivial only in degree $\mathrm{ind}(\mathscr{F}^{\pm},\mathbf{0})$, where it coincides with the coefficient vector space $\mathbb{F}$. Notice that the Morse index of $\mathscr{F}^{\pm}$ at $\mathbf{0}$ is precisely the Morse index of $\mathscr{F}$ at $\mathbf{0}$, and therefore Theorem A.5.2 readily gives the following fundamental result.

Theorem A.5.3 (Shifting). $\mathrm{C}_*(\mathscr{F},\mathbf{0})\simeq \mathrm{C}_{*-\mathrm{ind}(\mathscr{F},\mathbf{0})}(\mathscr{F}^{0},\mathbf{0})$. □

Corollary A.5.4. *The local homology group $\mathrm{C}_n(\mathscr{F},\mathbf{0})$ is trivial if $n<\mathrm{ind}(\mathscr{F},\mathbf{0})$ or $n>\mathrm{ind}(\mathscr{F},\mathbf{0})+\mathrm{nul}(\mathscr{F},\mathbf{0})$.* □

A.6 Minimax

On a finite-dimensional closed manifold, an elementary way to detect a critical value of a given function is to look for its global minima (or maxima). A more general class of critical values, that we are going to discuss in this section, is given by the so-called "minimax" values.

Let $\mathscr{M}$ be a complete Hilbert-Riemannian manifold, $\mathscr{F}:\mathscr{M}\to\mathbb{R}$ a C^1 function and $\mathfrak{U}$ a family of subsets of $\mathscr{M}$. We define the **minimax** of $\mathscr{F}$ over the family $\mathfrak{U}$ as

$$\operatorname*{minimax}_{\mathfrak{U}}\mathscr{F}:=\inf_{\mathscr{U}\in\mathfrak{U}}\ \sup_{p\in\mathscr{U}}\{\mathscr{F}(p)\}\ \in\mathbb{R}\cup\{\pm\infty\}.$$

The following statement guarantees that, under certain conditions, the minimax is a critical value of $\mathscr{F}$. We refer the reader to [HZ94, page 79] for a proof.

Theorem A.6.1 (Minimax Theorem). *Assume that the following conditions are satisfied:*

(i) *$\mathscr{F}$ satisfies the Palais-Smale condition,*

(ii) *there exists a pseudo-gradient $\mathscr{X}$ for $\mathscr{F}$ whose anti pseudo-gradient flow is defined for all the non-negative times, i.e.,*

$$\Phi_{\mathscr{X}}:[0,\infty)\times\mathscr{M}\setminus\mathrm{Crit}(\mathscr{F})\to\mathscr{M},$$

(iii) *the family $\mathscr{U}$ is positively invariant under the anti pseudo-gradient flow of $\mathscr{X}$, i.e., for each $\mathscr{U}\in\mathfrak{U}$ and $t\geq 0$ we have $\Phi_{\mathscr{X}}(t,\mathscr{U})\in\mathfrak{U}$,*

(iv) $\operatorname*{minimax}_{\mathfrak{U}}\mathscr{F}$ *is finite.*

Then $\operatorname*{minimax}_{\mathfrak{U}}\mathscr{F}$ *is a critical value of $\mathscr{F}$.* □

Remark A.6.1. If we take $\mathfrak{U}$ to be the family of singletons on $\mathscr{M}$, i.e.,

$$\mathfrak{U} = \{\{p\} \mid p \in \mathscr{M}\},$$

then condition (iii) above is trivially satisfied and

$$\operatorname*{minimax}_{\mathfrak{U}} \mathscr{F} = \min_{p \in \mathscr{M}} \{\mathscr{F}(p)\}. \qquad \square$$

Notice that condition (ii) above is automatically satisfied for any pseudo-gradient if the function $\mathscr{F}$ is bounded from below. As for the choice of a suitable family $\mathscr{U}$ satisfying conditions (iii) and (iv), there are several possibilities. In the following we discuss one of these, which leads to a homological version of the minimax theorem.

Let a be a nonzero homology class in $\mathrm{H}_d(\mathscr{M})$. We denote by $\mathfrak{S}_a$ the family consisting of the supports of the singular cycles in $\mathscr{M}$ representing a, i.e.,

$$\mathfrak{S}_a = \left\{ \bigcup_{v=1}^{V} \alpha_v(\Delta^d) \;\middle|\; \alpha = \sum_{v=1}^{V} \alpha_v,\ [\alpha] = a \right\},$$

where each α_v is a singular simplex of the cycle α, i.e., $\alpha_v : \Delta^d \to \mathscr{M}$. Notice that, by the homotopic invariance of singular homology, $\mathfrak{S}_a$ satisfies condition (iii) of the minimax theorem. Moreover, $\mathfrak{S}_a$ is a family of compact subsets of $\mathscr{M}$. Therefore, if the function $\mathscr{F}$ is bounded from below, its minimax over the family $\mathfrak{S}_a$ is finite. Hence, we have the following.

Theorem A.6.2 (Homological Minimax Theorem). *Let $\mathscr{M}$ be a complete Hilbert-Riemannian manifold, and $\mathscr{F} : \mathscr{M} \to \mathbb{R}$ a C^1 function that is bounded from below and satisfies the Palais-Smale condition. Then for each nonzero $a \in \mathrm{H}_*(\mathscr{M})$ the quantity $\operatorname{minimax}_{\mathfrak{S}_a} \mathscr{F}$ is finite and it is a critical value of $\mathscr{F}$.* $\square$

If $\mathscr{F}$ is also C^2, then by the following result due to Viterbo [Vit88] it is sometimes possible to estimate the Morse index and nullity pair of a critical point corresponding to the minimax critical value.

Theorem A.6.3. *Under the assumptions of Theorem A.6.2, if $\mathscr{F}$ is also C^2 and its critical points corresponding to the critical value $\operatorname{minimax}_{\mathfrak{S}_a} \mathscr{F}$ are isolated, then there exists a critical point p of $\mathscr{F}$ such that*

$$\mathscr{F}(p) = \operatorname*{minimax}_{\mathfrak{S}_a} \mathscr{F},$$
$$\operatorname{ind}(\mathscr{F}, p) \leq d \leq \operatorname{ind}(\mathscr{F}, p) + \operatorname{nul}(\mathscr{F}, p),$$

where d is the degree of a, i.e., $a \in \mathrm{H}_d(\mathscr{M})$. $\square$

Now, if two linearly independent homology classes are given, one might ask whether or not their associated minimax values (with respect to the function $\mathscr{F}$) are the same. Before answering this question, let us quickly recall some definitions from algebraic topology. Given a topological space X and two nonzero homology classes $a, b \in \mathrm{H}_*(X)$, we write $a \prec b$ when there exists $\omega \in \mathrm{H}^d(X)$, for some $d > 0$, such that $a = b \frown \omega$. Here, $\frown$: $\mathrm{H}_i(X) \otimes \mathrm{H}^j(X) \to \mathrm{H}_{i-j}(X)$ is the **cap product**, which is related to the **cup product** $\smile$: $\mathrm{H}^i(X) \otimes \mathrm{H}^j(X) \to \mathrm{H}^{i+j}(X)$ by

$$(\omega \smile \psi)(a) = \psi(a \frown \omega), \qquad \forall \omega, \psi \in \mathrm{H}^*(X),\ a \in \mathrm{H}_*(X). \tag{A.5}$$

The cup product gives an important homotopical invariant of X, the **cup-length** $\mathrm{CL}(X)$, defined as the maximum integer $l \geq 0$ such that there exist cohomology classes $\omega_1, \dots, \omega_l$, with $\omega_i \in \mathrm{H}^{d_i}(X)$ and $d_i > 0$ for each $i = 1, \dots, l$, satisfying

$$\omega_1 \smile \cdots \smile \omega_l \neq 0.$$

The relation "$\prec$" defined above can be used to give an alternative definition of the cup-length. In fact, if $\omega_1 \smile \cdots \smile \omega_l \neq 0$ in $\mathrm{H}^*(X)$, then there exists $a_l \in \mathrm{H}_*(X)$ such that $(\omega_1 \smile \cdots \smile \omega_l)(a_l) \neq 0$, and we can iteratively define $a_{i-1} := a_i \frown \omega_i$. Therefore, by (A.5), we have

$$\omega_1(a_1) = (\omega_2 \smile \omega_1)(a_2) = \cdots = (\omega_l \smile \cdots \smile \omega_1)(a_l) \neq 0, \tag{A.6}$$

which implies that $a_0 = a_1 \frown \omega_1$ is nonzero and, by construction, $a_0 \prec \cdots \prec a_l$. Conversely, if $a_0 \prec \cdots \prec a_l$, then there exists a nonzero $\omega_i \in \mathrm{H}^{d_i}(X)$ such that $d_i > 0$ and $a_{i-1} = a_i \frown \omega_i$ for each $i = 1, \dots, n$, so that (A.6) holds and we have that $\omega_1 \smile \cdots \smile \omega_l \neq 0$ in $\mathrm{H}^*(X)$. Summing up, we have obtained the following.

Proposition A.6.4. *The cup-length* $\mathrm{CL}(X)$ *is equal to the maximum* $l \in \mathbb{N} \cup \{0\}$ *such that there exist* $l + 1$ *nonzero homology classes* $a_0, \dots, a_l \in \mathrm{H}_*(X)$ *satisfying* $a_0 \prec \cdots \prec a_l$. □

Coming back to the question as to when two given homology classes have different minimax values with respect to some function, we have the following result.

Theorem A.6.5. *Let* $\mathscr{M}$ *be a complete Hilbert-Riemannian manifold, and* $\mathscr{F}: \mathscr{M} \to \mathbb{R}$ *a* C^1 *function that is bounded from below and satisfies the Palais-Smale condition. Consider non-zero homology classes* $a, b \in \mathrm{H}_*(\mathscr{M})$ *such that* $a \prec b$. *If* $\mathscr{F}$ *has only finitely many critical points corresponding to the critical value* $\mathrm{minimax}_{\mathfrak{S}_b} \mathscr{F}$, *then*

$$\underset{\mathfrak{S}_a}{\mathrm{minimax}}\, \mathscr{F} < \underset{\mathfrak{S}_b}{\mathrm{minimax}}\, \mathscr{F}. \qquad \square$$

This theorem, together with Proposition A.6.4, readily implies the following multiplicity result for the critical points of $\mathscr{F}$.

Corollary A.6.6. *Under the assumptions of Theorem* A.6.5, *the function* $\mathscr{F}$ *has at least* $\mathrm{CL}(\mathscr{M}) + 1$ *critical points.* □

Bibliography

[Abb01] A. Abbondandolo, *Morse theory for Hamiltonian systems*, Chapman & Hall/CRC Research Notes in Mathematics, vol. 425, Chapman & Hall/CRC, Boca Raton, FL, 2001.

[Abb03] ———, *On the Morse index of Lagrangian systems*, Nonlinear Anal. **53** (2003), no. 3-4, 551–566.

[AF03] R.A. Adams and J.J.F. Fournier, *Sobolev spaces*, second ed., Pure and Applied Mathematics (Amsterdam), vol. 140, Elsevier/Academic Press, Amsterdam, 2003.

[AF07] A. Abbondandolo and A. Figalli, *High action orbits for Tonelli Lagrangians and superlinear Hamiltonians on compact configuration spaces*, J. Differential Equations **234** (2007), no. 2, 626–653.

[AM78] R. Abraham and J.E. Marsden, *Foundations of mechanics*, Benjamin / Cummings Publishing Co. Inc. Advanced Book Program, 1978, Second edition, revised and enlarged, with the assistance of T. Raţiu and R. Cushman.

[AM06] A. Abbondandolo and P. Majer, *Lectures on the Morse complex for infinite-dimensional manifolds*, Morse theoretic methods in nonlinear analysis and in symplectic topology, NATO Sci. Ser. II Math. Phys. Chem., vol. 217, Springer, Dordrecht, 2006, pp. 1–74.

[Arn78] V.I. Arnold, *Mathematical methods of classical mechanics*, Springer-Verlag, New York, 1978.

[Arv02] W. Arveson, *A short course on spectral theory*, Graduate Texts in Mathematics, vol. 209, Springer-Verlag, New York, 2002.

[AS06] A. Abbondandolo and M. Schwarz, *On the Floer homology of cotangent bundles*, Comm. Pure Appl. Math. **59** (2006), no. 2, 254–316.

[AS09] ———, *A smooth pseudo-gradient for the Lagrangian action functional*, Adv. Nonlinear Stud. **9** (2009), no. 4, 597–623.

[Ban80] V. Bangert, *Closed geodesics on complete surfaces*, Math. Ann. **251** (1980), no. 1, 83–96.

[Ben86] V. Benci, *Periodic solutions of Lagrangian systems on a compact manifold*, J. Differential Equations **63** (1986), no. 2, 135–161.

[Bes66] C. Bessaga, *Every infinite-dimensional Hilbert space is diffeomorphic with its unit sphere*, Bull. Acad. Polon. Sci. Sér. Sci. Math. Astronom. Phys. **14** (1966), 27–31.

[BGH98] G. Buttazzo, M. Giaquinta, and S. Hildebrandt, *One-dimensional variational problems - an introduction*, Oxford Lecture Series in Mathematics and its Applications, vol. 15, The Clarendon Press Oxford University Press, New York, 1998.

[BH04] A. Banyaga and D. Hurtubise, *Lectures on Morse homology*, Kluwer Texts in the Mathematical Sciences, vol. 29, Kluwer Academic Publishers Group, Dordrecht, 2004.

[BK83] V. Bangert and W. Klingenberg, *Homology generated by iterated closed geodesics*, Topology **22** (1983), no. 4, 379–388.

[BL10] V. Bangert and Y. Long, *The existence of two closed geodesics on every Finsler 2-sphere*, Math. Ann. **346** (2010), no. 2, 335–366.

[BM85] J.M. Ball and V.J. Mizel, *One-dimensional variational problems whose minimizers do not satisfy the Euler-Lagrange equation*, Arch. Rational Mech. Anal. **90** (1985), no. 4, 325–388.

[Bot56] R. Bott, *On the iteration of closed geodesics and the Sturm intersection theory*, Comm. Pure Appl. Math. **9** (1956), 171–206.

[CdS01] A. Cannas da Silva, *Lectures on symplectic geometry*, Lecture Notes in Mathematics, vol. 1764, Springer-Verlag, Berlin, 2001.

[CG96] K.C. Chang and N. Ghoussoub, *The Conley index and the critical groups via an extension of Gromoll-Meyer theory*, Topol. Methods Nonlinear Anal. **7** (1996), no. 1, 77–93.

[Cha84] M. Chaperon, *Une idée du type "géodésiques brisées" pour les systèmes hamiltoniens*, C. R. Acad. Sci. Paris Sér. I Math. **298** (1984), no. 13, 293–296.

[Cha93] K.C. Chang, *Infinite-dimensional Morse theory and multiple solution problems*, Progress in Nonlinear Differential Equations and their Applications, 6, Birkhäuser Boston Inc., Boston, MA, 1993.

[CI99] G. Contreras and R. Iturriaga, *Global minimizers of autonomous Lagrangians*, 22° Colóquio Brasileiro de Matemática. [22nd Brazilian Mathematics Colloquium], Instituto de Matemática Pura e Aplicada (IMPA), Rio de Janeiro, 1999.

[Con78] C.C. Conley, *Isolated invariant sets and the Morse index*, CBMS Regional Conference Series in Mathematics, vol. 38, American Mathematical Society, Providence, R.I., 1978.

[Con84] ______, *Lecture at the University of Wisconsin*, April 6, 1984.

[Con06] G. Contreras, *The Palais-Smale condition on contact type energy levels for convex Lagrangian systems*, Calc. Var. Partial Differ. Equ. **27** (2006), no. 3, 321–395.

[CZ84] C.C. Conley and E. Zehnder, *Morse-type index theory for flows and periodic solutions for Hamiltonian equations*, Comm. Pure Appl. Math. **37** (1984), no. 2, 207–253.

[Dui76] J.J. Duistermaat, *On the Morse index in variational calculus*, Advances in Math. **21** (1976), no. 2, 173–195.

[Eke90] I. Ekeland, *Convexity methods in Hamiltonian mechanics*, Ergebnisse der Mathematik und ihrer Grenzgebiete (3) [Results in Mathematics and Related Areas (3)], vol. 19, Springer-Verlag, Berlin, 1990.

[Fat08] A. Fathi, *Weak KAM theorem in Lagrangian dynamics*, Cambridge Univ. Press, forthcoming, preliminary version number 10, 2008.

[GG10] V.L. Ginzburg and B.Z. Gürel, *Local Floer homology and the action gap*, J. Symplectic Geom. **8** (2010), no. 3, 323–357.

[GH96] M. Giaquinta and S. Hildebrandt, *Calculus of variations. I*, Grundlehren der Mathematischen Wissenschaften [Fundamental Principles of Mathematical Sciences], vol. 310, Springer-Verlag, Berlin, 1996, The Lagrangian formalism.

[Gin10] V.L. Ginzburg, *The Conley conjecture*, Ann. Math. **172** (2010), no. 2, 1127–1180.

[GL58] I.M. Gel'fand and V.B. Lidskiĭ, *On the structure of the regions of stability of linear canonical systems of differential equations with periodic coefficients*, Amer. Math. Soc. Transl. (2) **8** (1958), 143–181.

[GM69a] D. Gromoll and W. Meyer, *On differentiable functions with isolated critical points*, Topology **8** (1969), 361–369.

[GM69b] ———, *Periodic geodesics on compact Riemannian manifolds*, J. Differential Geometry **3** (1969), 493–510.

[Ham82] R.S. Hamilton, *The inverse function theorem of Nash and Moser*, Bull. Amer. Math. Soc. (N.S.) **7** (1982), no. 1, 65–222.

[Hat02] A. Hatcher, *Algebraic topology*, Cambridge University Press, 2002.

[Hei09] D. Hein, *The Conley conjecture for irrational symplectic manifolds*, to appear in J. Symplectic Geom., 2009.

[Hei11] ———, *The Conley conjecture for the cotangent bundle*, Archiv der Mathematik **96** (2011), no. 1, 85–100.

[Hin09] N. Hingston, *Subharmonic solutions of Hamiltonian equations on tori*, Ann. Math. **170** (2009), no. 2, 529–560.

[HZ94] H. Hofer and E. Zehnder, *Symplectic invariants and Hamiltonian dynamics*, Birkhäuser Advanced Texts: Basler Lehrbücher. [Birkhäuser Advanced Texts: Basel Textbooks], Birkhäuser Verlag, Basel, 1994.

[KH95] A. Katok and B. Hasselblatt, *Introduction to the modern theory of dynamical systems*, Encyclopedia of Mathematics and its Applications, vol. 54, Cambridge University Press, Cambridge, 1995, With a supplement by A. Katok and L. Mendoza.

[Kli78] W. Klingenberg, *Lectures on closed geodesics*, Springer-Verlag, Berlin, 1978, Grundlehren der Mathematischen Wissenschaften, Vol. 230.

[LA98] Y. Long and T. An, *Indexing domains of instability for Hamiltonian systems*, NoDEA Nonlinear Differential Equations Appl. **5** (1998), no. 4, 461–478.

[Lee03] J.M. Lee, *Introduction to smooth manifolds*, Graduate Texts in Mathematics, vol. 218, Springer-Verlag, New York, 2003.

[LL98] C. Liu and Y. Long, *An optimal increasing estimate of the iterated Maslov-type indices*, Chinese Sci. Bull. **43** (1998), no. 13, 1063–1066.

[LL00] ______, *Iteration inequalities of the Maslov-type index theory with applications*, J. Differential Equations **165** (2000), no. 2, 355–376.

[LL03] Y. Long and G. Lu, *Infinitely many periodic solution orbits of autonomous Lagrangian systems on tori*, J. Funct. Anal. **197** (2003), no. 2, 301–322.

[Lon90] Y. Long, *Maslov-type index, degenerate critical points, and asymptotically linear Hamiltonian systems*, Sci. China Ser. A **33** (1990), no. 12, 1409–1419.

[Lon00] ______, *Multiple periodic points of the Poincaré map of Lagrangian systems on tori*, Math. Z. **233** (2000), no. 3, 443–470.

[Lon02] ______, *Index theory for symplectic paths with applications*, Progress in Mathematics, vol. 207, Birkhäuser Verlag, Basel, 2002.

[LS85] F. Laudenbach and J.C. Sikorav, *Persistance d'intersection avec la section nulle au cours d'une isotopie hamiltonienne dans un fibré cotangent*, Invent. Math. **82** (1985), no. 2, 349–357.

[Lu09] G. Lu, *The Conley conjecture for Hamiltonian systems on the cotangent bundle and its analogue for Lagrangian systems*, J. Funct. Anal. **256** (2009), no. 9, 2967–3034.

[LZ90] Y. Long and E. Zehnder, *Morse-theory for forced oscillations of asymptotically linear Hamiltonian systems*, Stochastic processes, physics and geometry (Ascona and Locarno, 1988), World Sci. Publ., Teaneck, NJ, 1990, pp. 528–563.

[Mac09] B.D. MacCluer, *Elementary functional analysis*, Graduate Texts in Mathematics, vol. 253, Springer, New York, 2009.

[Mas72] V.P. Maslov, *Théorie des perturbations et méthodes asymptotiques*, Dunod, Paris, 1972.

[Mat91] J.N. Mather, *Action minimizing invariant measures for positive definite Lagrangian systems*, Math. Z. **207** (1991), no. 2, 169–207.

[Maz11] M. Mazzucchelli, *The Lagrangian Conley conjecture*, Comment. Math. Helv. **86** (2011), no. 1, 189–246.

[Mil56] J. Milnor, *Lectures on the h-cobordism theorem*, Princeton University Press, Princeton, NJ, 1956.

[Mil63] ———, *Morse theory*, Based on lecture notes by M. Spivak and R. Wells. Annals of Mathematics Studies, No. 51, Princeton University Press, Princeton, N.J., 1963.

[Mn91] R. Mañé, *Global variational methods in conservative dynamics*, 18° Colóquio Brasileiro de Matemática. [18th Brazilian Mathematics Colloquium], Instituto de Matemática Pura e Aplicada (IMPA), Rio de Janeiro, 1991.

[Mor25] M. Morse, *Relations between the critical points of a real function of n independent variables*, Trans. Amer. Math. Soc. **27** (1925), no. 3, 345–396.

[Mor96] ———, *The calculus of variations in the large*, American Mathematical Society Colloquium Publications, vol. 18, American Mathematical Society, Providence, RI, 1996, Reprint of the 1932 original.

[MS98] D. McDuff and D. Salamon, *Introduction to symplectic topology*, second ed., Oxford Mathematical Monographs, The Clarendon Press, Oxford University Press, New York, 1998.

[MW89] J. Mawhin and M. Willem, *Critical point theory and Hamiltonian systems*, Applied Mathematical Sciences, vol. 74, Springer-Verlag, New York, 1989.

[Pal63] R.S. Palais, *Morse theory on Hilbert manifolds*, Topology **2** (1963), 299–340.

[Pal66] ———, *Homotopy theory of infinite dimensional manifolds*, Topology **5** (1966), 1–16.

[PS64] R.S. Palais and S. Smale, *A generalized Morse theory*, Bull. Amer. Math. Soc. **70** (1964), 165–172.

[Rad92] H.B. Rademacher, *Morse-Theorie und geschlossene Geodätische*, Bonner Mathematische Schriften [Bonn Mathematical Publications], 229, Universität Bonn Mathematisches Institut, Bonn, 1992, Habilitationsschrift, Rheinische Friedrich-Wilhelms-Universität Bonn, Bonn, 1991.

[RS93a] J. Robbin and D. Salamon, *The Maslov index for paths*, Topology **32** (1993), no. 4, 827–844.

[RS93b] ———, *Phase functions and path integrals*, Symplectic geometry, London Math. Soc. Lecture Note Ser., vol. 192, Cambridge Univ. Press, Cambridge, 1993, pp. 203–226.

[Rud73] W. Rudin, *Functional analysis*, McGraw-Hill Book Co., New York, 1973, McGraw-Hill Series in Higher Mathematics.

[Sal99] D. Salamon, *Lectures on Floer homology*, Symplectic geometry and topology (Park City, UT, 1997), IAS/Park City Math. Ser., vol. 7, Amer. Math. Soc., Providence, RI, 1999, pp. 143–229.

[Sch93] M. Schwarz, *Morse homology*, Progress in Mathematics, vol. 111, Birkhäuser Verlag, Basel, 1993.

[Sma61] S. Smale, *Generalized Poincaré's conjecture in dimensions greater than four*, Ann. of Math. (2) **74** (1961), 391–406.

[SZ92] D. Salamon and E. Zehnder, *Morse theory for periodic solutions of Hamiltonian systems and the Maslov index*, Comm. Pure Appl. Math. **45** (1992), no. 10, 1303–1360.

[Tho49] R. Thom, *Sur une partition en cellules associée à une fonction sur une variété*, C. R. Acad. Sci. Paris **228** (1949), 973–975.

[Ton34] L. Tonelli, *Su gli integrali del calcolo delle variazioni in forma ordinaria*, Ann. Scuola Norm. Sup. Pisa Cl. Sci. (2) **3** (1934), no. 3-4, 401–450.

[Ver96] F. Verhulst, *Nonlinear differential equations and dynamical systems*, second ed., Universitext, Springer-Verlag, Berlin, 1996, Translated from the 1985 Dutch original.

[Vit88] C. Viterbo, *Indice de Morse des points critiques obtenus par minimax*, Ann. Inst. H. Poincaré Anal. Non Linéaire **5** (1988), no. 3, 221–225.

[Vit90] ______, *A new obstruction to embedding Lagrangian tori*, Invent. Math. **100** (1990), no. 2, 301–320.

[VPS76] M. Vigué-Poirrier and D. Sullivan, *The homology theory of the closed geodesic problem*, J. Differential Geometry **11** (1976), no. 4, 633–644.

[Web02] J. Weber, *Perturbed closed geodesics are periodic orbits: index and transversality*, Math. Z. **241** (2002), no. 1, 45–82.

[Wit82] E. Witten, *Supersymmetry and Morse theory*, J. Differential Geom. **17** (1982), no. 4, 661–692.

List of Symbols

Chapter 1

Chapter 2

Chapter 3

Chapter 4

$Q_{\mathscr{L}}$	Projection of the Euler-Lagrange flow onto M	85
$\mathbb{Z}_k$	Cyclic group of order k	86
$C_k^\infty(\mathbb{T};M)$	Space of continuous and k-broken smooth loops	86
Δ_k	Neighborhood of the diagonal of the k-fold product of M	87
λ_k	Embedding of Δ_k into $C_k^\infty(\mathbb{T};M)$	87
Λ_k	k-broken Euler-Lagrange loop space	87
$\gamma_{\boldsymbol{q}}$	Broken Euler-Lagrange loop between the k-uple of points $\boldsymbol{q}$	87
$\mathscr{A}_k$	Discrete action functional	89
L	Quadratic Lagrangian	93
A	Action functional associated to L	93
$\boldsymbol{E}$	Real Hilbert space	96
$\mathscr{B}$	Bounded symmetric bilinear form on $\boldsymbol{E}$	96
$\mathrm{ind}(\mathscr{B})$	Morse index of $\mathscr{B}$	96
$\{\boldsymbol{E}_n \mid n \in \mathbb{N}\}$	Family of Hilbert subspaces of $\boldsymbol{E}$	96
P_n	Orthogonal projector of $\boldsymbol{E}$ onto $\boldsymbol{E}_n$	96
$\boldsymbol{V}$	Finite-dimensional subspace of $\boldsymbol{E}$	96
$S(\boldsymbol{V})$	Sphere of $\boldsymbol{V}$	96
$\mathrm{Aff}_k(\mathbb{T};\mathbb{R}^m)$	k-broken affine loop space of $\mathbb{R}^m$	97
$\lfloor\cdot\rfloor$	Integer part	97
$(\mathscr{A})_c$	c-open sublevel of $\mathscr{A}$	100
$(\mathscr{A}_k)_c$	c-open sublevel of $\mathscr{A}_k$	100
r_k	Retraction of a sublevel of $\mathscr{A}$ onto a sublevel of $\mathscr{A}_k$	100
R_k	Deformation retraction from the identity to r_k	100
H_*	Singular homology functor	102
$\mathrm{C}_*(\mathscr{A}_k, \boldsymbol{q})$	Local homology of $\mathscr{A}_k$ at $\boldsymbol{q}$	102
$\mathrm{C}_*(\mathscr{A}, \gamma_{\boldsymbol{q}})$	Local homology of $\mathscr{A}$ at $\gamma_{\boldsymbol{q}}$	102
$\mathrm{H}_*(\lambda_k)$	Homology homomorphism induced by λ_k	102
$\mathrm{CL}(W^{1,2}(\mathbb{T};M))$	Cup-length of the free loop space $W^{1,2}(\mathbb{T};M)$	103
$\lambda_k^{[n]}$	Embedding of Δ_{nk} into $C_k^\infty(\mathbb{T}^{[n]};M)$	106
$\Lambda_k^{[n]}$	k-broken n-periodic Euler-Lagrange loop space	106
$\mathscr{A}_k^{[n]}$	Discrete mean action functional	107
$\psi_k^{[n]}$	Discrete nth-iteration map	107
$\boldsymbol{q}^{[n]}$	nth-iteration of $\boldsymbol{q}$	107

Chapter 5

$\boldsymbol{E}$	Real separable Hilbert space	110
$\mathscr{U}$	Open set in $\boldsymbol{E}$	110
$\mathscr{F}$	Function over $\mathscr{U}$	110
H	Operator on $\boldsymbol{E}$ associated to the Hessian of $\mathscr{F}$ at $\boldsymbol{x}$	110
$\boldsymbol{E}_\bullet$	Hilbert subspace of $\boldsymbol{E}$	110

Chapter 6

Appendix

Index